T0181858

Introduction to Advanced Electrodynamics

Kaushik Bhattacharya · Soumik Mukhopadhyay

Introduction to Advanced Electrodynamics

 Springer

Kaushik Bhattacharya
Department of Physics
Indian Institute of Technology Kanpur
Kanpur, India

Soumik Mukhopadhyay
Department of Physics
Indian Institute of Technology Kanpur
Kanpur, India

ISBN 978-981-16-7804-2   ISBN 978-981-16-7802-8   (eBook)
https://doi.org/10.1007/978-981-16-7802-8

This Springer imprint is published by the registered company Springer Nature Singapore Pte Ltd.
The registered company address is: 152 Beach Road, #21-01/04 Gateway East, Singapore 189721, Singapore

Preface

Classical electrodynamics is a core subject for budding physicists as well as serious students of Physics. It is, without doubt, one of the pillars of the grand edifice called modern physics. In the present book, we have tried to highlight some of the important aspects of advanced classical electrodynamics. In most of the cases, our points of view are not original as the subject has developed and matured over a long period of time and many have contributed with original papers and monographs on these topics. We claim our originality in the selection and in the manner of exposition of the topics which, we think, are essential for an overall understanding of classical electrodynamics.

Although there are various excellent and useful books on this subject, such as Introduction to electrodynamics by D. J. Griffiths, Classical Electrodynamics by J. D. Jackson, Classical electricity and magnetism by W. K. H. Panofsky and M. Philips, The classical theory of fields by L. D. Landau and E. M. Lifshitz and Electrodynamics of continuous media by L. D. Landau, E. M. Lifshitz and L. P. Pitaevskii, we think our endeavor to write another book on classical electrodynamics is still justified. As instructors, we have taught this subject for several years, to a large variety of students. What we inferred from our experience was that the subject required a book that would help instructors develop the course in a focussed manner keeping the content comprehensible to the graduate students with varying levels of mathematical abilities. In this book, we retain the basic ingredients of classical electrodynamics and elaborate on the various mathematical techniques employed to understand the complex scaffolding of the subject. It can be treated as a book on mathematical methods in physics as well, as we have meticulously explained most of the steps used to derive the results. We have tried to present the discussions on various topics in such a way that the reader will find the book self-contained although we do not claim that we have included all the topics relevant for a complete understanding of classical electrodynamics.

At this point, we can state what this book does not offer. It is not a research guide; it does not include topics on which one can get an idea about new research problems in classical electrodynamics. This book is not a compendium of results in classical electrodynamics. Many important topics are omitted not just due to the limitation of space, but with the intention to keep the material simple for beginners.

Many of our friends helped in our endevour. Some of them went through the manuscript patiently and encouraged us and others pointed out errors in the manuscript. We would specially like to thank Dipankar Chakrabarty, Subrata Sur, Soma Sanyal, Sudeshna Chattopadhyay and Suratna Das for their encouragement and advise.

In the end, we spell out whom this book is written for. The present book is ideally suited for the students who have credited a basic course of electromagnetism at the undergraduate level. This book also requires some preliminary knowledge of special relativity. Once these requirements are met, the reader can go through the materials presented without much difficulty. The material presented can be too heavy for a one-semester, fifty-lecture course. In the present form, the book can be used in two separate one-semester courses, complemented by additional materials from another textbook. The instructor is also free to design a one-semester course by omitting some of the chapters in the book.

Kanpur, India   Kaushik Bhattacharya
                                                                Soumik Mukhopadhyay

Contents

Chapter 1
Basic Laws of Electrodynamics

Largely developed by James Clerk Maxwell in the nineteenth century, classical non-relativistic electrodynamics formally unifies electricity and magnetism within a single framework. The story of magnetism, in fact, goes back a long way. The ancient Greeks knew of the strange properties of a naturally found mineral called loadstone which attracted iron. At least a thousand years back, the Chinese invented the compass which was later used by the Europeans for navigation. But it was not until the sixteenth century that a systematic exposition of the underlying laws governing magnetism began, primarily through Gilbert's treatise 'de Magnete', published in 1600. Magnetism became a subject of scientific enquiry. The ancient Greeks also knew how to generate static charges. However, it was in the eighteenth century that Charles Augustine de Coulomb measured the force of repulsion between like charges. In 1820, Oersted's experimental observation of the deflection of the compass needle in the vicinity of a current carrying wire paved the way for the unification of electricity and magnetism. Subsequent works of Gauss, Ampere, Faraday and many others led Maxwell to formulate the fundamental laws of electrodynamics.

In this chapter, we shall lay out the basic laws of non-relativistic electrodynamics in absence of material media, including four Maxwell's Field equations and their potential formulation.

1.1 Electric Field and the Principle of Superposition

The electromagnetic interaction is characterized by a quantity called 'electric charge' which has the following properties to start with: (1) The electric charge is invariant under change of reference frame; (2) The electric charges are available in two varieties, positive and negative; (3) The total charge is globally conserved, i.e., charges can neither be created nor annihilated. We begin with the simplest case of stationary

© Springer Nature Singapore Pte Ltd. 2021
K. Bhattacharya and S. Mukhopadhyay, *Introduction to Advanced Electrodynamics*,
https://doi.org/10.1007/978-981-16-7802-8_1

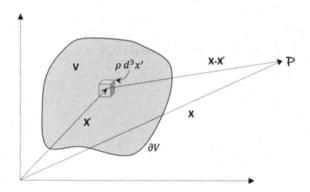

Fig. 1.1 A localized continuous charge distribution with volume V and boundary surface ∂V

charges. The interaction between the sources charges $\{q_i\}$ and the test charge Q is mediated by electric field $\mathbf{E}$ which is given by

$$\mathbf{E}(\mathbf{x}) = \sum_i q_i \frac{\mathbf{x} - \mathbf{x_i}'}{|\mathbf{x} - \mathbf{x_i}'|^3}. \tag{1.1}$$

Here, $\mathbf{x}$ is the position vector of the test charge Q, $\mathbf{x_i}'$ are the position vectors of the source charges q_i and $\frac{\mathbf{x} - \mathbf{x_i}'}{|\mathbf{x} - \mathbf{x_i}'|}$ is the unit vector along the direction connecting the position of the ith source charge and the test charge. The force exerted on the test charge Q, irrespective of whether the test charge is static or mobile, is given by $\mathbf{F} = q\mathbf{E}$. Also implicit in the definition of the electric field is the principle of superposition which states that the electric field due to a collection of test charges at a point is equal to the sum of the electric field generated by the individual source charges. Although charges are discrete, it is often convenient to substitute the discrete charge distribution q_i with a localized continuous charge distribution $\rho(\mathbf{x}')$ of volume V and boundary surface ∂V ($q_i \longrightarrow \rho\, d^3x'$). The electric field at any point $\mathbf{x}$ in that case is given by (Fig. 1.1)

$$\mathbf{E}(\mathbf{x}) = \int \rho(\mathbf{x}') \frac{\mathbf{x} - \mathbf{x}'}{|\mathbf{x} - \mathbf{x}'|^3}\, d^3x'. \tag{1.2}$$

Alternatively, given a discrete charge distribution $\{q_i(\mathbf{x_i}')\}$, one can, for convenience, define a volume charge density using the Dirac delta function in the following way:

$$\rho(\mathbf{x}') = \sum q_i \delta(\mathbf{x}' - \mathbf{x_i}'). \tag{1.3}$$

Based on Coulomb's law and the principle of superposition, we can find out the properties of the vector field $\mathbf{E}(\mathbf{x})$ related to its 'flux' across a closed surface and its 'circulation'. To understand electric flux, it is instructive to develop a visual representation of the electric field using the 'field lines'. The tangent to a point on a field line gives the direction of the electric field at that point, and the density of the

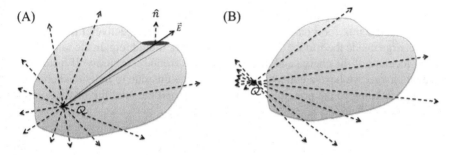

Fig. 1.2 A Electric flux across a closed surface enclosing a charge Q. The unit normal $\hat{\mathbf{n}}$ to the shaded region of area da is also shown. **B** If the charge is located outside the surface, the net electric flux is zero

field lines through an infinitesimal area around a point conveys the magnitude of the electric field at that point. For a given distribution of positive and negative charges, the field lines are directed away from the positive charge and toward the negative charge without intersecting each other.

The electric flux through an arbitrary surface is defined as

$$\phi_E = \int \mathbf{E} \cdot \hat{\mathbf{n}} \, da. \tag{1.4}$$

Here, $\hat{\mathbf{n}}$ is the unit normal to the surface element da and is customarily assumed to be directed outward for a closed surface. Now consider placing a single positive charge q within an arbitrary closed surface. The electric flux across the closed surface is non-zero as the value of $\mathbf{E} \cdot \hat{\mathbf{n}}$ for each area element on the surface has the same sign ($\mathbf{E}$ points away from the charge in every direction and $\hat{\mathbf{n}}$ is outward normal to the surface area element) and thus when integrated over, the closed surface, is bound to be non-zero and finite (unless the surface itself is pushed to infinity where the electric field is zero for all practical purposes). Interestingly, one can move the charge anywhere within the region enclosed by the surface and yet the electric flux remains the same! This of course does not mean that the electric field at each point on the surface remains unaltered. It so happens that the dot product $\mathbf{E} \cdot \hat{\mathbf{n}}$ sort of redistributes itself at each point on the surface so as to keep the total electric flux constant. On the other hand, if the charge is outside the closed surface, the total electric flux across the surface is zero since each field line penetrating into the surface manages to emerge out of the surface (Fig. 1.2B). Clearly, the electric flux through a closed surface depends on the charge enclosed by the surface.

Let us calculate, as a simple illustration, the total electric flux across a spherical surface enclosing a charge Q. Let us assume, for simplicity, that the charge is located at the center of the spherical surface. The electric field due to the charge Q on the surface of the sphere goes as $\frac{Q}{r^2}$ whereas the surface area of the sphere goes as $4\pi r^2$. Clearly, the net flux, which is a product of the electric field and the surface area, is independent of r and is equal to $4\pi Q$. This statement is true even if the surface is not spherical but an arbitrary closed surface. We have already convinced ourselves that the electric flux for a particular surface is the same irrespective of the location of the charge inside the surface. Thus, the electric flux should remain the same if we fix the location of the charge and make the surface around it arbitrary. If instead of a single charge, we have a collection of charges, the flux due to individual charges adds up according to the principle of superposition and the total flux is the sum of individual contribution.

$$\oint \mathbf{E} \cdot \hat{\mathbf{n}} \, da = \sum_i \oint \mathbf{E_i} \cdot \hat{\mathbf{n}} \, da. \tag{1.5}$$

Thus, Coulomb's inverse square law of electrostatics and the principle of superposition can be restated elegantly in the language of electric flux through a closed surface. It turns out that the electric flux through a closed surface is proportional to the total charge enclosed by the surface. This is the statement of Gauss's law, which, in integral form, is given by

$$\oint_{\partial V} \mathbf{E} \cdot \hat{\mathbf{n}} \, da = 4\pi \, Q_{\text{enc}}. \tag{1.6}$$

Here, ∂V is any closed surface and Q_{enc} is the total charge enclosed by ∂V.

We can further generalize the integral form of Gauss's law by assuming a continuous charge distribution $\rho(\mathbf{x}', t)$, which varies in space and time, instead of a discrete charge distribution. Then the total charge enclosed is given by the volume integral

$$Q_{\text{enc}} = \int_V \rho(\mathbf{x}', t) d^3 x'. \tag{1.7}$$

We can replace the primed variables with unprimed variables and extend the volume integral to all space as $\rho(\mathbf{x}) = 0$ for $x \neq x'$ anyway. On the other hand, we can convert the surface integral over electric field into a volume integral by applying divergence theorem as follows:

$$\oint_{\partial V} \mathbf{E} \cdot \hat{\mathbf{n}} \, da = \int_V \nabla \cdot \mathbf{E} \, d^3 x. \tag{1.8}$$

Here, V is the volume enclosed by the surface ∂V. It is then straightforward to show

$$\int_V \nabla \cdot \mathbf{E} \, d^3 x = 4\pi \int_V \rho(\mathbf{x}, t) d^3 x. \tag{1.9}$$

The equation is true for any surface enclosing the charge and consequently any volume V. It follows that the integrand must be equal. The differential form of Gauss's law is then given by

$$\nabla \cdot \mathbf{E}(\mathbf{x}, t) = 4\pi \rho(\mathbf{x}, t). \tag{1.10}$$

The divergence of the electric field at a point in space is equal to 4π times the charge density at that point. The point in space at which divergence of **E** is positive (negative) corresponds to the source (sink) of the field, i.e., the field lines emerge from the source and terminate at the sink.

1.1.1 Electric Scalar Potential

The electric potential is a scalar quantity which can be interpreted either as a scalar field or as electrostatic potential energy per unit charge. The negative gradient of the scalar field at a point in space is a measure of the electric field at that point. The electric potential at a point in space is also equal to the work done to bring a unit test charge from infinity to that point. Imagine a source charge q_i placed at $\mathbf{x_i}'$ which leads to an electric field **E** around it. The work done to bring a test charge Q from infinity to a point **x** in space is given by

$$W = -\int_{\infty}^{\mathbf{x}} \frac{q_i Q}{|\mathbf{x} - \mathbf{x_i}'|^3}(\mathbf{x} - \mathbf{x_i}') \cdot d\mathbf{l}. \tag{1.11}$$

Here, $d\mathbf{l}$ is the infinitesimal displacement along some path from infinity to **x**. Without loss of generality, we can place the source charge at the origin. Then we have

$$W = -\int_{\infty}^{\mathbf{r}} \frac{q_i Q}{|\mathbf{r}|^3} \mathbf{r} \cdot d\mathbf{l} = q_i Q \int_{\infty}^{r} \frac{d\tilde{r}}{\tilde{r}^2} = \frac{q_i Q}{r}. \tag{1.12}$$

Noting that $|\mathbf{r}|$ is simply the distance between the source charge and the test charge, we can write a more general expression for the work done per unit test charge or the electric potential at any point **x** due to a charge q_i placed at $\mathbf{x_i}'$, as follows:

$$\Phi = \frac{W}{Q} = \frac{q_i}{|\mathbf{x} - \mathbf{x_i}'|}. \tag{1.13}$$

For a collection of discrete charges $q_i(\mathbf{x_i}')$, the potential at any point in space is

$$\Phi(\mathbf{x}) = \sum_i \frac{q_i}{|\mathbf{x} - \mathbf{x_i}'|}. \tag{1.14}$$

For a continuous volume charge distribution $\rho(\mathbf{x_i}')$,

$$\Phi(\mathbf{x}) = \int \frac{\rho(\mathbf{x_i}')}{|\mathbf{x} - \mathbf{x_i}'|} d^3 x'. \tag{1.15}$$

The potential at any point in space due to a surface-charge distribution is given by

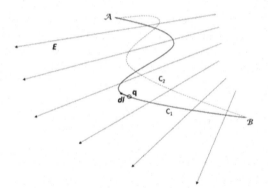

Fig. 1.3 The work done in moving charge q in an electric field from point $\mathcal{A}$ to point $\mathcal{B}$ is the same regardless of the paths taken. In this case, two different paths C_1 and C_2 are shown

$$\Phi(\mathbf{x}) = \int \frac{\sigma(\mathbf{x_i'})}{|\mathbf{x} - \mathbf{x_i'}|} da'. \tag{1.16}$$

The electric field can be expressed as gradient of the scalar potential. One can easily verify the statement now.

$$\mathbf{E} = -\nabla \sum_i \frac{q_i}{|\mathbf{x} - \mathbf{x_i'}|} = \sum_i q_i \frac{|\mathbf{x} - \mathbf{x_i'}|}{|\mathbf{x} - \mathbf{x_i'}|^3}. \tag{1.17}$$

The minus sign appearing before the gradient of the scalar potential is eventually eliminated when we calculate the electric field. Now that we are convinced that electric field can be expressed as the gradient of the scalar potential, it automatically follows that the curl of the electric field vanishes everywhere and that the line integral of the electric field between two points in space will be independent of the path taken (Fig. 1.3).

$$\nabla \times \mathbf{E} = 0. \tag{1.18}$$

Using Stokes' theorem, we have

$$\oint_C \mathbf{E} \cdot \mathbf{dl} = \int_S \nabla \times \mathbf{E} \cdot \hat{n} da = 0. \tag{1.19}$$

Alternatively,

$$\int_{\mathcal{A},C_1}^{\mathcal{B}} \mathbf{E} \cdot \mathbf{dl} = \int_{\mathcal{A},C_2}^{\mathcal{B}} \mathbf{E} \cdot \mathbf{dl}. \tag{1.20}$$

1.1.2 Poisson and Laplace Equations

Let us quickly summarize the important conclusions drawn so far. The electrostatic field is completely specified by the following two equations: $\nabla \cdot \mathbf{E}(\mathbf{x}) = 4\pi\rho(\mathbf{x})$ and

$\nabla \times \mathbf{E}(\mathbf{x}) = 0$. The latter suggests that electric field can be written as a gradient of scalar potential Φ. When the electric field is expressed in terms of potential, the differential form of Gauss's law is reduced to the following form.

$$\nabla^2 \Phi = -4\pi\rho. \tag{1.21}$$

The above equation is named the Poisson equation. We are already familiar with the formal solution of the Poisson equation, which is given below.

$$\Phi(\mathbf{x}) = \int_V \frac{\rho(\mathbf{x}_i')}{|\mathbf{x} - \mathbf{x}_i'|} d^3 x'. \tag{1.22}$$

Let us now verify that the above solution does indeed satisfy the Poisson equation. For simplicity, we consider a point charge q located at $\mathbf{x}'$. The charge density can be written as $\rho(\mathbf{x}) = q\delta(\mathbf{x} - \mathbf{x}')$. We are already familiar with the mathematical form of the potential due to a point charge. It follows that for $\mathbf{r} \neq \mathbf{r}'$, one can easily verify using Cartesian coordinates $\{x_i\}$, where $i = 1, 2, 3$,

$$\nabla^2 \frac{1}{|\mathbf{x} - \mathbf{x}'|} = \left(\frac{\partial^2}{\partial x_1^2} + \frac{\partial^2}{\partial x_2^2} + \frac{\partial^2}{\partial x_3^2}\right) \left[(x_1 - x_1')^2 + (x_2 - x_2')^2 + (x_3 - x_3')^2\right]^{-\frac{1}{2}}$$
$$= 0.$$

To deal with the singularity at $\mathbf{x} = \mathbf{x}'$, we redefine the potential by introducing a parameter a in the expression for potential and taking the $a \to 0$ limit in the following way.

$$\Phi(\mathbf{x}) = \lim_{a \to 0} \Phi_a(\mathbf{x}) = \lim_{a \to 0} \frac{q}{\left[(\mathbf{x} - \mathbf{x}')^2 + a^2\right]^{\frac{1}{2}}}.$$

Expressing the Laplacian in spherical coordinates and substituting $\mathbf{r} = \mathbf{x} - \mathbf{x}'$, we obtain

$$\nabla^2 \Phi_a(\mathbf{x}) = \nabla^2 \frac{q}{\left(r^2 + a^2\right)^{\frac{1}{2}}} = \frac{q}{r^2} \frac{\partial}{\partial r} \left(r^2 \frac{\partial}{\partial r} \frac{1}{\sqrt{r^2 + a^2}}\right) = -\frac{3qa^2}{\left(r^2 + a^2\right)^{\frac{5}{2}}}.$$

Clearly, as $a \to 0$, $\nabla^2 \Phi_a(\mathbf{x})$ becomes infinite at $r = 0$ or at $\mathbf{x} = \mathbf{x}'$ and reduces to zero elsewhere, exactly like a Delta function. Integrating over a volume not excluding $r = 0$, we get 4π.

$$\lim_{a \to 0} \int \frac{3a^2}{\left(r^2 + a^2\right)^{\frac{5}{2}}} r^2 \, d\Omega \, dr = 4\pi. \tag{1.23}$$

In other words, $\nabla^2(\frac{1}{r}) = 0$ for $r \neq 0$ and its volume integral is -4π. It immediately follows that

$$\nabla^2 \left(\frac{1}{|\mathbf{x} - \mathbf{x}'|}\right) = -4\pi\delta(\mathbf{x} - \mathbf{x}'). \tag{1.24}$$

We can now extend the argument for arbitrary continuous charge distribution $\rho(\mathbf{x}')$ and verify that Eq. (1.22) is the solution for the Poisson equation. We make use of Eq. (1.24) for the more general case.

$$\nabla^2 \Phi(\mathbf{x}) = \int \nabla^2 \left(\frac{1}{|\mathbf{x} - \mathbf{x}'|} \right) \rho(\mathbf{x}') d^3 x' = -4\pi \int \delta(\mathbf{x} - \mathbf{x}') \rho(\mathbf{x}') d^3 x' = -4\pi \rho(\mathbf{x}).$$
(1.25)

Thus, Eq. (1.22) is indeed the solution of the Poisson equation.

Often we encounter cases where we need to find out the potential in a limited region where there is no charge, i.e., $\rho = 0$. We are, of course, not interested in the trivial case where $\rho = 0$ everywhere! In such a non-trivial scenario, within that region, the Poisson equation reduces to the following form, called the Laplace equation.

$$\nabla^2 \Phi = 0.$$
(1.26)

1.1.3 Electrostatic Potential Energy for Discrete and Continuous Charge Distribution

Consider a localized charge distribution q_j with $j = 1, 2, ..., n - 1$ with corresponding position vectors $\mathbf{x_j}$. We have already defined the electrostatic potential $\Phi(\mathbf{x_i})$ as the work done against the electric field (due to the charge distribution q_j in that region) to bring a unit charge from infinity to any point $\mathbf{x_i}$. The potential $\Phi(\mathbf{x_i})$ vanishes at infinity. The potential energy of the charge q_i at the point $\mathbf{x_i}$ is the work done W_i to bring a charge q_i from infinity to $\mathbf{x_i}$ and is given by

$$W_i = q_i \Phi(\mathbf{x_i}).$$
(1.27)

We already know that the potential at the point $\mathbf{x_i}$ due to the collection of discrete charges q_j is given by

$$\Phi(\mathbf{x_i}) = \sum_{j \neq i}^{n-1} \frac{q_j}{|\mathbf{x_i} - \mathbf{x_j}|}.$$
(1.28)

Therefore, the potential energy W_i of the charge q_i can be rewritten as

$$W_i = q_i \sum_{j \neq i, j=1}^{n-1} \frac{q_j}{|\mathbf{x_i} - \mathbf{x_j}|}.$$
(1.29)

We now have a collection of n charges. The total potential energy of the collection is

$$W = \sum_i q_i \Phi(\mathbf{x_i}) = \sum_{i=1}^{n} \sum_{j<i} \frac{q_i q_j}{|\mathbf{x_i} - \mathbf{x_j}|} \tag{1.30}$$

$$= \frac{1}{2} \sum_{i,j=1, i\neq j}^{n} \frac{q_i q_j}{|\mathbf{x_i} - \mathbf{x_j}|}. \tag{1.31}$$

Two kinds of summations have been used here. In Eq. (1.30), the summation runs over two indices i and j separately with i running from 1 to n, and j is always less than i to avoid double counting. On the other hand, in Eq. (1.31), both i and j run from 1 to n while the factor $1/2$ takes care of the double counting. In both equations, i is never equal to j to exclude the infinite 'self energy' term, i.e., the charge interacting with itself.

The symmetric nature of the summations used in Eq. (1.31) is particularly useful in that it can be straightaway extended to the case of continuous charge distribution $\rho(\mathbf{x'})$. The constituent discrete charges in each pair $\{q_i q_j\}$ are replaced by continuous charge distributions using primed and unprimed variables in the following manner: $q_i \longrightarrow \rho(\mathbf{x})d^3x$ and $q_j \longrightarrow \rho(\mathbf{x'})d^3x'$. The potential energy of a continuous charge distribution or the work done to 'assemble' the charge distribution is given by

$$W = \frac{1}{2} \int \frac{\rho(\mathbf{x})\rho(\mathbf{x'})}{|\mathbf{x} - \mathbf{x'}|} \, d^3x \, d^3x'. \tag{1.32}$$

We replace one of the integrations over primed variables by the scalar potential term using Eq. (1.15) and get

$$W = \frac{1}{2} \int \rho(\mathbf{x})\Phi(\mathbf{x}) \, d^3x. \tag{1.33}$$

Then we eliminate the remaining charge density term altogether by making use of the Poisson Eq. (1.21) and obtain

$$W = -\frac{1}{8\pi} \int \Phi(\mathbf{x})\nabla^2\Phi(\mathbf{x}) \, d^3x. \tag{1.34}$$

We use the relation $\nabla \cdot (\Phi\nabla\Phi) = |\nabla\Phi|^2 + \Phi\nabla^2\Phi$ to rewrite Eq. (1.34) as follows:

$$W = -\frac{1}{8\pi} \int \left[\nabla \cdot (\Phi\nabla\Phi) - |\nabla\Phi|^2 \right] d^3x \tag{1.35}$$

$$= -\frac{1}{8\pi} \oint \Phi\nabla\Phi \cdot \hat{\mathbf{n}} \, da + \frac{1}{8\pi} \int |\nabla\Phi|^2 \, d^3x. \tag{1.36}$$

The first term is converted into closed surface integral using divergence theorem. The original volume integral containing charge density terms was over the space enclosing the localized charge distribution. Now that we have converted the first term in Eq. (1.36) into surface integral, we are at liberty to push the surface outward since ρ is zero there anyway and does not contribute to the original volume integral. We can eventually eliminate the first term by integrating over all space. This effectively places the surface at infinity where the potential Φ vanishes. Therefore, we are left with the following expression involving only field terms where the volume integral is over all space:

$$W = \frac{1}{8\pi} \int |\nabla \Phi|^2 \, d^3x = \frac{1}{8\pi} \int \mathbf{E}^2 \, d^3x. \tag{1.37}$$

Thus, the total electrostatic potential energy is finally expressed as an integral of the square of the electric field over all space. Note that according to Eq. (1.37), W is positive while according to Eq. (1.31), the potential energy of two opposite charges is negative. We emphasize that both equations are correct: they are actually calculating two different quantities. The 'field description' of potential energy, given by Eq. (1.37), includes not just the energy required to assemble the charges but also the self energy of each charge (energy of a charge due to its own field). This is problematic for a discrete point charge as the self energy of a point charge is infinite! However, for a continuous charge distribution, the charge at a point $\mathbf{x}'$ is $\rho(\mathbf{x}')d^3x'$ which approaches zero as $d^3x' \to 0$. Thus, the self energy is excluded automatically. The major difference between Eqs. (1.32) and (1.37) is that we can define an energy density (as if stored in the electric field over all space) for the latter. The electrostatic potential energy density u^1 is defined as follows:

$$u = \frac{1}{8\pi} \mathbf{E}^2. \tag{1.38}$$

Since the energy density is quadratic in the field, one has to be careful while adding the energies of the subsystems to calculate the total energy of the system. The total energy of the system will not be equal to the sum of the energies of the subsystems.

$$W_{\text{total}} = \frac{1}{8\pi} \int \left(\sum_i \mathbf{E}_i\right)^2 d^3x = \sum_i W_i + \frac{1}{4\pi} \int \sum_{i \neq j} \mathbf{E}_i \cdot \mathbf{E}_j \, d^3x. \tag{1.39}$$

1.2 Magnetic Field and the Principle of Superposition

In the previous section, we discussed how a stationary charge or collection of charges produces an electric field. A moving charge or a collection of charges produces a magnetic field in addition to the electric field. In this section, we shall discuss the simplest case of magnetostatics, that of the magnetic field produced by steady current, a uniform flow of charges which is constant in time. This is characterized by $\mathbf{J} = \rho\mathbf{v}$, the current density, where ρ is the charge density, $\mathbf{v}$ is the average velocity of each charge and $\frac{\partial J}{\partial t} = 0$. Clearly, a moving point charge does not constitute a steady current.

---

[1] Note that u is positive definite.

We note a fundamental point of departure from electrostatics: there is no magnetic charge. The basic unit of magnetostatics is the magnetic dipole. A magnetic dipole, described by the moment **m**, in presence of magnetic field **B** tends to align itself along the field direction due to the mechanical torque **N** exerted by the field, which is given below.

$$\mathbf{N} = \mathbf{m} \times \mathbf{B}. \tag{1.40}$$

A standard experiment to demonstrate the 'magnetic force' (attributed to Ampere) involves two current carrying wires. If the currents in the two wires are parallel to each other, meaning the currents are in the same direction, the wires attract each other while for currents in the opposite direction, the wires repel each other. Such a force cannot be of electrostatic origin. In fact, one can hold a compass needle at different positions near a current carrying wire and the tip of the needle circles around the wire, suggesting the existence of a 'magnetic field' which circles around the current direction. Thus, the directions of the current, the magnetic field and the force in the experiment involving the two current carrying wires give us a fairly good idea about the peculiar nature of the 'magnetic force'. In fact, the magnetic force on a charge q, moving with velocity v in a magnetic field **B**, is given by the 'Lorentz force law'.

$$\mathbf{F} = \frac{1}{c}(\mathbf{v} \times \mathbf{B}). \tag{1.41}$$

A famous example of the motion of a charged particle driven by the magnetic field is the 'cyclotron motion'. In cyclotron motion, the charged particle moves around the magnetic field in a circular path, with the magnetic force pointing inward at each point in the circular path, providing the centripetal acceleration. The equation of motion is $\frac{Q}{c}vB = \frac{mv^2}{R}$. Here, Q is the charge on the particle, v is the speed, B the magnetic field and R the radius of the circular path. From the equation, it is straightforward to understand that the linear momentum mv of the charged particle is proportional to the radius of the circular path. This is in fact the standard way to measure the momentum of an elementary particle experimentally: by measuring the radius of curvature of the particle trajectory in a region of known magnetic field.

Let us come back to the experiment on the parallel current carrying wires. The two current carrying wires produce magnetic fields separately circling around each wire. The charges flowing in the second wire perpendicular to the direction of the magnetic field produced by the first wire experience a magnetic force, according to the Lorentz force law, and vice versa. This leads to the force of attraction or repulsion depending on the direction of current in one wire relative to the other. We now extend the Lorentz force law to the case of force on a current carrying wire in a magnetic field. The magnetic force on a current carrying wire is given by

$$\mathbf{F} = \frac{1}{c}\int(\mathbf{v} \times \mathbf{B})\,dq = \frac{1}{c}\int(\mathbf{I} \times \mathbf{B})\,dl = \frac{1}{c}\int I\,(\mathbf{dl} \times \mathbf{B}). \tag{1.42}$$

Here, we have exchanged the vector sign in the end since both **I** and **dl** have the same direction.

The magnetic force on a volume current can be similarly written as

$$\mathbf{F} = \frac{1}{c} \int (\mathbf{v} \times \mathbf{B})\, \rho\, d^3x' = \frac{1}{c} \int (\mathbf{J} \times \mathbf{B})\, d^3x'. \tag{1.43}$$

Here, $\mathbf{J} = \rho \mathbf{v}$ is called the volume current density.

The total current (total charge per unit time) crossing any surface is given by the following surface integral:

$$I = \int \mathbf{J} \cdot \hat{\mathbf{n}}\, da. \tag{1.44}$$

Now consider a region of charge distribution $\rho(\mathbf{x}, t)$, within which a small volume V is bounded by a closed surface ∂V. The total charge per unit time leaving the volume V across the closed surface ∂V is given by

$$\oint_{\partial V} \mathbf{J} \cdot \hat{\mathbf{n}}\, da = \int_V (\nabla \cdot \mathbf{J})\, d^3x. \tag{1.45}$$

The closed surface integral is converted to a volume integral using the divergence theorem. We now give a precise mathematical description of the charge flow. We demand that the charge is conserved locally, and not just globally. Charges cannot simply disappear in space to reappear somewhere else! Whatever little volume we may consider, charges flowing out across the surface enclosing the volume will lead to a decrease of charges inside the volume

$$\oint_{\partial V} \mathbf{J} \cdot \hat{\mathbf{n}}\, da = -\frac{d}{dt} \int_V \rho\, d^3x = -\int_V \frac{\partial \rho}{\partial t}\, d^3x. \tag{1.46}$$

Note that the total time derivative becomes a partial when taken inside the integral. Using Eqs. (1.45) and (1.46), we obtain the equation of continuity describing local conservation of charge

$$\frac{\partial \rho}{\partial t}(\mathbf{x}, t) + \nabla \cdot \mathbf{J}(\mathbf{x}, t) = 0. \tag{1.47}$$

In magnetostatics, we deal with steady currents. This essentially means that charges do not pile up or diminish anywhere in space at any point of time: the charge density ρ remains constant in time. In that case, the equation of continuity reduces to the following form describing steady current flow:

$$\nabla \cdot \mathbf{J} = 0. \tag{1.48}$$

1.2.1 Biot-Savart Law of Magnetostatics

Equivalent to Coulomb's law in electrostatics is the Biot-Savart law of magnetostatics which is applicable not just for steady current but also for quasi stationary current. According to the Biot-Savart law, the magnetic field due to a current carrying wire is given by

$$\mathbf{B}(\mathbf{x}) = \frac{1}{c} \int I\mathbf{dl'} \times \frac{\mathbf{x} - \mathbf{x'}}{|\mathbf{x} - \mathbf{x'}|^3}. \tag{1.49}$$

Here, the integration is along the current path (Fig. 1.4).

The expression can be extended to surface current $\mathbf{K}(\mathbf{x'})$ and volume current $\mathbf{J}(\mathbf{x'})$ by replacing $I\mathbf{dl}$ with $\mathbf{K}(\mathbf{x'})da'$ and $\mathbf{J}(\mathbf{x'})d^3x'$, respectively, in Eq. (1.49).

What will be the magnetic field due to a point charge q moving with uniform velocity $\mathbf{v}$? A naive expectation, using the Biot-Savart law by pretending that a point charge in uniform motion does constitute a steady current (as pointed out earlier, this is not true), gives

$$\mathbf{B}(\mathbf{x}) = \frac{1}{c} q\mathbf{v} \times \frac{\mathbf{x} - \mathbf{x'}}{|\mathbf{x} - \mathbf{x'}|^3}. \tag{1.50}$$

It turns out that the expression is correct for charges with velocities that are small compared to the speed of light. However, a proper derivation of the expression involves using relativistic transformation of fields and then taking the non-relativistic limit. This will be discussed later in the book.

Similar to the electric field, the superposition principle applies to the magnetic field as well. It states that for a collection of source currents, the net magnetic field is the vector sum of fields due to each of the source currents measured separately.

As an illustration, let us calculate the force between two current loops (Fig. 1.5). This is instructive as we will have to use the Biot-Savart law to calculate the magnetic field at one current loop due to the other and additionally the force law of Eq. (1.42). We adopted the general convention that the position vectors of source elements are labeled by primed variables and the position vectors of the observation points by unprimed variables. However, in this case, we note that the position vector of the source element in the one loop is the position vector of the observation point for the other and vice versa.

Fig. 1.4 The current carrying wire with position vectors of the source current element $\mathbf{dl}$ and the observation point $\mathcal{P}$ are shown

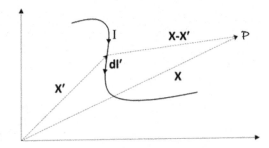

Fig. 1.5 Two current carrying loops with relevant position coordinates are shown. The origin is at $\mathcal{O}$

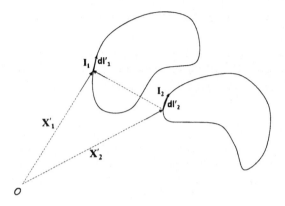

The magnetic field due to the second current loop carrying current I_2 at the position $\mathbf{x_1}'$ of the current element $\mathbf{dl}_1'$ in the first current loop is given by

$$\mathbf{B}_2(\mathbf{x}_1') = \frac{1}{c}\oint_2 I_2 \frac{\mathbf{dl_2}' \times (\mathbf{x_1}' - \mathbf{x_2}')}{|\mathbf{x_1}' - \mathbf{x_2}'|^3}. \tag{1.51}$$

Force on the first loop is then

$$\mathbf{F}_1 = \frac{1}{c}\oint_1 I_1\{\mathbf{dl_1}' \times \mathbf{B}_2(\mathbf{x_1}')\} = \frac{I_1 I_2}{c^2}\oint_1\oint_2 \frac{\mathbf{dl_1}' \times (\mathbf{dl_2}' \times \mathbf{x_{12}}')}{|\mathbf{x_{12}}'|^3}. \tag{1.52}$$

We now get rid of the primes as the primed and unprimed variables are interchangeable and thus redundant in this case. We evaluate the triple product in the numerator of the integrand as $\mathbf{dl}_1 \times (\mathbf{dl}_2 \times \mathbf{x}_{12}) = (\mathbf{dl}_1 \cdot \mathbf{x}_{12})\mathbf{dl}_2 - (\mathbf{dl}_1 \cdot \mathbf{dl}_2)\mathbf{x}_{12}$. When integrated over the closed loop, the first term goes to zero as $\mathbf{dl}_1 \cdot \mathbf{x}_{12}$ is a perfect integral with $|\mathbf{dl}_1| = |\mathbf{dx}_{12}|$. To be more explicit, $\mathbf{dl}_1 \cdot \mathbf{x}_{12} = \mathbf{dx}_{12} \cdot \mathbf{x}_{12} = \frac{1}{2}d(x_{12}{}^2)$. In the end, we are left with

$$\mathbf{F}_1 = -\frac{I_1 I_2}{c^2}\oint_1\oint_2 \frac{(\mathbf{dl}_1 \cdot \mathbf{dl}_2)\mathbf{x}_{12}}{|\mathbf{x}_{12}|^3}. \tag{1.53}$$

Force on the second loop is calculated just by interchanging the labels 1 and 2 and noting that $\mathbf{x}_{21} = -\mathbf{x}_{12}$.

$$\mathbf{F}_2 = \frac{I_1 I_2}{c^2}\oint_1\oint_2 \frac{(\mathbf{dl}_1 \cdot \mathbf{dl}_2)\mathbf{x}_{12}}{|\mathbf{x}_{12}|^3}. \tag{1.54}$$

Thus, $\mathbf{F}_1 = -\mathbf{F}_2$. The magnitude of the net mutual force between the two current loops can be finally written as

$$|\mathbf{F}_{12}| = \left| \frac{I_1 I_2}{c^2}\oint_1\oint_2 \frac{(\mathbf{dl}_1 \cdot \mathbf{dl}_2)\mathbf{x}_{12}}{|\mathbf{x}_{12}|^3} \right|. \tag{1.55}$$

1.2.2 General Derivation of Ampere's Law

We start from the Biot-Savart law for magnetic field due to a volume current distribution $\mathbf{J}(\mathbf{r}')$

$$\mathbf{B}(\mathbf{x}) = \frac{1}{c} \int \mathbf{J}(\mathbf{x}') \times \frac{\mathbf{x} - \mathbf{x}'}{|\mathbf{x} - \mathbf{x}'|^3} d^3 x'. \tag{1.56}$$

At this point, we ask a legitimate question. We know that the divergence of electric field at a point in space is non-zero if the charge density is non-zero at that point. What is the divergence of a magnetic field? For this, we take the divergence on both sides of Eq. (1.56). Before explicitly performing the calculation, we keep in mind that the integration is over the primed variables representing sources, while the operation ∇ is with respect to unprimed variables representing observation points and consequently

$$\nabla \cdot \mathbf{J}(\mathbf{x}') \times \frac{\mathbf{x} - \mathbf{x}'}{|\mathbf{x} - \mathbf{x}'|^3} = \frac{\mathbf{x} - \mathbf{x}'}{|\mathbf{x} - \mathbf{x}'|^3} \cdot (\nabla \times \mathbf{J}') - \mathbf{J}' \cdot \left(\nabla \times \frac{\mathbf{x} - \mathbf{x}'}{|\mathbf{x} - \mathbf{x}'|^3} \right). \tag{1.57}$$

The second term on the right-hand side of the equation vanishes as this is exactly similar to the case of electrostatic field. What about the first term? Here, $\mathbf{J}'$ is a function of the primed variable while the operation ∇ is over unprimed variables. Therefore, the first term vanishes as well. We can now express the Biot-Savart law in differential form as follows:

$$\nabla \cdot \mathbf{B}(\mathbf{x}) = 0. \tag{1.58}$$

Compare the equation with the corresponding one for the electric field. Unlike the electric field, the divergence of the magnetic field at each point in space vanishes. This tells us that there is no monopole term in the multipole expansion of magnetic sources or, in other words, there are no sources or sinks in the magnetic field lines.

We can also rewrite Eq. (1.56) in the following way:

$$\mathbf{B}(\mathbf{r}) = -\frac{1}{c} \int \mathbf{J}(\mathbf{x}') \times \nabla \frac{1}{|\mathbf{x} - \mathbf{x}'|} d^3 x' \tag{1.59}$$

$$= \frac{1}{c} \nabla \times \int \frac{\mathbf{J}(\mathbf{x}')}{|\mathbf{x} - \mathbf{x}'|} d^3 x'. \tag{1.60}$$

Here, we have used standard vector identity $\nabla \times (\phi \mathbf{G}) = \nabla \phi \times \mathbf{G} + \phi \nabla \times \mathbf{G}$. We have also used $\nabla \frac{1}{|\mathbf{x}-\mathbf{x}'|} = -\frac{\mathbf{x}-\mathbf{x}'}{|\mathbf{x}-\mathbf{x}'|^3}$ and $\nabla \times \mathbf{J}(\mathbf{x}') = 0$. Since ∇ is over unprimed variables, $\nabla \times \mathbf{J}(\mathbf{x}')$ vanishes.

We now define the 'vector potential' $\mathbf{A}$ for magnetostatics as

$$\mathbf{A}(\mathbf{x}) = \frac{1}{c} \int \frac{\mathbf{J}(\mathbf{x}')}{|\mathbf{x} - \mathbf{x}'|} d^3 x', \tag{1.61}$$

so that the magnetic field can be expressed as $\mathbf{B} = \nabla \times \mathbf{A}$.

Let us now proceed from here and compute $\nabla \times \mathbf{B}$, starting from Eq. (1.61) and using standard vector identities. We have

$$
\nabla \times \mathbf{B} = \nabla(\nabla \cdot \mathbf{A}) - \nabla^2 \mathbf{A}
$$
$$
= \frac{1}{c} \int \left[\nabla \left(\nabla \cdot \frac{\mathbf{J}(\mathbf{x}')}{|\mathbf{x} - \mathbf{x}'|} \right) - \nabla^2 \left(\frac{\mathbf{J}(\mathbf{x}')}{|\mathbf{x} - \mathbf{x}'|} \right) \right] d^3 x'.
$$

We use the standard vector identity $\nabla \cdot (\phi \mathbf{F}) = \mathbf{F} \cdot (\nabla \phi) + \phi(\nabla \cdot \mathbf{F})$ for the first term in the above integrand. Then we replace the operator ∇ with ∇' in the gradient term, which is accompanied by the negative sign as follows:

$$
\nabla \times \mathbf{B} = \frac{1}{c} \int \left[\nabla \left(-\nabla' \frac{1}{|\mathbf{x} - \mathbf{x}'|} \cdot \mathbf{J}(\mathbf{x}') \right) + \nabla \left(\frac{1}{|\mathbf{x} - \mathbf{x}'|} \nabla \cdot \mathbf{J}(\mathbf{x}') \right) \right] d^3 x'
$$
$$
- \frac{1}{c} \int \mathbf{J}(\mathbf{x}') \nabla^2 \frac{1}{|\mathbf{x} - \mathbf{x}'|} d^3 x'.
$$

The second term in the integrand vanishes since $\mathbf{J}$ is a function of primed variables and the derivative is over unprimed variables. For the same reason, $\mathbf{J}$ is taken outside the Laplacian in the third term. We are already familiar with the Laplacian in the third term from electrostatics. At this stage, we are left with

$$
\nabla \times \mathbf{B} = \frac{1}{c} \int \left[\nabla \left(-\nabla' \frac{1}{|\mathbf{x} - \mathbf{x}'|} \cdot \mathbf{J}(\mathbf{x}') \right) + \mathbf{J}(\mathbf{x}') 4\pi \delta(\mathbf{x} - \mathbf{x}') \right] d^3 x'. \quad (1.62)
$$

Let us take up the first term. Since the integration is over primed variables, we can place the unprimed ∇ operator outside the integral and again use the vector identity to bring back the divergence term. We concentrate only on the integral for the time being and write

$$
\int \left(-\nabla' \frac{1}{|\mathbf{x} - \mathbf{x}'|} \cdot \mathbf{J}(\mathbf{x}') \right) d^3 x' = \int \left[-\nabla' \cdot \frac{\mathbf{J}(\mathbf{x}')}{|\mathbf{x} - \mathbf{x}'|} + \frac{1}{|\mathbf{x} - \mathbf{x}'|} \nabla' \cdot \mathbf{J}(\mathbf{x}') \right] d^3 x'.
$$

The first term is a volume integral over divergence term and can be made to vanish by turning it into a surface integral. Now we start from Eq. (1.62) and rewrite it as follows:

$$
\nabla \times \mathbf{B} = \frac{1}{c} \nabla \int \left[\frac{1}{|\mathbf{x} - \mathbf{x}'|} \nabla' \cdot \mathbf{J}(\mathbf{x}') \right] d^3 x' + \frac{1}{c} \int \left[\mathbf{J}(\mathbf{x}') 4\pi \delta(\mathbf{x} - \mathbf{x}') \right] d^3 x'.
$$
$$
\quad (1.63)
$$

The first term vanishes since $\nabla' \cdot \mathbf{J}' = 0$ as we are dealing with steady current distribution and the second term can be easily evaluated using the delta function, so that we finally obtain Ampere's law in differential form

$$
\nabla \times \mathbf{B}(\mathbf{x}) = \frac{4\pi}{c} \mathbf{J}(\mathbf{x}). \quad (1.64)
$$

Ampere's law states that, unlike the electrostatic field, the curl of the magnetic field is non-zero. Physically, this means that magnetic field lines circulate around current paths. One can easily express Ampere's law in integral form by taking the surface integral on both sides and using Stokes' theorem. The integral form of Ampere's law is useful for solving problems with certain symmetry of the current distribution in space and is given by

$$\oint_C \mathbf{B} \cdot \mathbf{dl} = \frac{4\pi}{c} \int_{\partial V} \mathbf{J} \cdot \hat{\mathbf{n}} \, da. \tag{1.65}$$

It is worth repeating the statement of Ampere's law based on the integral form: The line integral of the magnetic field over a closed loop C is equal to the current through any open surface ∂V attached to the loop. This statement regarding the arbitrariness of the Amperian surface, in fact, provides the motivation for adding a correction term to Ampere's law for non-steady current.

1.2.3   Vector Potential

We have already defined the vector potential $\mathbf{A}(\mathbf{x})$ in Eq. (1.61). Such a definition is not completely free from arbitrariness. Since the magnetic field $\mathbf{B}$ is divergence-free everywhere, it automatically follows that $\mathbf{B}$ can be expressed as curl of some vector field $\mathbf{A}$ as

$$\mathbf{B} = \nabla \times \mathbf{A}. \tag{1.66}$$

The most general expression of $\mathbf{A}$, therefore, includes an added term which is a gradient of an arbitrary scalar function $\chi(\mathbf{x})$, as curl of gradient of any scalar function goes to zero:

$$\mathbf{A}(\mathbf{x}) = \frac{1}{c} \int \frac{\mathbf{J}(\mathbf{x}')}{|\mathbf{x} - \mathbf{x}'|} + \nabla \chi(\mathbf{x}). \tag{1.67}$$

For a given magnetic field $\mathbf{B}$, the choice of vector potential is not unique. It can undergo the following transformation, without leading to any change in $\mathbf{B}$:

$$\mathbf{A}' \longrightarrow \mathbf{A} + \nabla \chi. \tag{1.68}$$

Such a transformation is called gauge transformation. We shall discuss gauge transformation in greater detail later on. Here, we have a limited purpose: we choose a particular gauge so that each Cartesian component of the vector potential satisfies the Poisson equation, the solutions for which we are already familiar with. We use Eq. (1.66) to replace $\mathbf{B}$ in Eq. (1.64) and get

$$\nabla \times (\nabla \times \mathbf{A}) = \nabla(\nabla \cdot \mathbf{A}) - \nabla^2 \mathbf{A} = \frac{4\pi}{c} \mathbf{J}. \tag{1.69}$$

Clearly, if we choose $\nabla \cdot \mathbf{A} = 0$, the above equation reduces to the following form, with each orthogonal component of $\mathbf{A}$ satisfying the Poisson equation:

$$\nabla^2 \mathbf{A} = -\frac{4\pi}{c}\mathbf{J}. \tag{1.70}$$

The solutions for the vector potential readily turn out to be

$$\mathbf{A}(\mathbf{x}) = \frac{1}{c} \int \frac{\mathbf{J}(\mathbf{x}')}{|\mathbf{x} - \mathbf{x}'|} d^3 x'. \tag{1.71}$$

1.3 Faraday's Law of Electromagnetic Induction

Apart from being the first to introduce the concept of 'magnetic field lines', Michael Faraday conducted numerous experiments on the response of electrical circuits in presence of time-varying magnetic fields. The time-varying magnetic fields would be produced either by moving a magnet or abruptly turning off steady current in a wire or by moving a current carrying wire near a 'pick-up' coil. The transient response in the 'pick-up' coil would be detected by a galvanometer connected to the coil. Faraday observed that transient current flowed in the pick-up coil whenever there was a time-varying magnetic field either due to change of source current or due to relative motion between the sources producing non-uniform magnetic field (such as magnets or current carrying wires) and the pick-up coil. He concluded that the transient current flow was due to changing magnetic flux linked with the pick-up coil. More precisely, the time-varying magnetic flux generates an 'electromotive force' $\mathcal{E}$ (the line integral of the electric field around the coil circuit) which makes the current flow. Faraday's conclusion can be neatly expressed using the following mathematical description:

$$\mathcal{E} = \oint_C \mathbf{E}' \cdot \mathbf{dl} = -k\frac{d}{dt}\int_{\partial V} \mathbf{B} \cdot \hat{\mathbf{n}}\, da. \tag{1.72}$$

The left-hand side of the above equation is the electromotive force $\mathcal{E}$ expressed as the closed line integral around the pick-up coil $\oint_C \mathbf{E}' \cdot \mathbf{dl}$, while the right-hand side is the total time derivative of the magnetic flux through any open surface attached to the coil (Fig. 1.6). The minus sign is due to Lenz's law which states that the induced current is in such a direction as to oppose the change of magnetic flux, while k is a constant of proportionality depending on the choice of units.

In order to evaluate the constant of proportionality k, we consider a situation in which the loop C moves with uniform velocity $\mathbf{v}$ ($v/c \ll 1$) in the lab frame. The most crucial aspect of Faraday's observation is that it is the relative motion between the source (or the primary circuit) and the pick-up coil (the secondary circuit) that decides the 'induced' electromotive force or emf. Since $v/c \ll 1$, relativistic effects can be ignored and a Galilean transformation would suffice. As a result of uniform

Fig. 1.6 Magnetic flux linked through surface ∂V attached to the loop $\mathcal{C}$

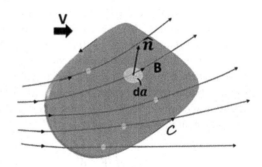

motion, the flux linked with the secondary coil may change either due to change of magnetic field with time at a point in space or due to the change in location of the boundary of the coil in a non-uniform magnetic field although the direction of the unit normal $\hat{\mathbf{n}}$ to the open surface attached to loop remains the same. The total time derivative of magnetic flux through the moving circuit is given by

$$\frac{d}{dt}\int_{\partial V}\mathbf{B}\cdot\hat{\mathbf{n}}\,da = \int\left(\frac{\partial}{\partial t}+\mathbf{v}\cdot\nabla\right)\mathbf{B}\cdot\hat{\mathbf{n}}\,da = \int\left[\frac{\partial\mathbf{B}}{\partial t}+(\mathbf{v}\cdot\nabla)\mathbf{B}\right]\cdot\hat{\mathbf{n}}\,da$$

$$= \int_{\partial V}\frac{\partial\mathbf{B}}{\partial t}\cdot\hat{\mathbf{n}}\,da + \int_{\partial V}\nabla\times(\mathbf{B}\times\mathbf{v})\cdot\hat{\mathbf{n}}\,da\,.$$

Here, we used $\nabla\times(\mathbf{B}\times\mathbf{v}) = (\mathbf{v}\cdot\nabla)\mathbf{B} - \mathbf{v}(\nabla\cdot\mathbf{B})$. Since $\nabla\cdot\mathbf{B} = 0$, we eventually have $(\mathbf{v}\cdot\nabla)\mathbf{B} = \nabla\times(\mathbf{B}\times\mathbf{v})$. Equation (1.72) can now be written as follows:

$$\oint_{\mathcal{C}}\mathbf{E}'\cdot\mathbf{dl} = -k\left[\int_{\partial V}\frac{\partial\mathbf{B}}{\partial t}\cdot\hat{\mathbf{n}}\,da + \int_{\partial V}\nabla\times(\mathbf{B}\times\mathbf{v})\cdot\hat{\mathbf{n}}\,da\right]$$

$$= -k\int_{\partial V}\frac{\partial\mathbf{B}}{\partial t}\cdot\hat{\mathbf{n}}\,da - k\oint_{\mathcal{C}}\mathbf{B}\times\mathbf{v}\cdot\mathbf{dl}\,,$$

which gives

$$\oint_{\mathcal{C}}\left[\mathbf{E}' - k(\mathbf{v}\times\mathbf{B})\right]\cdot\mathbf{dl} = -k\int_{\partial V}\frac{\partial\mathbf{B}}{\partial t}\cdot\hat{\mathbf{n}}\,da\,. \qquad (1.73)$$

We need to keep in mind that $\mathbf{E}'$ is the electric field in the moving frame attached to the loop. Application of Faraday's law in the laboratory frame gives

$$\oint_{\mathcal{C}}\mathbf{E}\cdot\mathbf{dl} = -k\int_{\partial V}\frac{\partial\mathbf{B}}{\partial t}\cdot\hat{\mathbf{n}}\,da. \qquad (1.74)$$

The principle of Galilean invariance demands that

$$\mathbf{E} = \mathbf{E}' - k(\mathbf{v}\times\mathbf{B}). \qquad (1.75)$$

Just a cursory glance at Eq. (1.75) gives us the impression that the $\mathbf{v} \times \mathbf{B}$ term must be associated with the Lorentz force law. In the frame attached to the loop, the force on a charged particle is $q\mathbf{E}'$, as the charged particle is at rest in that frame. There will be an additional force on the charges in the moving loop as observed from the laboratory frame which should be absent in the frame attached to the loop. The moving charge which constitutes a current in the laboratory frame experiences a Lorentz force provided the constant k is equal to $1/c$ in CGS units. With this choice of the constant k, we finally write down Faraday's law

$$\oint_C \mathbf{E}' \cdot \mathbf{dl} = -\frac{1}{c} \frac{d}{dt} \int_{\partial V} \mathbf{B} \cdot \hat{\mathbf{n}} \, da \, . \tag{1.76}$$

Note that we are now back to the total time derivative on the right-hand side, as we initially started with. The line integral is over $\mathbf{E}' = \mathbf{E} + \frac{1}{c}(\mathbf{v} \times \mathbf{B})$, the electric field $\mathbf{E}'$ being measured in the rest frame of the loop moving with velocity $\mathbf{v}$ with respect to the laboratory frame.

For a circuit held fixed in the laboratory frame so that both $\mathbf{E}$ and $\mathbf{B}$ are measured in the same frame, $\mathbf{v} = 0$ and the line integral is converted into surface integral using Stokes' theorem. We then have

$$\int_{\partial V} \left(\nabla \times \mathbf{E} + \frac{1}{c} \frac{\partial \mathbf{B}}{\partial t} \right) \cdot \hat{\mathbf{n}} \, da = 0. \tag{1.77}$$

As the open surface ∂V attached to C is arbitrary, the integrand must vanish everywhere in space. The differential form of Faraday's law is given by

$$\nabla \times \mathbf{E} = -\frac{1}{c} \frac{\partial \mathbf{B}}{\partial t}. \tag{1.78}$$

Unlike the electrostatic field, the electric field generated by time-varying magnetic field has a non-zero curl. The line integral of such an electric field will depend on the path, which can also be verified experimentally.

One more interesting point. Recall that the magnetic field in Ampere's law $\nabla \times \mathbf{B} = \frac{4\pi}{c}\mathbf{J}$ in differential form corresponds to that in the Biot-Savart law in integral form. Faraday's law in differential form is structurally similar to Ampere's law in that $\mathbf{B}$ is now replaced by $\mathbf{E}$ and $4\pi\mathbf{J}$ by $-\frac{\partial \mathbf{B}}{\partial t}$. Using this analogy, we can readily write down the solution for $\mathbf{E}$ as follows:

$$\mathbf{E} = -\frac{1}{4\pi c} \int \frac{\frac{\partial \mathbf{B}}{\partial t} \times (\mathbf{x} - \mathbf{x}')}{|\mathbf{x} - \mathbf{x}'|^3}. \tag{1.79}$$

1.3.1 Energy Stored in Magnetic Field

A steady current distribution produces a magnetic field, according to Ampere's law. On the other hand, Faraday's law says that when a magnetic field is changed (or the current distribution which produces the magnetic field), it sets up an electromotive force that tries to oppose the change. Whenever a steady current is being set up, it has to go through a transient state starting from zero current to reach the final steady state. In the process, the source establishing the steady current has to do some work against the 'induced' electromotive force $\mathcal{E}$ (also called the back emf), which will be stored as energy in the steady current distribution, or, alternatively, in the magnetic field produced. The total energy stored in the magnetic field is the total work done to establish it.

The work done against the back emf to get a steady current I going in a circuit is given by

$$\frac{dW}{dt} = -\mathcal{E}I = \frac{I}{c}\frac{d\phi}{dt}. \qquad (1.80)$$

Here, $\phi = \int_S \mathbf{B} \cdot \mathbf{da}$ is the magnetic flux linked with the circuit. The flux ϕ is proportional to the current so that $\phi = LI$. The constant of proportionality L is called the self inductance, which depends solely on the geometry of the circuit. The total work done W, starting from zero initial current to final current I, is obtained on integrating Eq. (1.80)

$$W = \frac{1}{2c}LI^2. \qquad (1.81)$$

Now let us generalize the expression dealing with a line current to surface and volume currents. We note that for line current I,

$$\phi = LI = \int_S \mathbf{B} \cdot \hat{\mathbf{n}}\,da = \int_{\partial V} \nabla \times \mathbf{A} \cdot \hat{\mathbf{n}}\,da = \oint \mathbf{A} \cdot \mathbf{dl}. \qquad (1.82)$$

Therefore,

$$W = \frac{1}{2c}I(LI) = \frac{1}{2c}I \oint \mathbf{A} \cdot \mathbf{dl} = \frac{1}{2c} \oint (\mathbf{A} \cdot \mathbf{I})\,dl. \qquad (1.83)$$

The above expression readily generalizes to volume current $\mathbf{J}$ as

$$W = \frac{1}{2c} \int_V (\mathbf{A} \cdot \mathbf{J})\,d^3x. \qquad (1.84)$$

The integration is over the volume V occupied by the current distribution. Now we replace $\mathbf{J}$ by magnetic field using Ampere's law, $\nabla \times \mathbf{B} = \frac{4\pi}{c}\mathbf{J}$.

$$W = \frac{1}{8\pi} \int_V \mathbf{A} \cdot (\nabla \times \mathbf{B})\,d^3x. \qquad (1.85)$$

The derivative can be transferred from **B** to **A** using the following relation:

$$\nabla \cdot (\mathbf{A} \times \mathbf{B}) = \mathbf{B} \cdot (\nabla \times \mathbf{A}) - \mathbf{A} \cdot (\nabla \times \mathbf{B})$$
$$= B^2 - \mathbf{A} \cdot (\nabla \times \mathbf{B}).$$

Finally, we have

$$W = \frac{1}{8\pi} \int_V \left[B^2 - \nabla \cdot (\mathbf{A} \times \mathbf{B}) \right] d^3x$$
$$= \frac{1}{8\pi} \left[\int_V B^2 \, d^3x - \oint_{\partial V} (\mathbf{A} \times \mathbf{B}) \cdot \hat{\mathbf{n}} \, da \right]. \qquad (1.86)$$

A part of the volume integral has been replaced by a closed surface integral using the divergence theorem. As the original volume integral in Eq. (1.84) was over the region occupied by the current distribution, the surface ∂V encloses the localized current distribution completely. We could have integrated over a larger volume of space as well. Outside the region of current distribution, $\mathbf{J} = 0$ and it would not change the final result of the integration if the volume is increased further. However, so far as Eq. (1.86) is concerned, for a given volume of current distribution, an increase of the region of integration does change both the integrals over V and ∂V keeping the total value of W unchanged. This is understandable as the integrals now involve fields that extend beyond the region of current distribution. The volume integral would increase as more region in space is integrated over. Since W should remain unchanged in the process, this means that the surface integral should decrease. This is expected: the fields fall off faster than $\frac{1}{|\mathbf{x}-\mathbf{x}'|}$ as the surface moves away from the region of current distribution. Eventually, if we integrate over all space, the surface integral goes to zero. The energy can then be interpreted as being distributed and stored in the magnetic field. The final expression for the total magnetic energy stored in the field is the following volume integral over all space:

$$W = \frac{1}{8\pi} \int B^2 \, d^3x. \qquad (1.87)$$

Similar to the electrostatic potential energy density $\frac{1}{8\pi} E^2$ for a charge distribution given by Eq. (1.38), we can define a magnetic energy density $\frac{1}{8\pi} B^2$. We need to keep in mind that the magnetic field does not do any work. It is the work done against the induced electric field in the process of setting up the magnetic field, excluding the ohmic losses, that is stored as energy in the field.

1.4   Maxwell's Correction to Ampere's Law: The Displacement Current

So long as we dealt with steady-state situations, electric and magnetic phenomena appeared to be two independent categories. Faraday's law, however, suggested that there could be a connection between the two as time-varying magnetic field produced electric field. Does time-varying electric field produce a magnetic field? Maxwell answered the question by introducing a correction term to Ampere's law which not only unified electricity and magnetism, but opened up a new frontier area in science, that of electromagnetic radiation.

Let us write down the set of differential equations involving electric and magnetic fields, as discussed so far:

$$\nabla \cdot \mathbf{E} = 4\pi\rho. \tag{1.88}$$

$$\nabla \cdot \mathbf{B} = 0. \tag{1.89}$$

$$\nabla \times \mathbf{E} = -\frac{1}{c}\frac{\partial \mathbf{B}}{\partial t}. \tag{1.90}$$

$$\nabla \times \mathbf{B} = \frac{4\pi}{c}\mathbf{J}. \tag{1.91}$$

The first three equations are valid for non-steady-state phenomena involving time-varying sources $\rho(\mathbf{x}', t)$ and $\mathbf{J}(\mathbf{x}', t)$. However, the fourth equation, namely Ampere's law, is only valid for steady current under the condition $\nabla \cdot \mathbf{B} = 0$ and, consequently, encounters a serious inconsistency in the case of non-steady currents. To demonstrate this explicitly, let us take divergence on both sides of the equation.

$$\nabla \cdot (\nabla \times \mathbf{B}) = \frac{4\pi}{c}\nabla \cdot \mathbf{J}. \tag{1.92}$$

Clearly, while the left-hand side of the equation goes to zero, the right-hand side does not. For non-steady currents, equation of continuity demands that $\nabla \cdot \mathbf{J} = -\frac{\partial\rho}{\partial t}$. The situation can be restored by noting that $\rho = \frac{1}{4\pi}\nabla \cdot \mathbf{E}$ and replacing ρ in the equation of continuity as follows:

$$\nabla \cdot \mathbf{J} = -\frac{1}{4\pi}\frac{\partial(\nabla \cdot \mathbf{E})}{\partial t}. \tag{1.93}$$

We can now easily find out the quantity with vanishing divergence for non-steady currents.

$$\nabla \cdot \left(\mathbf{J} + \frac{1}{4\pi}\frac{\partial \mathbf{E}}{\partial t}\right) = 0. \tag{1.94}$$

Maxwell replaced $\mathbf{J}$ by the quantity $\mathbf{J} + \frac{1}{4\pi}\frac{\partial \mathbf{E}}{\partial t}$ in Ampere's law for time-varying fields and called the extra term the 'displacement current'. Incorporating Maxwell's correction in Ampere's law, we finally write down the correct equation.

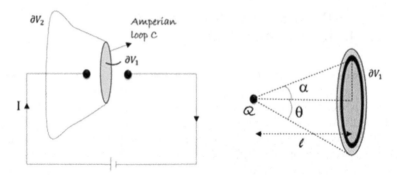

Fig. 1.7 Applying Ampere's law on a 'spark-gap' circuit

$$\nabla \times \mathbf{B} = \frac{4\pi}{c} \left(\mathbf{J} + \frac{1}{4\pi} \frac{\partial \mathbf{E}}{\partial t} \right)$$

$$= \frac{4\pi}{c} \mathbf{J} + \frac{1}{c} \frac{\partial \mathbf{E}}{\partial t}. \tag{1.95}$$

In one stroke, the inclusion of displacement current achieves the following: (1) Consistency with the equation of continuity for time-varying fields; (2) Ampere's original law for magnetostatics is automatically recovered when $\mathbf{E}$ is constant; (3) For $\mathbf{J} = 0$, there appears a striking symmetry between $\mathbf{E}$ and $\mathbf{B}$ fields in Faraday's law and modified Ampere's law. While time-varying magnetic fields induce electric fields, time-varying electric fields produce magnetic fields. In fact, if we put $\rho = 0$, along with $\mathbf{J} = 0$, all four equations, Eqs. (1.88), (1.89), (1.90) and (1.95), look remarkably symmetric under interchange of $\mathbf{E}$ and $\mathbf{B}$. Of course, there is a minus sign in Faraday's law! We shall see how useful it is when we encounter electromagnetic waves later on!

Coming back to displacement current, we present a classic scenario where Ampere's law for magnetostatics invariably fails. Let us consider the example of charging up a metal blob in a spark-gap circuit using a battery (Fig. 1.7) and apply Ampere's law, without the displacement current term, in integral form: $\oint_C \mathbf{B} \cdot \mathbf{dl} = \frac{4\pi}{c} \int_{\partial V} \mathbf{J} \cdot \hat{\mathbf{n}} \, da = \frac{4\pi}{c} I_{\text{enc}}$. Here, I_{enc} is the current enclosed by the Amperian loop C or, more precisely, the total current through any open surface ∂V attached to the loop. Now, consider the two surfaces ∂V_1 and ∂V_2 attached to the Amperian loop in Fig. 1.7. While the shaded flat surface ∂V_1 does not enclose a current I, giving $I_{\text{enc}} = 0$, the unshaded curved surface ∂V_2 does and gives $I_{\text{enc}} = I$. However, as discussed earlier, $\oint \mathbf{B} \cdot \mathbf{dl}$ should be the same regardless of the shape of the open surface so long as it is attached to the loop.

In this example, charges accumulate on the metal blobs leading to time-varying electric field in the region between the two blobs. Now if we include the displacement current term and accordingly modify the integral form, we should be able to see that $\oint \mathbf{B} \cdot \mathbf{dl}$ turns out to be the same for the two surfaces. Let us verify this explicitly. The termination of the wire leads to a build-up of charge $\mathcal{Q}(t)$ at the metal blob with time (so that $\frac{dQ}{dt} = I$). Consider the circular loop shown in Fig. 1.7 which has radius R and subtends an angle 2θ with respect to the charge $+Q$. The electric flux through the flat surface attached to the loop is given by

$$
\begin{aligned}
\int_{\partial V_1} \mathbf{E} \cdot \hat{\mathbf{n}} \, da &= \int_{\partial V_1} E \cos \alpha \, da \\
&= Q \int_0^\theta \frac{1}{(l/\cos\alpha)^2} \cos\alpha \, 2\pi \, l \tan\alpha \, d(l \tan\alpha) \\
&= 2\pi Q \int_0^\theta \sin\alpha \, d\alpha = 2\pi Q(1 - \cos\theta).
\end{aligned}
\tag{1.96}
$$

Here, 2α is the angle subtended by the metal blob with charge $+Q$ to the elementary ring shown in the figure. The component of the electric field due to the charge $+Q$ on the elementary ring and along the area vector is $E \cos\alpha$. The electric field E is calculated assuming inverse square law with the distance of the charge $+Q$ from the elementary ring being $(l/\cos\alpha)$. The radius of the ring is $l \tan\alpha$. Therefore, the elementary area is $2\pi \, l \tan\alpha \, d(l \tan\alpha)$.

Note that the total electric flux through the closed surface, consisting of the flat and the curved surface together, will be equal to $4\pi Q$ from Gauss's law. Hence, electric flux through the curved surface ∂V_2 is given by

$$
\begin{aligned}
\int_{\partial V_2} \mathbf{E} \cdot \hat{\mathbf{n}} \, da &= 4\pi Q - 2\pi Q(1 - \cos\theta) \\
&= 2\pi Q(1 + \cos\theta).
\end{aligned}
\tag{1.97}
$$

For ∂V_1, $I_{\text{enc}} = 0$. Hence,

$$
\begin{aligned}
\oint \mathbf{B} \cdot \mathbf{dl} &= \frac{1}{c} \frac{d}{dt} \int_{\partial V_1} \mathbf{E} \cdot \hat{\mathbf{n}} da \\
&= \frac{2\pi}{c} \frac{dQ}{dt}(1 - \cos\theta) = \frac{2\pi}{c} I(1 - \cos\theta).
\end{aligned}
\tag{1.98}
$$

On the other hand, for ∂V_2, $I_{\text{enc}} = I$. Hence,

$$
\begin{aligned}
\oint \mathbf{B} \cdot \mathbf{dl} &= \frac{4\pi}{c} I + \frac{1}{c} \frac{d}{dt} \int_{\partial V_2} \mathbf{E} \cdot \hat{\mathbf{n}} \, da \\
&= \frac{4\pi}{c} I - 2\pi \frac{dQ}{dt}(1 + \cos\theta) \\
&= \frac{2\pi}{c} I(1 - \cos\theta).
\end{aligned}
\tag{1.99}
$$

Thus, for both surfaces, $\oint \mathbf{B} \cdot \mathbf{dl}$ turns out to be the same once we include the contribution due to the displacement current. Note the minus sign for displacement current in the calculation for ∂V_2. The surface area vector is outward normal to ∂V_2 and opposite to the direction of the current, hence the minus sign. We must add that the calculation is oversimplified. We have assumed inverse square law for the charge $+Q(t)$ on one metal blob and have not taken into consideration the effect of the other metal blob with charge $-Q(t)$. However, this is just for illustrative purposes and we can always keep the surface ∂V_2 sufficiently close to $+Q$, so that the effect of the other blob can be neglected.

In practice, it was not so easy to detect displacement current using such a setup in Maxwell's time. One needed to apply extremely high voltage (typically few 10's of kilovolts) across a tiny gap (typically of the order of 1 cm) to have any appreciable effect of displacement current. Heinrich Hertz developed the primitive radio wave transmitter (called the spark-gap transmitter) based on the same principle in the late nineteenth century.

1.5   Maxwell Equations; Scalar and Vector Potentials

Let us now write down all four equations in final form, called the Maxwell equations (in absence of material media) as

$$\nabla \cdot \mathbf{E} = 4\pi\rho. \qquad \textbf{Gauss's law} \qquad (1.100)$$

$$\nabla \cdot \mathbf{B} = 0. \qquad \textbf{No monopoles} \qquad (1.101)$$

$$\nabla \times \mathbf{E} = -\frac{1}{c}\frac{\partial \mathbf{B}}{\partial t}. \qquad \textbf{Faraday's law} \qquad (1.102)$$

$$\nabla \times \mathbf{B} = \frac{4\pi}{c}\mathbf{J} + \frac{1}{c}\frac{\partial \mathbf{E}}{\partial t}. \qquad \textbf{Generalized Ampere's law} \qquad (1.103)$$

We now need to revisit the definitions of the scalar and vector potentials in electrostatics and magnetostatics. Since $\nabla \cdot \mathbf{B} = 0$ is still valid for time-varying fields, we can express the magnetic field $\mathbf{B}$ as curl of vector potential $\mathbf{A}$, as before

$$\mathbf{B} = \nabla \times \mathbf{A}. \qquad (1.104)$$

On the other hand, although Gauss's law is still applicable, curl of electric field is no longer zero for time-varying magnetic fields. Therefore, the electric field cannot be expressed as gradient of scalar potential alone. Replacing $\mathbf{B} = \nabla \times \mathbf{A}$ in Faraday's law, we get

$$\nabla \times \mathbf{E} = -\frac{1}{c}\frac{\partial}{\partial t}(\nabla \times \mathbf{A}). \qquad (1.105)$$

Therefore,

$$\nabla \times \left(\mathbf{E} + \frac{1}{c}\frac{\partial \mathbf{A}}{\partial t}\right) = 0. \qquad (1.106)$$

Since the curl of the quantity within the bracket vanishes, it can be defined as the gradient of scalar potential Φ. It follows that

$$\mathbf{E} = -\nabla\Phi - \frac{1}{c}\frac{\partial\mathbf{A}}{\partial t}. \tag{1.107}$$

What happens to the Poisson equation we built up for electrostatics and magnetostatics (each orthogonal component of vector potential satisfied the Poisson equation.)? Making use of Eq. (1.107) in Gauss's law, we have

$$\nabla^2\Phi + \frac{1}{c}\frac{\partial}{\partial t}(\nabla\cdot\mathbf{A}) = -4\pi\rho. \tag{1.108}$$

Replacing the fields with potentials in generalized Ampere's law, we obtain

$$\nabla\times(\nabla\times\mathbf{A}) = \frac{4\pi}{c}\mathbf{J} + \frac{1}{c}\frac{\partial}{\partial t}\left(-\nabla\Phi - \frac{1}{c}\frac{\partial\mathbf{A}}{\partial t}\right). \tag{1.109}$$

Using the vector identity, $\nabla\times(\nabla\times\mathbf{A}) = \nabla(\nabla\cdot\mathbf{A}) - \nabla^2\mathbf{A}$ on the left-hand side of the above equation, and rearranging the terms, we finally arrive at the following equation:

$$\left(\nabla^2\mathbf{A} - \frac{1}{c^2}\frac{\partial^2\mathbf{A}}{\partial t^2}\right) - \nabla\left(\nabla\cdot\mathbf{A} + \frac{1}{c}\frac{\partial\Phi}{\partial t}\right) = -\frac{4\pi}{c}\mathbf{J}. \tag{1.110}$$

We have already converted the Maxwell equations involving $\mathbf{E}$ and $\mathbf{B}$ fields with three components each into two coupled inhomogeneous equations, Eqs. (1.108) and (1.110), involving scalar potential Φ and the three components of $\mathbf{A}$. But we can do better! We use the arbitrariness associated with the definition of Φ and $\mathbf{A}$ to decouple the two equations. We note that the fields $\mathbf{E}$ and $\mathbf{B}$ remain unchanged under the following simultaneous transformations:

$$\mathbf{A}' = \mathbf{A} - \nabla\chi. \tag{1.111}$$

$$\Phi' = \Phi + \frac{1}{c}\frac{\partial\chi}{\partial t}. \tag{1.112}$$

Here, $\chi = \chi(\mathbf{x}, t)$ is any well-behaved scalar function. We assume that Φ and $\mathbf{A}$ can be adjusted in such a way that the following condition is satisfied:

$$\nabla\cdot\mathbf{A} + \frac{1}{c}\frac{\partial\Phi}{\partial t} = 0. \tag{1.113}$$

This is the so-called 'Lorenz gauge condition'. Under this condition, the two coupled inhomogeneous equations are decoupled, leading to the following inhomogeneous wave equations: inhomogeneous for Φ, and each Cartesian components of $\mathbf{A}$, separately

$$\nabla^2 \Phi - \frac{1}{c^2} \frac{\partial^2 \Phi}{\partial t^2} = -4\pi\rho. \tag{1.114}$$

$$\nabla^2 \mathbf{A} - \frac{1}{c^2} \frac{\partial^2 \mathbf{A}}{\partial t^2} = -\frac{4\pi}{c} \mathbf{J}. \tag{1.115}$$

The striking structural similarity between Eqs. (1.114) and (1.115) suggests that the solution of these equations, namely the potentials Φ and the three Cartesian components of $\mathbf{A}$, will have the same mathematical form. We shall come back to the discussion on the Lorenz gauge later in the book.

1.6   A Short Note on Units and Dimensions

The system of units in classical electrodynamics is determined by the way we write down the fundamental laws of electrodynamics such as Coulomb's law, the Biot-Savart law, Faraday's law and the Lorentz force law. There are primarily two systems of units that we shall discuss in this section: the SI (Systéme International) and the Gaussian system. Although the choice of unit systems is, to a large extent, a matter of convenience nowadays, it is also an outcome of the way the field of electromagnetism progressed over time in history. Throughout the book, we will be using the Gaussian system. However, it is instructive to be accustomed to the SI system as well, particularly because of its growing popularity and extensive use in everyday life.

The units and dimensions of all the physical quantities we encounter in electromagnetism are derived, in the sense that they are defined, both in magnitude and dimension, in terms of the basic dimensions, viz., mass (m), length (l) and time (t).

The unit of charge in the Gaussian system is called the 'esu' or statcoulomb and in the SI system, Coulomb. The Coulomb force law between two charges q_1 and q_2 separated by a distance r in Gaussian system is given by

$$|\mathbf{F}| = \frac{q_1 q_2}{r^2}. \tag{1.116}$$

In terms of basic dimensions, the dimension of the square of charge is $ml^3 t^{-2}$. The unit of charge 1 esu in the Gaussian system is so defined that when two charges of 1 esu each are at a distance 1 cm apart, the charges will exert a force of 1 dyne on one another, i.e., $(1 \text{ esu})^2 = 1 \text{ dyne} \times \text{cm}^2$.

In SI system, the charges are measured in Coulomb (C). The inverse square force law in the SI system is given by

$$|\mathbf{F}| = \frac{1}{4\pi\epsilon_0} \frac{q_1 q_2}{r^2}. \tag{1.117}$$

Here, $\epsilon_0 = 8.854 \times 10^{-12} \, C^2/Nm^2$ is a fundamental constant called the vacuum permittivity. We can easily find out the conversion factor between the Gaussian and SI units of charges by calculating the two equal charges in Coulomb that leads to a force of 1 dyne if the charges are 1cm apart. It turns out that 1 esu $= 3.336 \times 10^{-10} \, C$ or, conversely, 1 C $= 2.998 \times 10^9$ esu.

Similarly, the unit of current, defined as the time rate of charge flow ($I = \frac{dq}{dt}$), is also determined by the force between two infinite current carrying wires unit distance apart as given by Ampere's law or the Biot-Savart law. The unit of current in the Gaussian system is esu/second or statampere. The same in the SI system is Ampere (A) where 1 A $= 3 \times 10^9$ esu/sec or statampere.

The electric field is defined as the force per unit charge and the dimensions can be directly determined from Coulomb's law itself in either system of units. In the Gaussian system, the unit of the electric field is dyne/esu, while the same is Newton/C in the SI system.

The Lorentz force law suggests that the magnetic field has also the same dimension as the electric field in the Gaussian system

$$\mathbf{F} = q \left[\mathbf{E} + \frac{1}{c} (\mathbf{v} \times \mathbf{B}) \right]. \tag{1.118}$$

This is one major advantage of the Gaussian (or the CGS) system over the SI system. The unit of the magnetic field in the Gaussian system is called the Gauss (G), where 1 G $= 1$ dyne/esu. In the SI system, the Lorentz force law is written as

$$\mathbf{F} = q(\mathbf{E} + \mathbf{v} \times \mathbf{B}). \tag{1.119}$$

Therefore, in the SI system, E/B has the dimension of speed. In other words, while the electric field has the same dimension in both systems, the magnetic field is dimensionally different. The magnetic field in the two systems obeys the conversion relation $[\frac{1}{c} B]_G \equiv [B]_{SI}$. The unit of magnetic field in SI system is Tesla, where 1 Tesla $\equiv 10^4$ Gauss. Note that we have deliberately used the 'equivalent' sign instead of 'equality' here since the two units are dimensionally different.

The advantage of the Gaussian system vis-a-vis SI is more apparent when we compare the total energy stored in the electromagnetic field in the two units.

$$U_{em} = \frac{1}{8\pi} \int (E^2 + B^2) \, d^3x. \qquad \text{(Gaussian)} \tag{1.120}$$

$$U_{em} = \frac{1}{2} \int \left(\epsilon_0 E^2 + \frac{1}{\mu_0} B^2 \right) d^3x. \quad \text{(SI)} \tag{1.121}$$

Clearly, the symmetric nature of electromagnetic energy density as observed in the case of Gaussian units is lost in SI units.

Is there a recipe to convert the equations from Gaussian to SI system and vice versa?[2] We present a simple illustration of how to switch from one system to another. Take, for example, the Biot-Savart law in the Gaussian system (the quantities with dimensions are shown within brackets):

$$\frac{1}{[c]}[\mathbf{dB}]_G = \frac{1}{[c]^2}[I]\frac{[\mathbf{dl}] \times [\mathbf{r}]}{[r]^3}. \tag{1.122}$$

The same law in the SI system takes the form

$$[\mathbf{dB}]_{SI} = \frac{[\mu_0]}{4\pi}[I]\frac{[\mathbf{dl}] \times [\mathbf{r}]}{[r]^3}. \tag{1.123}$$

Here, $\mu_0 = 4\pi \times 10^{-7}\,\mathrm{N/A^2}$ is another fundamental constant, called the vacuum permeability. Now let us write down the Biot-Savart law, independent of the two systems, as follows:

$$k_1[\mathbf{dB}] = \frac{k_2}{[c]^2}[I]\frac{[\mathbf{dl}] \times [\mathbf{r}]}{[r]^3}. \tag{1.124}$$

Here, k_1 and k_2 are quantities to be determined for each system separately. By inspection, it is easy to see that $k_1 = 1/c$, $k_2 = 1$ for Gaussian system and $k_1 = 1$, $k_2 = 1/4\pi\epsilon_0$ for SI system, respectively. Here, we have used the fact that in SI system, $\frac{1}{c^2} = \mu_0\epsilon_0$. Thus, given an equation in the Gaussian system, we can move over to the SI system, or vice versa, using the transformation of the fields $(\mathbf{E}, \mathbf{B})$ and the sources $(\rho, \mathbf{J})$, as prescribed below

$$[\mathbf{E}]_G \longrightarrow [\mathbf{E}]_{SI}. \tag{1.125}$$

$$\frac{1}{c}[\mathbf{B}]_G \longrightarrow [\mathbf{B}]_{SI}. \tag{1.126}$$

$$[\rho, \mathbf{J}]_G \longrightarrow \frac{1}{4\pi\epsilon_0}[\rho, \mathbf{J}]_{SI}. \tag{1.127}$$

Let us verify this with the help of an example. We start with generalized Ampere's law in Gaussian units

$$\nabla \times [\mathbf{B}]_G = \frac{4\pi}{c}[\mathbf{J}]_G + \frac{1}{c}\frac{\partial[\mathbf{E}]_G}{\partial t}.$$

We first divide both sides of the above equation by c and then use the transformation for the fields given in Eqs. (1.125) and (1.126) to rewrite the equation in terms of $[\mathbf{B}]_{SI}$ and $[\mathbf{E}]_{SI}$ as an intermediate step. We then replace $[\mathbf{J}]_G$ by $\frac{1}{4\pi\epsilon_0}[\mathbf{J}]_{SI}$ and use the relation $\frac{1}{c^2} = \mu_0\epsilon_0$ to write down the equation in final form in SI system

---

[2] For the interested reader, a generalized treatment regarding the choice of units and dimensions across different systems is given in Jackson's book Classical Electrodynamics, third edition.

$$\nabla \times [\mathbf{B}]_{\text{SI}} = \mu_0 [\mathbf{J}]_{\text{SI}} + \mu_0 \epsilon_0 \frac{\partial [\mathbf{E}]_{\text{SI}}}{\partial t} .$$

In this manner, we can convert all other Maxwell's equations involving fields and sources from the Gaussian system to SI and vice versa. The other equations involving potentials and sources can be transformed similarly, keeping in mind that Φ transforms the same way as $\mathbf{E}$ and $\mathbf{A}$ transforms the same way as $\mathbf{B}$. We list some of the important equations involving fields and potentials in Gaussian and SI systems in Table 1.1. The units of various relevant physical quantities in Gaussian and SI systems along with conversion factors are given in Table 1.2.

Table 1.1 Equations of electrodynamics in Gaussian and SI systems

	(Gaussian)	(SI)
Maxwell's equations	$\nabla \cdot \mathbf{E} = 4\pi\rho$	$\nabla \cdot \mathbf{E} = \frac{\rho}{\epsilon_0}$
	$\nabla \cdot \mathbf{B} = 0$	$\nabla \cdot \mathbf{B} = 0$
	$\nabla \times \mathbf{E} = -\frac{1}{c}\frac{\partial \mathbf{B}}{\partial t}$	$\nabla \times \mathbf{E} = -\frac{\partial \mathbf{B}}{\partial t}$
	$\nabla \times \mathbf{B} = \frac{4\pi}{c}\mathbf{J} + \frac{1}{c}\frac{\partial \mathbf{E}}{\partial t}$	$\nabla \times \mathbf{B} = \mu_0\mathbf{J} + \mu_0\epsilon_0\frac{\partial \mathbf{E}}{\partial t}$
Inhomogeneous wave equations	$\nabla^2\Phi - \frac{1}{c^2}\frac{\partial^2\Phi}{\partial t^2} = -4\pi\rho$	$\nabla^2\Phi - \frac{1}{c^2}\frac{\partial^2\Phi}{\partial t^2} = -\frac{\rho}{\epsilon_0}$
	$\nabla^2\mathbf{A} - \frac{1}{c^2}\frac{\partial^2\mathbf{A}}{\partial t^2} = -\frac{4\pi}{c}\mathbf{J}$	$\nabla^2\mathbf{A} - \frac{1}{c^2}\frac{\partial^2\mathbf{A}}{\partial t^2} = -\mu_0\mathbf{J}$
Equation of continuity	$\nabla \cdot \mathbf{J} + \frac{\partial \rho}{\partial t} = 0$	$\nabla \cdot \mathbf{J} + \frac{\partial \rho}{\partial t} = 0$
Lorentz force law	$\mathbf{F} = q\left[\mathbf{E} + \frac{1}{c}(\mathbf{v} \times \mathbf{B})\right]$	$\mathbf{F} = q(\mathbf{E} + \mathbf{v} \times \mathbf{B})$

Table 1.2 The units and conversion factors of important physical quantities in Gaussian and SI systems

Physical quantity	(SI)	Conversion factor	(Gaussian)
Length (l)	1 meter (m)	10^2	centimeters (cm)
Mass (m)	1 kilogram (kg)	10^3	grams (g)
Time (t)	1 Second (s)	1	Second (s)
Work, energy (W, U)	1 Joule (J)	10^7	ergs
Charge (q)	1 Coulomb	3×10^9	statcoulombs
Charge density (ρ)	1 Coulomb m^{-3}	3×10^3	statcoulomb cm^{-3}
Current (I)	1 Ampere	3×10^9	statamperes
Current density (J)	1 Ampere m^{-2}	3×10^5	statampere cm^{-2}
Potential (Φ)	1 Volt	$\frac{1}{300}$	statvolt
Electric field (E)	1 Volt m^{-1}	$\frac{1}{3} \times 10^{-4}$	statvolt cm^{-1}
Capacitance (C)	1 Farad (F)	9×10^{11}	cm
Magnetic flux (ϕ)	1 Weber	10^8	Gauss cm^{-2}
Magnetic field (B)	1 Tesla	10^4	Gauss
Magnetic H field	1 Amp m^{-1}	$4\pi \times 10^{-3}$	Oersted (Oe)

1.7 Problems

1. The time-averaged potential of a neutral hydrogen atom is given by

$$\Phi = q \frac{e^{-2r/a_0}}{r} \left(1 + \frac{r}{a_0} \right).$$

   Here, q is the magnitude of the electronic charge, and a_0 is the Bohr radius. Find the distribution of charge (both continuous and discrete) that will give this potential and interpret your result physically.

2. Consider three charged spheres of radius R: first, conducting; second, having a uniform charge density within its volume and the third, having a spherically symmetric charge density that varies radially as r^n ($n > -3$). Each has a total charge Q. Use the Gauss law to obtain the electric fields both inside and outside each sphere. Sketch the behavior of the fields as a function of radius for the first two spheres, and for the third with $n = -2, +2$.

3. A long cylindrical wire with inner radius $r = a$ and outer radius $r = b$ carries a uniform current I. Find the magnetic field (i) inside the hollow region ($r < a$); (ii) inside the conducting region ($a < r < b$); (iii) outside the conductor ($r > a$).

4. A large parallel plate capacitor with uniform surface charge σ on the upper plate and $-\sigma$ on the lower plate moves with a constant speed v. Calculate the net magnetic field in the regions above the two plates, in between the two plates and below the two plates. Find out the magnitude and direction of the magnetic force per unit area on the upper plate. At what speed v would the magnetic force balance the electrical force?

5. Find the magnetic vector potential everywhere for the following cases: (i) An infinitely long solenoid of radius R with a constant surface current density $\mathbf{K}$, which is perpendicular to the axis of the solenoid. (ii) An infinite sheet with a uniform surface current $K\hat{\mathbf{x}}$. (iii) A long conducting wire of radius R carrying uniform current I along its axis. (iv) A spherical shell of radius R carrying a uniform surface charge σ and spinning at constant angular velocity ω.

6. A metallic sphere of radius R is moving with a constant velocity $v_0\hat{\mathbf{x}}$ in an uniform magnetic field $B_0\hat{\mathbf{z}}$. Find (i) the electric field inside the sphere; (ii) the induced charge density (surface or bulk) in the sphere; (iii) the electric dipole moment of the charge distribution; (iv) the potential difference between points $y = \pm R$ on the sphere.

7. A conducting rod of mass m slides freely along a rail made of two parallel conducting wires (along $\hat{\mathbf{x}}$) connected together by a resistance R. The two wires are separated along the y-direction by a distance l. The resistor R, the two parallel wires and the rod form a closed rectangular loop, and the whole thing is immersed in a constant magnetic field $-B\hat{\mathbf{z}}$ (pointing into the page). The rod is given an initial velocity $v_0\hat{\mathbf{x}}$ such that the area of the loop increases. Determine the following: (i) Compute the electric current in the loop at any time t. (ii) Compute the rod's displacement $x(t)$ and velocity $v(t)$. (iii) Compute the power

lost through the Joule heating and plot it as a function of time. What is the total power dissipated?

8. Two long coaxial cylindrical shells of radii a and b $(b > a)$ are placed with their axis along the z-direction. A current I goes through the inner shell in the z-direction and returns through the outer shell, the current being uniformly distributed on the surface. If $I = I_0 \exp{-t/\tau}$, find the induced electric field everywhere.

9. Two large plates at $z = \pm\frac{d}{2}$ carry slow time-varying surface currents $K(t)\hat{x}$ and $-K(t)\hat{x}$, respectively. (i) Find the magnetic field everywhere. Justify the use of the laws of magnetostatics in this context. (ii) Find the induced electric field everywhere.

10. A parallel plate capacitor is made of two circular sheets of radius R with a separation $d \ll R$. The capacitor is getting charged at a very slow rate with charge $Q(t) = Q_0\{1 - \exp(-t/t_0)\}$. (i) Plot the charge as a function of time. (ii) Determine and plot the displacement current as a function of time. (iii) Determine the magnetic field in between the plates. (iv) What is the origin of the magnetic field? When and where would the above analysis represent the true fields in between the plates?

11. A parallel plate capacitor with circular plates (radius R) is being charged by a constant current I_C as shown in the figure below. Assume that the electric field inside the capacitor is uniform and neglect fringing fields near the edges. The instantaneous total charge on the plates is $\pm Q(t)$. Calculate the magnetic field **B** for the closed loop C with radius $r < R$ for the two surfaces S_1 (shaded flat surface) and S_2 (the unshaded surface which does not cut any capacitor plate) attached to C as shown in the figure, using the generalized form of Ampere's law (including the displacement current). Calculate the distribution of surface current density flowing in the capacitor plates by applying the same law on surface S_3, which cuts the capacitor plate.

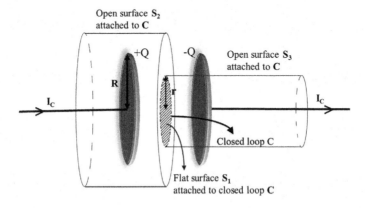

Open surface S_2 attached to C

Open surface S_3 attached to C

+Q -Q

R

r

I_C I_C

Closed loop C

Flat surface S_1 attached to closed loop C

Chapter 2
Boundary Value Problems in Electrostatics

Normally, if the charge distribution $\rho(\mathbf{x}')$ or the current distribution $\mathbf{J}(\mathbf{x}')$ is specified, the problem is essentially solved using the integral solutions for the scalar or the vector potentials given in Chap. 1, respectively. However, there are boundaries present in the form of material media such as conductors, dielectrics or magnetized materials. In some cases, rather than specifying the source charges or currents, the potentials or the fields are given at the boundaries. Thus, it is necessary to introduce new mathematical tools to incorporate such boundary conditions into the solutions of the Poisson or Laplace equations.

In the present chapter, we shall limit our discussion to the problems of electrostatics (with point charges fixed in space or source charge densities having no time variation). We shall be primarily interested in problems with conductors as boundaries, where either the potential Φ or the normal derivative $\frac{\partial \Phi}{\partial n}$ is specified on the boundaries. The normal derivative $\frac{\partial \Phi}{\partial n}$ or component of the electric field normal to the boundary surface E_n is proportional to the surface-charge density σ. In this context, the solution of the Poisson equation using the Green function and method of images will be discussed with illustrative examples. In the end, solutions of the Laplace equation, for a given region with the potential Φ specified on the boundary surface, will be discussed for different coordinate systems.[1]

---

[1] Readers should have elementary knowledge of the Legendre polynomials and the Bessel functions for certain parts of the chapter.

© Springer Nature Singapore Pte Ltd. 2021
K. Bhattacharya and S. Mukhopadhyay, *Introduction to Advanced Electrodynamics*,
https://doi.org/10.1007/978-981-16-7802-8_2

2.1 Boundary Conditions in Electrostatics

The differential forms of Maxwell's equations for electrostatics, $\nabla \cdot \mathbf{E} = 4\pi\rho$ and $\nabla \times \mathbf{E} = 0$, are valid locally at each point in space. As we have seen earlier, these equations can be written in integral form using divergence theorem and Stokes' theorem, respectively, as follows:

$$\oint_{\partial V} \mathbf{E} \cdot \hat{\mathbf{n}}\, da = 4\pi \int_V \rho(\mathbf{x})\, d^3x. \tag{2.1}$$

$$\oint_C \mathbf{E} \cdot \mathbf{dl} = 0 \,. \tag{2.2}$$

Our immediate task is now to apply these equations to find out how the electric fields are related to each other on either side of an idealized surface-charge density σ. The electric field on one side of the surface charge is labeled as $\mathbf{E}_1$ and on the other side as $\mathbf{E}_2$ (Fig. 2.1).

We apply Eq. (2.1) on the infinitesimal Gaussian pillbox as shown in Fig. 2.1. In the limit of the height of the pillbox going to zero, the curved surface does not contribute to the electric flux. Only the top and bottom flat surfaces contribute. In that case, the total electric flux is given by $\oint_{\partial V} \mathbf{E} \cdot \hat{\mathbf{n}}\, da = (\mathbf{E}_2 - \mathbf{E}_1) \cdot \hat{\mathbf{n}}\, \Delta a$ where Δa is the area of the top and bottom surfaces. The total charge enclosed by the Gaussian pillbox is contributed by the idealized surface charge such that $\int_V \rho(\mathbf{x})\, d^3x = \sigma\, \Delta a$. From Eq. (2.1), we have the following relation for the normal component of the electric field on either side of the surface charge:

$$(\mathbf{E}_2 - \mathbf{E}_1) \cdot \hat{\mathbf{n}} = 4\pi\sigma \,. \tag{2.3}$$

Thus, the normal component of the electric field is discontinuous at the surface and the discontinuity is proportional to the surface-charge density σ.

We now apply Eq. (2.2) over the infinitesimal rectangular loop with its longer arms parallel to the interface and on either side of the boundary as shown in Fig. 2.1. The flat surface attached to the loop has the area vector tangential to the boundary. In the limit of the length of smaller arms approaching zero with the length of the

Fig. 2.1 The boundary surface carries idealized surface-charge density σ. The electric field on either side of the boundary is labeled as $\mathbf{E}_1$ and $\mathbf{E}_2$

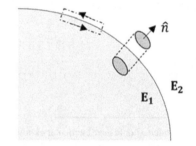

larger arm being Δl, $\oint_C \mathbf{E} \cdot \mathbf{dl} = (E_2{}^t - E_1{}^t)\,\Delta l$, where the superscript t stands for components tangential to the boundary surface. Equation (2.2) then gives $E_2{}^t = E_1{}^t$. A more formal notation using the unit vector $\hat{\mathbf{n}}$ normal to the boundary surface leads to the following relation connecting the tangential components of electric field on either side of the boundary surface:

$$\hat{\mathbf{n}} \times (\mathbf{E}_2 - \mathbf{E}_1) = 0. \tag{2.4}$$

We finally write down the boundary condition for the electrostatic field by combining Eqs. (2.3) and (2.4) as

$$\mathbf{E}_2 - \mathbf{E}_1 = 4\pi\sigma\hat{\mathbf{n}}. \tag{2.5}$$

Here, $\hat{\mathbf{n}}$ is the unit vector normal to the boundary surface, directed from region 1 to region 2.

Let us now find out the corresponding boundary conditions for the scalar potential Φ as it is generally advantageous to calculate the potential first, before finding the electric field for a given charge distribution. The potential difference between two points on either side of the surface charge, but very close to it, is given by

$$\Phi_2 - \Phi_1 = -\int_1^2 \mathbf{E} \cdot \mathbf{dl}. \tag{2.6}$$

As the path length of the line integral approaches zero, the integral itself vanishes. It follows that $\Phi_2 = \Phi_1$. The scalar potential Φ is, thus, continuous across the boundary surface. However, since $\mathbf{E} = -\nabla\Phi$, the gradient of potential Φ will follow the same boundary condition as the electric field $\mathbf{E}$. Substituting the electric fields in Eq. (2.5), we have

$$\nabla\Phi_2 - \nabla\Phi_1 = -4\pi\sigma\hat{\mathbf{n}}. \tag{2.7}$$

We note that $\nabla\Phi \cdot \hat{\mathbf{n}} = \frac{\partial\Phi}{\partial n}$ is the normal derivative of Φ, describing change of Φ along $\hat{\mathbf{n}}$, i.e., the direction normal to the boundary surface. Using this notation, we write down the boundary condition for the normal derivative of the scalar potential as follows:

$$\frac{\partial\Phi_2}{\partial n} - \frac{\partial\Phi_1}{\partial n} = -4\pi\sigma. \tag{2.8}$$

As an illustrative example, we take the case of a conducting sphere (or of any other shape). The region 1 in Fig. 2.1 is now imagined as a conductor. The conductor has charge density σ on the surface, and inside the conductor, the electric field is zero. We want to find out the normal component of the electric field just outside the conductor. We apply Gauss's law in integral form over the tiny pillbox as before, half of which is outside the conductor and the other half inside. Again, in the limit of the height of the pillbox approaching zero, the electric flux across the curved surface is zero. The flux through the bottom flat surface is also zero as $\mathbf{E} = 0$ inside the conductor. If Δa is the area of the top surface of the pillbox, which is infinitesimally close to but outside the conductor, then

$$\oint_{\partial V} \mathbf{E} \cdot \hat{\mathbf{n}} \, da = E_n \, \Delta a = 4\pi \sigma \, \Delta a.$$

Here, E_n is the normal component of the electric field just outside the conductor. We are finally left with the following expression for E_n or, alternatively, the normal derivative of potential Φ, $\frac{\partial \Phi}{\partial n}$, just outside the conductor with surface-charge density σ:

$$E_n = -\frac{\partial \Phi}{\partial n} = 4\pi \sigma. \tag{2.9}$$

Thus, the normal derivative of potential just outside a conductor is proportional to the surface-charge density.

In the next section, we start developing the technique of solving boundary value problems of electrostatics based on the Green function.

2.2 Green's Theorem

The divergence theorem for a well-behaved vector field $\mathcal{A}$ within volume V, enclosed by surface ∂V, is given by

$$\int_V \nabla \cdot \mathcal{A} \, d^3x = \oint_{\partial V} \mathcal{A} \cdot \hat{\mathbf{n}} \, da. \tag{2.10}$$

We define $\mathcal{A}$ in terms of two arbitrary scalar fields ξ and ψ as follows:

$$\mathcal{A} = \xi \nabla \psi. \tag{2.11}$$

Using standard vector identities,

$$\nabla \cdot \mathcal{A} = \nabla \xi \cdot \nabla \psi + \xi \nabla^2 \psi, \tag{2.12}$$

$$\mathcal{A} \cdot \hat{\mathbf{n}} = \xi \nabla \psi \cdot \hat{\mathbf{n}} = \xi \frac{\partial \psi}{\partial n}. \tag{2.13}$$

Here, $\hat{\mathbf{n}}$ is the unit vector outward normal to surface ∂V and $\frac{\partial \psi}{\partial n}$ is the derivative outward normal to the same surface. Substituting Eqs. (2.12) and (2.13) back in Eq. (2.10), we obtain

$$\int_V \left[\nabla \xi \cdot \nabla \psi + \xi \nabla^2 \psi \right] d^3x = \oint_{\partial V} \xi \frac{\partial \psi}{\partial n} \, da. \tag{2.14}$$

We write down the same equation with ξ and ψ interchanged. Then we eliminate the $\nabla \xi \cdot \nabla \psi$ subtracting the equation with interchanged ξ and ψ from Eq. (2.14).

Finally, we arrive at the following equation:

$$\int_V \left[\xi \nabla^2 \psi - \psi \nabla^2 \xi \right] d^3x = \oint_{\partial V} \left[\xi \frac{\partial \psi}{\partial n} - \psi \frac{\partial \xi}{\partial n} \right] da. \tag{2.15}$$

Equations (2.14) and (2.15) are called Green's first and second identities, respectively. Green's second identity is also known as Green's theorem.

Suppose we choose two particular solutions of Poisson's equations $\xi = \Phi$, where Φ is the scalar potential, and $\psi = \frac{1}{|\mathbf{x} - \mathbf{x}'|}$, such that $\nabla^2 \Phi = -4\pi\rho(\mathbf{x})$ and $\nabla^2 \psi = -4\pi\delta(\mathbf{x} - \mathbf{x}')$. As usual, $\mathbf{x}$ is the field point or the observation point and $\mathbf{x}'$ is the position coordinate of the source charges. Replacing these solutions in Green's second identity, we get

$$\int_V \left[\Phi \left\{ -4\pi\delta(\mathbf{x} - \mathbf{x}') \right\} - \frac{1}{|\mathbf{x} - \mathbf{x}'|} \{ -4\pi\rho(\mathbf{x}) \} \right] d^3x =$$
$$\oint_{\partial V} \left[\Phi \frac{\partial}{\partial n} \frac{1}{|\mathbf{x} - \mathbf{x}'|} - \frac{1}{|\mathbf{x} - \mathbf{x}'|} \frac{\partial \Phi}{\partial n} \right] da. \tag{2.16}$$

Now interchanging primed and unprimed variables and evaluating the integral involving delta function, we have

$$\Phi(\mathbf{x}) = \int_V \frac{\rho(\mathbf{x}')}{|\mathbf{x} - \mathbf{x}'|} d^3x' + \frac{1}{4\pi} \oint_{\partial V} \left[\frac{1}{|\mathbf{x} - \mathbf{x}'|} \frac{\partial \Phi}{\partial n'} - \Phi(\mathbf{x}') \frac{\partial}{\partial n'} \frac{1}{|\mathbf{x} - \mathbf{x}'|} \right] da'. \tag{2.17}$$

This simple mathematical exercise leads to the following important observations.

1. The first term on the right-hand side of Eq. (2.17) is the solution of the Poisson equation in absence of boundaries. If $\rho(\mathbf{x}') \neq 0$ and $\frac{\partial \Phi}{\partial n'}$ on the surface falls off faster that $\frac{1}{|\mathbf{x} - \mathbf{x}'|}$, then the surface integral vanishes at infinity and we are left with the familiar solution:

$$\Phi(\mathbf{x}) = \int_V \frac{\rho(\mathbf{x}')}{|\mathbf{x} - \mathbf{x}'|} d^3x'. \tag{2.18}$$

2. For $\rho(\mathbf{x}') = 0$ everywhere within V, the problem reduces to that of solving Laplace's equation and the solution for the potential anywhere inside the volume is readily expressed in terms of a surface integral involving the potential Φ and its normal derivative $\frac{\partial \Phi}{\partial n'}$ (equivalently, the electric fields $-\frac{\partial \Phi}{\partial n'}$ or the surface-charge densities $-\frac{1}{4\pi} \frac{\partial \Phi}{\partial n'}$) specified at the boundary surface as

$$\Phi(\mathbf{x}) = \frac{1}{4\pi} \oint_{\partial V} \left[\frac{1}{|\mathbf{x} - \mathbf{x}'|} \frac{\partial \Phi}{\partial n'} - \Phi(\mathbf{x}') \frac{\partial}{\partial n'} \frac{1}{|\mathbf{x} - \mathbf{x}'|} \right] da'. \tag{2.19}$$

However, simultaneous use of both boundary conditions (Φ and $\frac{\partial \Phi}{\partial n'}$ specified on the boundary surface simultaneously) turns out to be an over specification of the problem, leading to no solution as we shall find out now.

2.3   Uniqueness of Solutions: Dirichlet and Neumann Boundary Conditions

How does one specify the boundary conditions so that a unique solution of the Poisson or Laplace equation is obtained? It turns out that there are two such boundary conditions for which a unique and physically reasonable solution for the potential inside the boundary exists. The boundary conditions are described in the following way.

1. Dirichlet conditions: Here, the potential Φ is specified on the boundary surface (Fig. 2.2 A).
2. Neumann conditions: In this case, the normal derivative of the potential on the boundary surface $\frac{\partial \Phi}{\partial n'}$ or the normal component of the electric field $-\frac{\partial \Phi}{\partial n'}$ on the boundary surface is specified (Fig. 2.2 B).

As we shall see, the specification of either one or the other condition is enough for a unique solution. Simultaneous specification of both boundary conditions does not yield any solution.

We suppose that there exist two solutions of the Poisson equation for the same boundary conditions: Φ_1 and Φ_2. We define $\mathcal{U} = \Phi_1 - \Phi_2$. Since Φ_1 and Φ_2 satisfy Poisson's equation and $\Phi_1 = \Phi_2$ or $\frac{\partial \mathcal{U}}{\partial n} = 0$, according to the Dirichlet or the Neumann boundary conditions, respectively, we immediately obtain $\nabla^2 \mathcal{U} = 0$ inside the volume V and $\mathcal{U} = 0$ on the boundary surface ∂V. Now using Green's first identity for any two arbitrary scalar function (Eq. 2.14) and putting $\xi = \psi = \mathcal{U}$, we obtain

$$\int_V \left[\nabla \mathcal{U} \cdot \nabla \mathcal{U} + U \nabla^2 \mathcal{U} \right] d^3x = \oint_{\partial V} \mathcal{U} \frac{\partial \mathcal{U}}{\partial n} \, da. \tag{2.20}$$

The closed surface integral on the right-hand side goes to zero under the Dirichlet ($\mathcal{U} = 0$) and the Neumann ($\frac{\partial \mathcal{U}}{\partial n} = 0$) boundary conditions, separately. On the left-hand side, we have $\nabla^2 \mathcal{U} = 0$ everywhere within the volume V. Therefore, we finally have, for both boundary conditions,

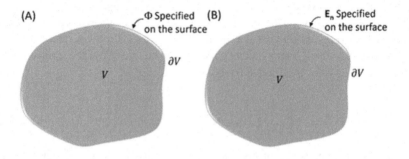

Fig. 2.2   **A** Dirichlet and **B** Neumann boundary conditions

$$\int_V |\nabla \mathcal{U}|^2 \, d^3x = 0. \tag{2.21}$$

Consequently, $\nabla \mathcal{U} = 0$. This suggests that $\mathcal{U}$ is constant throughout the volume V. For the Dirichlet condition, we have already specified $\mathcal{U} = 0$ on the boundary surface ∂V. It follows that $\mathcal{U} = 0$ everywhere inside the volume V. Hence, $\Phi_1 = \Phi_2$ everywhere and the solution is unique. Similarly, for the Neumann condition, the solution will be unique everywhere as well, apart from an arbitrary additive constant. There will be a unique solution even in the case of mixed boundary condition, with the Dirichlet and the Neumann conditions specified exclusively over parts of the boundary surface, i.e., the Dirichlet condition specified on a subsurface ∂V_1 and the Neumann condition specified on the subsurface ∂V_2 (with the subsurfaces ∂V_1 and ∂V_2 constituting the closed surface ∂V).

Now we consider the so-called Cauchy boundary condition where both Φ and $\frac{\partial \Phi}{\partial n}$ are simultaneously and arbitrarily specified on the boundary surface ∂V. Let us, for the time being, ignore the Neumann boundary condition specified on ∂V and solve the corresponding Poisson equation subject to the Dirichlet boundary conditions. As we have already convinced ourselves, the resulting solution is unique. From this unique solution, one can evaluate $\frac{\partial \Phi}{\partial n}$ everywhere in V and take the limiting value of $\frac{\partial \Phi}{\partial n}$ at the surface ∂V. However, since $\frac{\partial \Phi}{\partial n}$ is already specified on the surface, we should not, in general, expect the limiting values to be consistent with the original Neumann boundary conditions. We conclude that the problem is over-specified or the solution is over-determined so that no unique solution exists for the Poisson equation with the Cauchy boundary conditions on the closed surface ∂V.

2.4 Formal Solution of the Poisson Equation Using the Green Function

As discussed earlier, the potential of a unit point charge is $\frac{1}{|\mathbf{x}-\mathbf{x}'|}$ for a system with no boundaries. It satisfies the following equation:

$$\nabla^2 \frac{1}{|\mathbf{x} - \mathbf{x}'|} = -4\pi \delta (\mathbf{x} - \mathbf{x}') . \tag{2.22}$$

Now consider the following expression for the potential at $\mathbf{x}$ due to a continuous charge distribution $\rho(\mathbf{x}')$:

$$\Phi(\mathbf{x}) = \int \frac{1}{|\mathbf{x} - \mathbf{x}'|} \rho(\mathbf{x}') \, d^3x' . \tag{2.23}$$

In this case, $\frac{1}{|\mathbf{x}-\mathbf{x}'|}$ acts like a 'response function', describing how the source charge $\rho(\mathbf{x}') \, d^3x'$ at $\mathbf{x}'$ generates a 'response' at $\mathbf{x}$, by contributing to the potential $\Phi(\mathbf{x})$.

Alternatively, the right-hand side of the equation can be viewed as an integral operator converting $\rho(\mathbf{x}')$ into $\Phi(\mathbf{x})$.

The function $\frac{1}{|\mathbf{x}-\mathbf{x}'|}$ is only one of a class of functions, whose Laplacian is a delta function. Such functions are called the Green functions $G(\mathbf{x}, \mathbf{x}')$ and are defined as follows:

$$\nabla^2 G(\mathbf{x}, \mathbf{x}') = -4\pi\delta(\mathbf{x} - \mathbf{x}'). \tag{2.24}$$

The solution to the Poisson equation for a unit point source at $\mathbf{x}'$ is a Green function. Now if we sum the solution of the Poisson equation due to many such point sources acting simultaneously, then in the continuous limit, we have the following integral expression for the potential at $\mathbf{x}$:

$$\Phi(\mathbf{x}) = \int G(\mathbf{x}, \mathbf{x}')\,\rho(\mathbf{x}')\,d^3x'. \tag{2.25}$$

For solving boundary value problems, we need to be a bit more elaborate in our definition of the Green function. We now define

$$G(\mathbf{x}, \mathbf{x}') = \frac{1}{|\mathbf{x} - \mathbf{x}'|} + F(\mathbf{x}, \mathbf{x}'), \tag{2.26}$$

where the function $F(\mathbf{x}, \mathbf{x}')$ satisfies the Laplace equation inside the volume V, $\nabla^2 F(\mathbf{x}, \mathbf{x}') = 0$.

Green's second identity or Green's theorem described by Eq. (2.15) or Eq. (2.17) is not a solution satisfying the correct boundary conditions as both Φ (Dirichlet) and $\frac{\partial \Phi}{\partial n}$ (Neumann) terms are present in the surface integral. Our task is now to utilize the Green function $G(\mathbf{x}, \mathbf{x}')$, especially the restricted freedom associated with the function $F(\mathbf{x}, \mathbf{x}')$ to eliminate either the Dirichlet or the Neumann term in the surface integral. In this sense, the function $F(\mathbf{x}, \mathbf{x}')$ represents the required rearrangements of charges on the boundary to arrive at the desired boundary condition at ∂V and thereby affect the potential within V. We can now derive an equation that gives $\Phi(\mathbf{x})$ for a given proper boundary condition.

We replace $\psi = G(\mathbf{x}, \mathbf{x}')$ in Eq. (2.15) or Eq. (2.17) and write

$$\Phi(\mathbf{x}) = \int_V G(\mathbf{x}, \mathbf{x}')\rho(\mathbf{x}')\,d^3x' + \frac{1}{4\pi} \oint_{\partial V} \left[G(\mathbf{x}, \mathbf{x}')\frac{\partial \Phi}{\partial n'} - \Phi(\mathbf{x}')\frac{\partial}{\partial n'}G(\mathbf{x}, \mathbf{x}') \right] da'. \tag{2.27}$$

The restriction on the Green function will depend on the nature of the boundary condition that needs to be satisfied. Alternatively, we redefine $G(\mathbf{x}, \mathbf{x}')$ in Eq. (2.27) so as to satisfy either the Dirichlet or the Neumann boundary condition.

In order to satisfy the Dirichlet condition, we need to eliminate the $\frac{\partial \Phi}{\partial n'}$ term in the surface integral in Eq. (2.27). This is easily achieved if we impose the condition

$$G(\mathbf{x}, \mathbf{x}') = 0 \tag{2.28}$$

on the surface ∂V and force the solution $\Phi(\mathbf{x})$ to only depend on the $\Phi(\mathbf{x}')$ values at the surface ∂V, as required by the Dirichlet boundary condition

$$\Phi(\mathbf{x}) = \int_V G(\mathbf{x}, \mathbf{x}')\rho(\mathbf{x}')\, d^3x' - \frac{1}{4\pi} \oint_{\partial V} \Phi(\mathbf{x}')\frac{\partial}{\partial n'}G(\mathbf{x}, \mathbf{x}')\, da'. \qquad (2.29)$$

For the Neumann boundary condition to be satisfied, we require that the surface integral in Eq. (2.27) only depends on the $\frac{\partial\Phi}{\partial n'}$ term. However, it turns out that we cannot have $\frac{\partial}{\partial n'}G(\mathbf{x}, \mathbf{x}') = 0$ on the surface ∂V. Let us verify the statement first. We have, within volume V,

$$\nabla^2 G(\mathbf{x}, \mathbf{x}') = -4\pi\,\delta(\mathbf{x} - \mathbf{x}').$$

Taking the volume integral on both sides, we obtain

$$\int_V \nabla^2 G(\mathbf{x}, \mathbf{x}')\, d^3x' = -4\pi \int_V \delta(\mathbf{x} - \mathbf{x}')\, d^3x' = -4\pi.$$

Finally, we apply Gauss's divergence theorem to convert the volume integral into an integral over surface ∂V, enclosing the volume V and write

$$\int_V \nabla^2 G(\mathbf{x}, \mathbf{x}')\, d^3x' = \oint_{\partial V} \nabla G(\mathbf{x}, \mathbf{x}') \cdot \hat{\mathbf{n}}'\, da = \oint_{\partial V} \frac{\partial}{\partial n'}G(\mathbf{x}, \mathbf{x}')\, da = -4\pi.$$

Clearly, $\frac{\partial}{\partial n'}G(\mathbf{x}, \mathbf{x}') \neq 0$. However, if the surface area is S, we are free to define an average value

$$\frac{\partial}{\partial n'}G(\mathbf{x}, \mathbf{x}') = \frac{-4\pi}{S}, \qquad (2.30)$$

for all $\mathbf{x}'$ over the surface ∂V. Plugging the average value of $\frac{\partial}{\partial n'}G(\mathbf{x}, \mathbf{x}')$ in Eq. (2.27), we arrive at the following expression of the potential under the Neumann boundary condition

$$\Phi(\mathbf{x}) = \int_V G(\mathbf{x}, \mathbf{x}')\rho(\mathbf{x}')\, d^3x' + \frac{1}{4\pi} \oint_{\partial V} G(\mathbf{x}, \mathbf{x}')\frac{\partial\Phi}{\partial n'}\, da' + \langle\Phi\rangle_{\partial V}. \qquad (2.31)$$

Here, $\langle\Phi\rangle_{\partial V} = \frac{1}{S}\oint_{\partial V} \Phi(\mathbf{x}')\, da'$ is the average value of Φ over the surface ∂V, acting as a trivial additive constant in the overall solution. If the region of interest V is bounded by two surfaces with one closed surface having finite surface area and the other pushed to infinity, the effective area S of the surface enclosing the region becomes infinite and $\langle\Phi\rangle_{\partial V}$ vanishes.

Without even knowing the concrete form of the Dirichlet or the Neumann boundary conditions, such as $\rho(\mathbf{x}')$, $\Phi(\mathbf{x}')$ or $\frac{\partial\Phi}{\partial n'}$, we have reduced the problem into simple boundary conditions given by Eq. (2.28) and Eq. (2.30), respectively, using the Green function. Thus. the solution for Φ can now be obtained, for a given geometry and boundary condition, with a suitable choice of the Green function, for any source charge distribution $\rho(\mathbf{x}')$ within volume V and knowledge of potential $\Phi(\mathbf{x}')$ on the corresponding closed surface ∂V (or the component of electric field normal to the surface). However, it turns out that finding the Green function for a given geometry is not a simple task.

What is the Physical meaning of $F(\mathbf{x}, \mathbf{x}')$? It is the solution of the Laplace equation within volume V, and hence can be interpreted as potential due to the charge distribution external to V. This charge distribution outside V is chosen such that either the boundary condition given by Eq. (2.28) (zero potential) or Eq. (2.30) (zero normal derivative) is satisfied when combined with the potential due to the actual source charge distribution within V. Finding $F(\mathbf{x}, \mathbf{x}')$ is, thus, equivalent to finding the so-called 'image charges' in the method of images, to be discussed in the next section.

2.5 Method of Images

We have discussed the formal solution of the scalar potential in the Poisson equation in presence of boundaries in terms of constructing a suitable Green function obeying specific boundary conditions (either the Dirichlet or the Neumann boundary conditions). The choice of the Green function will depend crucially on the geometry of the problem. The method of images is closely related to the Green function technique. It makes use of 'image charges' to reproduce the desired boundary condition and allows the construction of the Green function by physical reasoning.

Consider a region of interest V having a collection of point charges. The boundary conditions are specified on the surface ∂V enclosing the charges. For example, the boundary surfaces could be conductors either grounded or held at constant potential (the Dirichlet condition). The method of images consists of two following generic components:

1. Construct a collection of 'image charges' suitably placed outside the region of interest V and adjusted in sign and magnitude such that the required boundary condition on ∂V is satisfied. The potential due to the image charges placed outside the region of interest satisfies the Laplace equation inside the region of interest.
2. Now remove the boundary in the original problem to include the 'image charges' and calculate the potential within the region of interest due to the 'actual' and the 'image' charges combined.

Let us illustrate the method with a simple example: a point charge near a grounded conducting plane (Fig. 2.3A). The conductor is at a potential $\Phi = 0$. By symmetry, we place an image charge $-q$ at exactly the same distance from the conducting plane as the original charge in the region of interest and get rid of the conductor altogether (Fig. 2.3B). Clearly, the potential at the boundary between the conductor (in the original problem) and the vacuum is zero, as required by the boundary condition

(A)   (B)

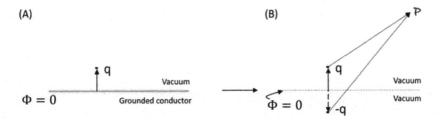

Fig. 2.3 **A** Original problem and **B** Image problem

given. Now we can calculate the potential at any point in the region of interest (the vacuum), due to the charges $+q$ and $-q$ (Fig. 2.3B), and the problem is essentially solved. Once we calculate the potential Φ in the region of interest, it is straightforward to calculate other quantities such as electric field $\mathbf{E} = -\nabla\Phi$ anywhere in the region of interest, the surface-charge density on the conductor $\sigma = -\frac{1}{4\pi}\frac{\partial\Phi}{\partial n}$, which is at the boundary of the region of interest, and consequently, the total charge 'induced' on the conducting plane.

2.5.1 Point Charge Near a Conducting Sphere

Let us now explicitly solve a slightly more difficult problem of a point charge near a 'grounded' conducting sphere of radius R, centered at the origin (Fig. 2.4). Suppose the charge q is placed at a point $\mathbf{y}$ outside the sphere. The image charge needs to be placed outside the region of interest, i.e., anywhere inside the sphere. The sign and magnitude of the image charge q' and its location $\mathbf{y}'$ should be such that the potential $\Phi(\mathbf{x})$ is zero on the surface of the sphere, i.e., for $|\mathbf{x}| = R$. By symmetry, we should place the image charge q' somewhere inside the sphere on the straight line connecting the origin and the charge q. Or in other words, the position vectors $\mathbf{y}$ and $\mathbf{y}'$ should be parallel to each other.

Fig. 2.4 The image problem of a charge near a conducting sphere

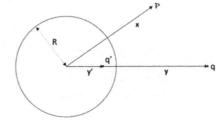

The potential due to the charges q and q' anywhere outside the sphere is given by

$$\Phi(\mathbf{x}) = \frac{q}{|\mathbf{x} - \mathbf{y}|} + \frac{q'}{|\mathbf{x} - \mathbf{y}'|}. \tag{2.32}$$

We now choose q' such that the required boundary condition $\Phi(\mathbf{x}) = 0$ for $|\mathbf{x}| = R$ is satisfied

$$\frac{q'}{|\mathbf{x} - \mathbf{y}'|} = -\frac{q}{|\mathbf{x} - \mathbf{y}|}. \tag{2.33}$$

Therefore,

$$\frac{q'^2}{q} = \frac{(\mathbf{x} - \mathbf{y}')^2}{(\mathbf{x} - \mathbf{y})^2}. \tag{2.34}$$

If θ is the angle between $\mathbf{x}$ and $\mathbf{y}$ (or $\mathbf{y}'$), then we have for $|\mathbf{x}| = R$,

$$2R\left[y' - \left(\frac{q'}{q}\right)^2 y\right]\cos\theta + \left(\frac{q'}{q}\right)^2 (R^2 + y^2) - (R^2 + y'^2) = 0. \tag{2.35}$$

The equation is satisfied for arbitrary values of θ only if the following relations are satisfied simultaneously:

$$\frac{y'}{y} = \left(\frac{q'}{q}\right)^2. \tag{2.36}$$

$$\left(\frac{q'}{q}\right)^2 (R^2 + y^2) = (R^2 + y'^2). \tag{2.37}$$

From Eq. (2.36), we obtain (keeping in mind that image charge should be negative)

$$q' = -q\sqrt{\frac{y'}{y}}. \tag{2.38}$$

Equation (2.37) is quadratic in y' and can be easily factored out as follows:

$$\left(y' - \frac{R^2}{y}\right)(y' - y) = 0. \tag{2.39}$$

It follows that $y' \neq y$, as the image charge and the actual charge cannot generally coincide. Therefore, the position of the image charge is given by

$$y' = \frac{R^2}{y}. \tag{2.40}$$

From Eqs. (2.38) and (2.40), we finally obtain the image charge

$$q' = -q\frac{R}{y}. \tag{2.41}$$

Clearly, if the charge q is brought closer to the sphere or if y is reduced, the image charge q' grows in magnitude and moves away from the center of the sphere toward the surface. When the charge q is just outside the sphere, i.e., $y \simeq R$, the magnitude of image charge is equal to q, and the location, just inside the sphere. Here, the region of interest is any point outside the sphere with boundaries on the surface of the sphere and at infinity. The potential at any point outside the sphere and on the surface of the sphere is simply that due to the image charge q' at $\mathbf{y}'$ and the charge q at $\mathbf{y}$, and its form is

$$\Phi(\mathbf{x}) = \frac{q}{\sqrt{x^2 + y^2 - 2xy\cos\theta}} + \frac{q'}{\sqrt{x^2 + y'^2 - 2xy'\cos\theta}}. \tag{2.42}$$

The total 'induced' surface charge on the sphere must be q'. The charge density σ on the surface of the sphere is given by the normal derivative of Φ on the surface $|\mathbf{x}| = R$. Here, the unit normal to the surface is $\hat{\mathbf{n}} \equiv \hat{\mathbf{x}}$. Therefore,

$$\sigma = -\frac{1}{4\pi}\left(\frac{\partial\Phi}{\partial x}\right)_{|\mathbf{x}|=R} \tag{2.43}$$

$$= \frac{1}{4\pi}\left[\frac{q(R - y\cos\theta)}{(R^2 + y^2 - 2Ry\cos\theta)^{3/2}} + \frac{q'(R - y'\cos\theta)}{(R^2 + y'^2 - 2Ry'\cos\theta)^{3/2}}\right]. \tag{2.44}$$

Here θ is the angle between $\mathbf{x}$ and $\mathbf{y}$ (or $\mathbf{y}'$). Now we replace q' and y' in Eq. (2.44), using Eqs. (2.40) and (2.41). After simple algebraic manipulation, we obtain

$$\sigma = -\frac{q}{4\pi R^2}\left(\frac{R}{y}\right)\frac{\left(1 - \frac{R^2}{y^2}\right)}{\left(1 + \frac{R^2}{y^2} - 2\frac{R}{y}\cos\theta\right)^{3/2}}. \tag{2.45}$$

Since $y \geq R$, the surface-charge density is negative, as expected (Fig. 2.5).

Now it is easy to extend the problem to the case of point charge q near a charged conducting sphere of total charge Q. The potential on the surface of the sphere is now non-zero but constant. The way to tackle this problem using the method of images is as follows. First, find the image charge q' such that the potential on the surface of the sphere is zero. Second, find another image charge q'' which leads to the constant potential on the surface. We have already solved the first part of the problem. The location and magnitude of q' are given by Eq. (2.40) and Eq. (2.41), respectively. It is straightforward to find the image charge q''. By symmetry, it should be located at the center of the sphere in the original problem. Moreover, since q' in effect leads to an induced total surface charge on the sphere of the same amount, the magnitude of

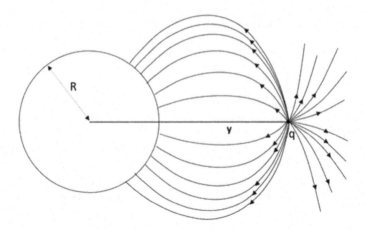

Fig. 2.5 Electric field lines for the positive charge q near a grounded conducting sphere

q'' should be such that the net charge on the boundary of what was a sphere in the original problem remains Q. Clearly, $q'' = Q - q'$. We finally use the principle of superposition to find the potential, at any point outside the sphere and on the spherical boundary surface ($|\mathbf{x}| \geq R$), as

$$\Phi(\mathbf{x}) = \frac{q}{|\mathbf{x} - \mathbf{y}|} + \frac{q'}{|\mathbf{x} - \mathbf{y}'|} + \frac{q''}{\mathbf{x}} \tag{2.46}$$

$$= \frac{q}{|\mathbf{x} - \mathbf{y}|} - \frac{q\frac{R}{y}}{|\mathbf{x} - \frac{R^2}{y^2}\mathbf{y}|} + \frac{Q + q\frac{R}{y}}{\mathbf{x}} . \tag{2.47}$$

The total 'induced' surface-charge density is that due to the image charge q', which we have already calculated in Eq. (2.45), plus that due to q'', which is $\frac{q''}{4\pi R^2}$, uniformly distributed on the sphere

$$\sigma = -\frac{q}{4\pi R^2}\left(\frac{R}{y}\right) \frac{\left(1 - \frac{R^2}{y^2}\right)}{\left(1 + \frac{R^2}{y^2} - 2\frac{R}{y}\cos\theta\right)^{3/2}} + \frac{Q + q\frac{R}{y}}{4\pi R^2} . \tag{2.48}$$

The force on q is calculated by applying the Coulomb force law on the pairs q, q' and q, q'', respectively. Note that both forces are along $\mathbf{y}$-direction and they add up as

$$\mathbf{F} = \left[\frac{q \cdot \left(-q\frac{R}{y}\right)}{\left(y - \frac{R^2}{y}\right)^2} + \frac{q \cdot \left(Q + q\frac{R}{y}\right)}{y^2}\right]\hat{\mathbf{y}} = \frac{q^2}{y^2}\left[\frac{Q}{q} - \left(\frac{R}{y}\right)^3 \frac{2 - \frac{R^2}{y^2}}{\left(1 - \frac{R^2}{y^2}\right)^2}\right]\hat{\mathbf{y}} . \tag{2.49}$$

Fig. 2.6 Force between a point charge and a charged conducting sphere. The dimensionless force F_r as a function of dimensionless distance y_r between the charge and the sphere. At small distances between the point charge and the sphere, the force is attractive

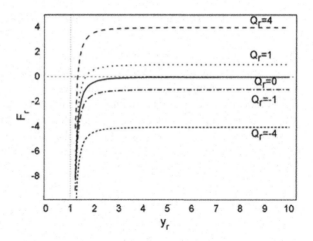

The force follows inverse square law for $y \gg R$, as at large distances, the interaction is effectively between two point charges q and Q. As y becomes smaller, the induced surface-charge distribution on the sphere comes into play. We plot the dimensionless force F_r in unit of $\frac{q^2}{y^2}$ as a function of dimensionless distance $y_r = \frac{y}{R}$ of the charge q from the sphere for different values of $\frac{Q}{q}$ (Fig. 2.6). Several interesting features are observed as follows:

1. If Q is negative, the force is negative (attractive) for all distances.
2. If Q is zero, we recover the case of charge in front of a grounded conducting sphere.
3. If $Q > 0$, there is always a crossover from positive, repulsive force at large distances to negative, attractive force at small distances.

Thus, whatever may be the value of Q, the force is always attractive at small distances. The unstable fixed point or the point of unstable equilibrium corresponding to $F_r = 0$ shifts closer to the sphere as Q is increased although the asymptotic approach to the fixed point starts at roughly the same distance from the sphere, regardless of the charge Q.

The work function of a metal can be partly attributed to the same phenomenon discussed above. As soon as a free charge is sought to be removed from the metal, the problem reduces to the same category as that of a point charge near a conductor with a non-zero net charge. As a result, there will be an attractive force on the charge at short range due to the image charge induced on the metal. Work needs to be done against this attractive force to remove the free charge on the metal.

2.5.2 Conducting Sphere in Uniform Electric Field

The problem of conducting sphere in a uniform electric field can be solved very easily now with the method of images. We assume that the uniform electric field is produced by an appropriate combination of two charges with opposite signs at an appropriate distance. Let us try to find out the magnitude and location of the charges which lead to a uniform electric field around the conducting sphere.

We place a charge $-Q$ at $\mathbf{y}$ and another charge $+Q$ at $-\mathbf{y}$ outside the spherical region as shown in Fig. 2.7. The electric field at an arbitrary point $\mathcal{P}$ due to the combination of the two charges is given by

$$\mathbf{E}(\mathbf{x}) = Q \left[\frac{\mathbf{x} + \mathbf{y}}{|\mathbf{x} + \mathbf{y}|^3} - \frac{\mathbf{x} - \mathbf{y}}{|\mathbf{x} - \mathbf{y}|^3} \right]. \tag{2.50}$$

If the charges are far away from the spherical region and from the observation point, i.e., if $|\mathbf{y}| \gg R, |\mathbf{x}|$, we have $\frac{1}{|\mathbf{x} \pm \mathbf{y}|^3} = \frac{1}{(x^2 + y^2 \pm 2\mathbf{x} \cdot \mathbf{y})^{\frac{3}{2}}} = \frac{1}{y^3} \frac{1}{(1 + \frac{x^2}{y^2} \pm 2\frac{x}{y}\hat{\mathbf{x}} \cdot \hat{\mathbf{y}})^{\frac{3}{2}}} \approx \frac{1}{y^3}$. As a result, the electric field becomes effectively independent of $\mathbf{x}$ in a limited region where $\mathbf{E}(\mathbf{x}) = \frac{2Q}{y^2}\hat{\mathbf{y}}$.[2] Now suppose a conducting sphere is placed at the origin. The image charges corresponding to the point charges $\pm Q$ located at $\mathbf{y} = \mp r'\hat{\mathbf{y}}$ are $\mp \frac{QR}{r'}$, located at $\mp y = \frac{R^2}{r'}$, respectively. The potential due to all four charges anywhere in the region of interest at a distance $|\mathbf{x}| = r$ from the origin is given by

$$\Phi(\mathbf{r}) = \frac{Q}{\sqrt{r^2 + r'^2 + 2rr'\cos\theta}} - \frac{Q}{\sqrt{r^2 + r'^2 - 2rr'\cos\theta}}$$
$$+ \frac{QR/r'}{\sqrt{r^2 + \left(\frac{R^2}{r'}\right)^2 - 2r\frac{R^2}{r'}\cos\theta}} - \frac{QR/r'}{\sqrt{r^2 + \left(\frac{R^2}{r'}\right)^2 + 2r\frac{R^2}{r'}\cos\theta}}. \tag{2.51}$$

The charges $\pm Q$ are far away from the sphere so that $\frac{R}{r'} \ll 1$. We also assume that the observation point is sufficiently close to the sphere such that $\frac{r}{r'} \ll 1$. Keeping this in mind, we factor out r'^2 from the denominators of the first two terms and r^2 from the denominators of the last two terms on the right-hand side of Eq. (2.51). After expanding each of them and keeping only the lower order terms (so that terms with factors Q/r'^n, where $n > 2$, are excluded), we arrive at the following expression:

$$\Phi(\mathbf{r}) = \frac{Q}{r'}\left(1 - \frac{r}{r'}\cos\theta\right) - \frac{Q}{r'}\left(1 + \frac{r}{r'}\cos\theta\right)$$
$$+ \frac{QR}{rr'}\left(1 + \frac{R^2}{rr'}\cos\theta\right) - \frac{QR}{rr'}\left(1 - \frac{R^2}{rr'}\cos\theta\right)$$
$$= -\frac{2Q}{r'^2}\left(r - \frac{R^3}{r^2}\right)\cos\theta. \tag{2.52}$$

---

[2] The result is exact in the limit $Q, |\mathbf{y}| \to \infty$.

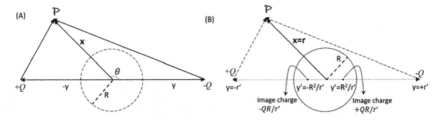

Fig. 2.7 **A** Two charges $\pm Q$ placed far away from the dashed circular region. **B** In presence of a conducting sphere placed in the dashed circular region shown in **A**, image charges are produced

As discussed earlier, in the limit $r' \gg R$, we identify $\frac{2Q}{r'^2}$ as the magnitude of the uniform electric field $E_0\hat{\mathbf{y}}$, around the region where the sphere is located. Therefore,

$$\Phi(\mathbf{r}) = -E_0 \left(r - \frac{R^3}{r^2} \right) \cos \theta. \tag{2.53}$$

The first term $-E_0 r \cos \theta$ is the potential corresponding to the uniform electric field E_0, while the second term is the potential due to the image charges. The induced surface-charge density on the sphere is proportional to the normal derivative of the potential on the surface and is given as

$$\sigma = -\frac{1}{4\pi} \left(\frac{\partial \Phi}{\partial r} \right)_{r=R} = \frac{E_0}{4\pi} \left[1 + \left(\frac{2R^3}{r^3} \right)_{r=R} \right] \cos \theta = \frac{3E_0}{4\pi} \cos \theta. \tag{2.54}$$

Note that the image charges $\pm \frac{QR}{r'}$ separated by a distance $\frac{2R^2}{r'}$ constitute an electric dipole of moment $p = \frac{QR}{r'} \cdot \frac{2R^2}{r'} = E_0 R^3$ and the potential due to the dipole is $\frac{p \cos \theta}{r^2}$. The total induced surface charge (obtained by integrating σ over the surface of the sphere) is zero, although the sphere is not necessarily grounded.

2.6 Construction of the Green Function from Images

Let us now try to understand the image problems of Sect. 2.5 in terms of the Green functions. The potential due to the unit point charge and its images are, in fact, the Green function satisfying either one of the boundary conditions given by Eqs. (2.28) and (2.30). We consider the example of a unit point charge at $\mathbf{y}$ near a grounded conducting sphere of radius R, with the center of the sphere at the origin and the image charge q' located at $\mathbf{y}'$ inside the sphere, similar to what is shown in Fig. 2.4. Clearly, the potential $\Phi(\mathbf{x})$ anywhere outside the sphere satisfies the following equation:

$$\nabla^2 \Phi(\mathbf{x}) = -4\pi \delta(\mathbf{x} - \mathbf{y}), \tag{2.55}$$

Fig. 2.8 The center of the
conducting sphere is at the
origin. The angle between
the position vector of charge
$q = 1$ given by $\mathbf{x}'$ and that of
the observation point $\mathcal{P}$
given by $\mathbf{x}$ is γ, whereas the
angles subtended by $\mathbf{x}$ and $\mathbf{x}'$
on the z-axis are θ and θ',
respectively (not shown in
the figure)

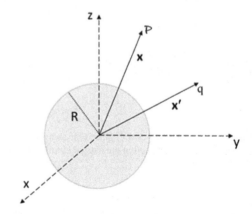

with the boundary condition on the surface of the sphere ($|\mathbf{x}| = R$) being $\Phi(\mathbf{x}) = 0$.
These are, in fact, the same conditions that define the Green function for this problem.
We now refer back to the potential in Eq. (2.32) for the image problem of similar
geometry discussed in Sect. 2.5 and use the notation developed in Sect. 2.4 (replace
y by x' and y' by x'') to write down the Green function for the present problem:

$$G(\mathbf{x}, \mathbf{x}') = \frac{1}{|\mathbf{x} - \mathbf{x}'|} + \frac{q'}{|\mathbf{x} - \mathbf{x}''|}. \tag{2.56}$$

Here, $\mathbf{x}'$ is the location of the unit source charge, and $\mathbf{x}$ is the observation point
(Fig. 2.8). On the other hand, x'' is the location of the image charge q'. We also have
$q' = -\frac{R}{|x'|} = -\frac{R}{x'}$ and $x'' = \frac{R^2}{x'^2}\mathbf{x}'$. Therefore,

$$G(\mathbf{x}, \mathbf{x}') = \frac{1}{|\mathbf{x} - \mathbf{x}'|} - \frac{R/x'}{|\mathbf{x} - \frac{R^2}{x'^2}\mathbf{x}'|}. \tag{2.57}$$

The Green function $G(\mathbf{x}, \mathbf{x}')$ gives the response $\Phi(\mathbf{x})$ observed at $\mathbf{x}$ due to the unit
source charge at $\mathbf{x}'$ near the grounded conducting sphere. It can be easily verified
that $G(\mathbf{x}, \mathbf{x}')$ satisfies the required Dirichlet boundary condition on the surface of the
sphere: $G(\mathbf{x}, \mathbf{x}')_{|\mathbf{x}|=R} = 0$. Note that the Green function is symmetric under inter-
change of $\mathbf{x}$ and $\mathbf{x}'$[3] and hence we could as well put $|\mathbf{x}'| = R$ to verify the boundary
condition.

The solution for Φ for the Green function satisfying the Dirichlet condition is
given by Eq. (2.29). In general, if we are given a different boundary condition, such
as, for example, the potential Φ on the surface of the sphere being non-zero or not
necessarily constant, then we need the same Green function to calculate the normal
derivative $\frac{\partial G}{\partial n'}$ on the surface of the sphere. Here, $\mathbf{n}'$ is outward drawn unit normal
to the boundary surface. Since the region of interest is outside the sphere and the

---

[3] This is a generic property of the Green function.

boundary is on the surface of the sphere, $\mathbf{n}'$ is directed radially toward the center of the sphere.

Let the angle between $\mathbf{x}$ and $\mathbf{x}'$ be γ. Then Eq. (2.57) can be restated as follows:

$$G(\mathbf{x}, \mathbf{x}') = \frac{1}{\sqrt{x^2 + x'^2 - 2xx'\cos\gamma}} - \frac{1}{\sqrt{(\frac{x^2 x'^2}{R^2} + R^2 - 2xx'\cos\gamma)}}. \tag{2.58}$$

Also,

$$\left(\frac{\partial G}{\partial n'}\right)_{x'=R} = -\left(\frac{\partial G}{\partial x'}\right)_{x'=R} \tag{2.59}$$

$$= \frac{R - x\cos\gamma}{(x^2 + R^2 - 2xR\cos\gamma)^{3/2}} - \frac{\frac{x^2}{R} - x\cos\gamma}{(x^2 + R^2 - 2xR\cos\gamma)^{3/2}}$$

$$= -\frac{x^2 - R^2}{R(x^2 + R^2 - 2xR\cos\gamma)^{3/2}}. \tag{2.60}$$

Note that $\frac{\partial G}{\partial n'}$ is negative in this case as $x > R$.

The geometry of the problem demands use of spherical polar coordinates-ordinates, $\hat{\mathbf{x}} = (\sin\theta\sin\varphi, \ \sin\theta\cos\varphi, \ \cos\theta)$ and $\hat{\mathbf{x}}' = (\sin\theta'\sin\varphi', \ \sin\theta'\cos\varphi', \ \cos\theta')$. Now, replacing $\frac{\partial G}{\partial n'}$ from Eq. (2.60) in Eq. (2.29), we finally obtain the solution of the Laplace equation outside the sphere for a given Dirichlet condition, i.e., with the potential Φ being specified on the surface

$$\Phi(\mathbf{x}) = \frac{1}{4\pi}\oint_{\partial V} \Phi(R, \theta, \varphi)\frac{x^2 - R^2}{R(x^2 + R^2 - 2xR\cos\gamma)^{3/2}} R^2\, d\Omega', \tag{2.61}$$

where $\cos\gamma = \hat{\mathbf{x}} \cdot \hat{\mathbf{x}}' = \sin\theta\sin\theta'\cos(\varphi - \varphi') + \cos\theta\cos\theta'$ and $d\Omega = d(\cos\theta)d\varphi$.

For a given charge distribution $\rho(\mathbf{x}')$ outside the sphere, and the potential being specified on the surface, we have the following solution of the Poisson equation:

$$\Phi(\mathbf{x}) = \int_V G(\mathbf{x}, \mathbf{x}')\rho(\mathbf{x}')\, d^3x' + \frac{1}{4\pi}\oint_{\partial V} \Phi(R, \theta, \varphi)\frac{x^2 - R^2}{R(x^2 + R^2 - 2xR\cos\gamma)^{3/2}} R^2\, d\Omega'. \tag{2.62}$$

2.6.1 Application of the Green Function to a Pair of Conducting Hemispherical Shells at Different Fixed Potentials

Let us now apply Eq. (2.61) on a conducting sphere consisting of two hemispherical shells, insulated from each other and kept at different constant potentials: $+V$ on the upper hemisphere and $-V$ on the lower hemisphere. The insulating ring, which

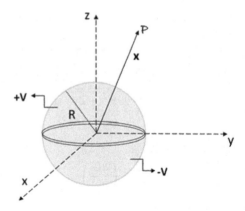

Fig. 2.9 Two hemispherical shells isolated from each other by an insulating ring on the $z = 0$ plane. The upper hemisphere is kept at potential $+V$ and the lower hemisphere is kept at potential $-V$

electrically isolates the two hemispheres, lies in the $z = 0$ plane as shown in Fig. 2.9. We assume that there are no charges outside the sphere and $\rho(\mathbf{x}') = 0$. We want to find the potential Φ anywhere outside the sphere subject to the given boundary condition on the surface of the sphere: $\Phi(R, \theta', \varphi') = +V$ when $0 \leq \theta' \leq \frac{\pi}{2}$ and $\Phi(R, \theta', \varphi') = -V$ when $\frac{\pi}{2} \leq \theta' \leq \pi$. Equation (2.62) then gives

$$\Phi(\mathbf{x}) = \frac{R(x^2 - R^2)}{4\pi} \left[\int_0^{2\pi} d\varphi' \int_0^1 d(\cos\theta') \frac{V}{(x^2 + R^2 - 2xR\cos\gamma)^{3/2}} \right.$$
$$\left. + \int_0^{2\pi} d\varphi' \int_{-1}^0 d(\cos\theta') \frac{(-V)}{(x^2 + R^2 - 2xR\cos\gamma)^{3/2}} \right]. \qquad (2.63)$$

Here, $\cos\gamma$ is dependent on the primed coordinates-ordinates. We consider first a special case where we want to evaluate Φ only on the z-axis. For observation points $\mathbf{x}$ constrained to be on the positive z-axis, $\theta = 0$ and $\cos\theta = +1$. Then, $\cos\gamma = \cos\theta'$. The resulting integral from Eq. (2.63) is easy to evaluate. Using the substitutions $\cos\theta' = u$ and $k = \frac{2Rx}{x^2 + R^2}$, we have

$$\int d(\cos\theta') \frac{V}{(x^2 + R^2 - 2xR\cos\theta')^{3/2}} = \frac{1}{(x^2 + R^2)^{3/2}} \int \frac{du}{(1 - ku)^{3/2}}$$
$$= \frac{1}{(x^2 + R^2)^{3/2}} \frac{2}{k} \frac{1}{\sqrt{1 - ku}}$$
$$= \frac{1}{ax\sqrt{x^2 + R^2 - 2ax\cos\theta'}}.$$

Therefore,

$$\int_{-1}^0 \frac{d(\cos\theta')}{(x^2 + R^2 - 2xR\cos\theta')^{3/2}} = \frac{1}{Rx}\left(\frac{1}{\sqrt{x^2 + a^2}} - \frac{1}{x + R} \right), \qquad (2.64)$$

$$\int_0^1 \frac{d(\cos\theta')}{(x^2 + R^2 - 2xR\cos\theta')^{3/2}} = \frac{1}{Rx}\left(\frac{1}{x - R} - \frac{1}{\sqrt{x^2 + a^2}} \right). \qquad (2.65)$$

Plugging the results from Eqs. (2.64) and (2.65) in Eq. (2.63), and putting $x = z$, we write down the final expression for Φ on the z-axis as follows:

$$\Phi(z) = V\left(1 - \frac{z^2 - R^2}{z\sqrt{z^2 + R^2}}\right). \tag{2.66}$$

The second integral in Eq. (2.63) can be suitably modified using the change of primed variables: $\varphi' \to \pi + \varphi'$ and $\theta' \to \pi - \theta'$, which lead to $\cos\gamma \to -\cos\gamma$. Therefore,

$$\Phi(\mathbf{x}) = \frac{VR(x^2 - R^2)}{4\pi} \int_0^1 d(\cos\theta')$$

$$\left[\int_0^{2\pi} d\varphi' \frac{1}{(x^2 + R^2 - 2xR\cos\gamma)^{3/2}} - \int_{\pi}^{3\pi} d\varphi' \frac{1}{(x^2 + R^2 + 2xR\cos\gamma)^{3/2}}\right].$$

Note that the problem has azimuthal symmetry as the potential is independent of φ. Therefore, we can independently reset the limits of φ in the second integral to $0 \le \varphi \le 2\pi$ and write

$$\Phi(\mathbf{x}) = \frac{VR(x^2 - R^2)}{4\pi} \int_0^{2\pi} d\varphi' \int_0^1 d(\cos\theta')$$

$$\left[\frac{1}{(x^2 + R^2 - 2xR\cos\gamma)^{3/2}} - \frac{1}{(x^2 + R^2 + 2xR\cos\gamma)^{3/2}}\right]. \tag{2.67}$$

It is not possible to solve the integral exactly. However, we can expand the denominators in the integrand in the power series as follows:

$$(x^2 + R^2 - 2xR\cos\gamma)^{-3/2} = \frac{1}{x^3}\left(1 + \frac{R^2}{x^2} - 2\frac{R}{x}\cos\gamma\right)^{-3/2}$$

$$= \frac{1}{x^3}\left[1 - \frac{3}{2}\left(\frac{R^2}{x^2} - 2\frac{R}{x}\cos\gamma\right)\right.$$

$$+ \frac{1}{2!}\left(\frac{3}{2}\right)\left(\frac{5}{2}\right)\left(\frac{R^2}{x^2} - 2\frac{R}{x}\cos\gamma\right)^2$$

$$- \frac{1}{3!}\left(\frac{3}{2}\right)\left(\frac{5}{2}\right)\left(\frac{7}{2}\right)\left(\frac{R^2}{x^2} - 2\frac{R}{x}\cos\gamma\right)^3 + \cdots\right]$$

$$= \frac{1}{x^3}\left[1 + \frac{R}{x}(3\cos\gamma) + \left(\frac{R}{x}\right)^2\left(-\frac{3}{2} + \frac{15}{2}\cos^2\gamma\right)\right.$$

$$+ \left(\frac{R}{x}\right)^3\left(\frac{-15}{2}\cos\gamma + \frac{35}{2}\cos^3\gamma\right) + \cdots\right].$$

We expand the other term $(x^2 + R^2 + 2xR\cos\gamma)^{-3/2}$ similarly as

$$(x^2 + R^2 - 2xR\cos\gamma)^{-3/2} = \frac{1}{x^3}\left[1 - \frac{R}{x}(3\cos\gamma) + \left(\frac{R}{x}\right)^2\left(-\frac{3}{2} + \frac{15}{2}\cos^2\gamma\right)\right.$$
$$\left. + \left(\frac{R}{x}\right)^3\left(\frac{15}{2}\cos\gamma - \frac{35}{2}\cos^3\gamma\right) + \cdots\right].$$

It turns out that terms with even powers of $\cos\gamma$ cancel each other out in the integrand in Eq. (2.67) and we are left with only odd powers of $\cos\gamma$

$$\Phi(\mathbf{x}) = \frac{VR^2}{2\pi x^2}\left(1 - \frac{R^2}{x^2}\right)\int_0^{2\pi}d\varphi'\int_0^1 d(\cos\theta')$$
$$\left[3\cos\gamma + \left(\frac{-15}{2}\cos\gamma + \frac{35}{2}\cos^3\gamma\right)\frac{R^2}{x^2} + \cdots\right]. \qquad (2.68)$$

Keeping in mind that the terms having odd powers of $\cos(\varphi - \varphi')$ integrates to zero, the integrals involving $\cos\gamma$ and $\cos^3\gamma$ are not difficult to perform and are done in the following way:

$$\int_0^{2\pi}d\varphi'\int_0^1 d(\cos\theta')\cos\gamma = \pi\cos\theta, \qquad (2.69)$$

$$\int_0^{2\pi}d\varphi'\int_0^1 d(\cos\theta')\cos^3\gamma = \frac{\pi}{4}\cos\theta\left(3 - \cos^2\theta\right). \qquad (2.70)$$

We plug the results of the integrals of odd powers of $\cos\gamma$ in Eqs. (2.69) and (2.70) in Eq. (2.68) and collect the coefficients of $(\frac{R}{x})^2$, $(\frac{R}{x})^4$, etc., to arrive at the series solution:

$$\Phi(\mathbf{x}) = \frac{3VR^2}{2x^2}\left(1 - \frac{R^2}{x^2}\right)\left[\cos\theta + \left(\frac{-5}{2}\cos\theta + \frac{35}{24}\cos\theta(3 - \cos^2\theta)\right)\frac{R^2}{x^2} + \cdots\right].$$
$$= \frac{3VR^2}{2x^2}\left[\cos\theta + \left(-\cos\theta - \frac{5}{2}\cos\theta + \frac{35}{8}\cos\theta - \frac{35}{24}\cos^3\theta\right)\frac{R^2}{x^2} + \cdots\right].$$
$$= \frac{3VR^2}{2x^2}\left[\cos\theta - \frac{7R^2}{12x^2}\left(\frac{5}{2}\cos^3\theta - \frac{3}{2}\cos\theta\right) + \cdots\right]. \qquad (2.71)$$

As a special case, we can now calculate the potential on the z-axis, where $\cos\theta = 1$. At large distances, the potential on the z-axis varies as $\frac{1}{z^2}$. The angular factors in the series in Eq. (2.71), such as $\cos\theta$ and $\left(\frac{5}{2}\cos^3\theta - \frac{3}{2}\cos\theta\right)$ are the Legendre polynomials of odd order. In other words, the solution is a series expansion in terms of the orthogonal Legendre polynomials of odd order.

2.7 Laplace Equation as Boundary Value Problem

A systematic approach to solving boundary value problems is to find the solution of the Laplace equation in some region V (where $\rho = 0$) as an expansion in terms of complete sets of orthogonal functions.[4] The coefficients of the orthogonal functions in the expansion are obtained by requiring that the given boundary condition (potential Φ or charge density σ specified on the boundaries) is satisfied.

Without going into a generalized mathematical introduction, we shall solve the Laplace equation using the method of separation of variables for some specific geometries by appropriately choosing the coordinate system. Let us start with a simple example. We assume a 3D rectangular box-like region of sides a, b and c. Suppose we need to solve the Laplace equation inside the rectangular box with the potential given everywhere on the surface (six faces of the box). We can divide the problem into six distinct boundary value problems and superpose the corresponding six solutions to get the final solution. For example, we take the boundary condition in the original problem for one face and assume the potential to be zero everywhere else on the surface to get the first solution. Similarly, we set up the problem for the second face to get to the second solution and so on. Eventually, we construct a linear combination of the six solutions to reach the final solution for the potential inside.

Therefore, we solve just one such problem where the potential everywhere on the boundaries is zero except for one side where the potential is specified. Then it is trivial to extend the problem to the case where the potential everywhere on the surface is non-zero. Let us choose a Cartesian coordinate system. The origin is chosen as shown in Fig. 2.10. The boundary conditions are as follows: (1) $\Phi = 0$ for $x = 0, a$; $y = 0, b$; $z = 0$ and (2) $\Phi = \Phi_0(x, y)$ for $z = c$. The Laplace equation in Cartesian coordinates-ordinates is given by

$$\frac{\partial^2 \Phi}{\partial x^2} + \frac{\partial^2 \Phi}{\partial y^2} + \frac{\partial^2 \Phi}{\partial z^2} = 0. \tag{2.72}$$

We look for solution in the form of product of functions of x, y, z, separately as

$$\Phi(\mathbf{x}) = X(x)Y(y)Z(z). \tag{2.73}$$

Substituting Eq. (2.73) in Eq. (2.72), we obtain

$$\frac{1}{X}\frac{d^2 X}{dx^2} + \frac{1}{Y}\frac{d^2 Y}{dy^2} + \frac{1}{Z}\frac{d^2 Z}{dz^2} = 0. \tag{2.74}$$

Each term on the left-hand side of the equation is a function of a single variable. Consequently, we may write

---

[4] For an introduction to the solution of the Laplace equation by separation of variables, readers at this point may consult any standard textbook on Mathematical Physics.

Fig. 2.10 The potential on
all sides of the box is zero
except for the top side (dark
shade). The potential on the
top side is $\Phi_0(x, y)$

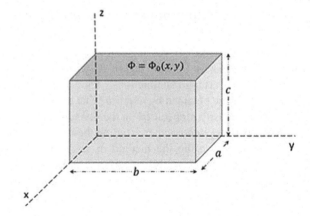

Fig. 2.10 The potential on all sides of the box is zero except for the top side (dark shade). The potential on the top side is $\Phi_0(x, y)$

$$\frac{1}{X}\frac{d^2X}{dx^2} = -\alpha^2, \tag{2.75}$$

$$\frac{1}{Y}\frac{d^2Y}{dy^2} = -\beta^2, \tag{2.76}$$

$$\frac{1}{Z}\frac{d^2Z}{dz^2} = \alpha^2 + \beta^2 = -\gamma^2, \tag{2.77}$$

where α^2, β^2 are positive constants and $\gamma = i\sqrt{\alpha^2 + \beta^2}$. The solution for Φ is the product of solutions of Eq. (2.75), Eq. (2.76) and Eq. (2.77), respectively. It is given as

$$\Phi(\mathbf{x}) = (d_{x1}e^{+i\alpha x} + d_{x2}e^{-i\alpha x})(d_{y1}e^{+i\beta y} + d_{y2}e^{-i\beta y})(d_{z1}e^{+i\gamma z} + d_{z2}e^{-i\gamma z}). \tag{2.78}$$

Keeping in mind the boundary conditions, we choose $d_{x1} = -d_{x2}$, $d_{y1} = -d_{y2}$ and $d_{z1} = -d_{z2}$, such that the solution can be written as

$$\Phi(\mathbf{x}) = \sin \alpha x \, \sin \beta y \, \sinh(\sqrt{\alpha^2 + \beta^2}z). \tag{2.79}$$

Note that, although, in principle, we could have used cosine functions as well, we could not have used hyperbolic cosine instead of the hyperbolic sine as the potential must vanish at $z = 0$. The boundary condition for Φ requires that $\Phi = 0$ for $x = a$ and $y = b$. This gives $\alpha a = n\pi$ (where $n = 0, 1, 2...$) and $\beta b = m\pi$ (where $m = 0, 1, 2...$). Therefore,

$$\Phi_{nm}(\mathbf{x}) = \sin \frac{n\pi}{a}x \, \sin \frac{m\pi}{b}y \, \sinh\left(\sqrt{\frac{n^2\pi^2}{a^2} + \frac{m^2\pi^2}{b^2}}z\right). \tag{2.80}$$

The solution meets the boundary conditions everywhere except at the top surface $z = c$. We now construct a linear combination of the Φ_{nm}'s to write down the following expansion:

$$\Phi(\mathbf{x}) = \sum_{n,m=1}^{\infty} \mathcal{A}_{nm}\Phi_{nm}(\mathbf{x}), \tag{2.81}$$

where $\mathcal{A}_{nm}$ are the coefficients to be determined using the boundary condition at the top surface. The boundary condition at $z = c$ gives

$$\Phi(x, y, c) = \Phi_0(x, y) = \sum_{nm} \mathcal{A}_{nm} \sin\frac{n\pi}{a}x \, \sin\frac{m\pi}{b}y \, \sinh\left(\sqrt{\frac{n^2}{a^2}+\frac{m^2}{b^2}}\pi c\right).$$

$$\tag{2.82}$$

We multiply both sides of Eq. (2.82) by $\sin\frac{n'\pi}{a}x \, \sin\frac{m'\pi}{b}y$ and integrate over the top surface of the box as

$$\int_0^a\int_0^b \Phi_0(x, y)\sin\frac{n'\pi}{a}x \, \sin\frac{m'\pi}{b}y \, dx \, dy = \sum_{nm} \mathcal{A}_{nm}\sinh(\sqrt{\frac{n^2}{a^2}+\frac{m^2}{b^2}}\pi c)$$

$$\int_0^a\int_0^b \sin\frac{n\pi}{a}x \, \sin\frac{m\pi}{b}y \, \sin\frac{n'\pi}{a}x \, \sin\frac{m'\pi}{b}y \, dx \, dy. \tag{2.83}$$

Note that $\sin\frac{n\pi}{a}x$, where $n = 0, 1, 2...$, are a set of orthogonal functions. In general, the orthogonality relation is written as

$$\int_0^L \sin\left(\frac{n\pi}{L}x\right)\sin\left(\frac{n'\pi}{L}x\right) dx = \frac{L}{2}\delta_{nn'}. \tag{2.84}$$

Using the orthogonality relation in Eq. (2.84) on the right-hand side of Eq. (2.83), we get

$$\int_0^{a,b} \Phi_0(x, y)\sin\left(\frac{n'\pi}{a}x\right)\sin\left(\frac{m'\pi}{b}y\right) dx \, dy$$

$$= \sum_{nm} \mathcal{A}_{nm}\sinh\left(\sqrt{\frac{n^2}{a^2}+\frac{m^2}{b^2}}\pi c\right)\frac{ab}{4}\delta_{nn'}\delta_{mm'} = \mathcal{A}_{n'm'}\sinh\left(\sqrt{\frac{n'^2}{a^2}+\frac{m'^2}{b^2}}\pi c\right)\frac{ab}{4}.$$

Therefore,

$$\mathcal{A}_{nm} = \frac{4}{ab}\frac{1}{\sinh\left(\sqrt{\frac{n'^2}{a^2}+\frac{m'^2}{b^2}}\pi c\right)}\int_0^{a,b} \Phi_0(x, y)\sin\left(\frac{n'\pi}{a}x\right)\sin\left(\frac{m'\pi}{b}y\right) dx \, dy.$$

$$\tag{2.85}$$

Thus, the solution of the Laplace equation inside the box with the given non-zero potential $\Phi_0(x, y)$ only on the top surface is given by Eq. (2.81) with the coefficients

$\mathcal{A}_{nm}$ given by Eq. (2.85). As discussed earlier, we can now construct a solution for the potential being non-zero everywhere on the surface of the box, using a linear combination of the solutions in Eq. (2.81, 2.85) for each face separately, assuming in each case that the potential in other faces is zero.

2.8   Laplace Equation in Spherical Coordinates

The choice of coordinate system for solving the Laplace equation depends on the geometry or the symmetry of the problem at hand. For planar boundary surfaces, it was appropriate to use Cartesian system. For problems involving spherical symmetry, it is convenient to use spherical coordinates. Let us first write down the Laplace equation in spherical coordinates as

$$\frac{1}{r^2}\frac{\partial}{\partial r}\left(r^2\frac{\partial \Phi}{\partial r}\right) + \frac{1}{r^2 \sin\theta}\frac{\partial}{\partial \theta}\left(\sin\theta\frac{\partial \Phi}{\partial \theta}\right) + \frac{1}{r^2 \sin^2\theta}\frac{\partial^2 \Phi}{\partial \varphi^2} = 0. \tag{2.86}$$

At the outset, we assume that the problems we are going to deal with will have azimuthal symmetry so that the potential Φ is independent of φ. Equation (2.86) then immediately reduces to

$$\frac{\partial}{\partial r}\left(r^2\frac{\partial \Phi}{\partial r}\right) + \frac{1}{\sin\theta}\frac{\partial}{\partial \theta}\left(\sin\theta\frac{\partial \Phi}{\partial \theta}\right) = 0. \tag{2.87}$$

We now need to look for solutions in the following product form:

$$\Phi(r, \theta) = R(r)\Theta(\theta). \tag{2.88}$$

Substituting Φ given by Eq. (2.88) into Eq. (2.87) and dividing by Φ, we have

$$\frac{1}{R}\frac{d}{dr}\left(r^2\frac{dR}{dr}\right) + \frac{1}{\Theta \sin\theta}\frac{d}{d\theta}\left(\sin\theta\frac{d\Theta}{d\theta}\right) = 0. \tag{2.89}$$

The Laplace equation is now easily separable. We write down the separated ordinary differential equations as

$$\frac{1}{R}\frac{d}{dr}\left(r^2\frac{dR}{dr}\right) = l(l+1), \tag{2.90}$$

$$\frac{1}{\Theta \sin\theta}\frac{d}{d\theta}\left(\sin\theta\frac{d\Theta}{d\theta}\right) = -l(l+1), \tag{2.91}$$

where $l(l+1)$ is a constant. It is easy to verify that radial Eq. (2.90) has the general solution

$$R(r) = A_l \, r^l + B_l \, r^{-(l+1)} \,. \tag{2.92}$$

Here, A_l and B_l are the arbitrary constants. The solutions of the angular equation, Eq. (2.91), are the Legendre polynomials given as

$$\Theta(\theta) = P_l(\cos\theta) \,. \tag{2.93}$$

The general solution is then given by

$$\Phi(r, \theta) = \sum_{l=0}^{\infty} \left[A_l \, r^l + B_l \, r^{-(l+1)} \right] p_l(\cos\theta) \,. \tag{2.94}$$

The Legendre polynomials $P_l(x)$ can be defined by the Rodrigues formula

$$P_l(x) = \frac{1}{2^l l!} \left(\frac{d}{dx} \right)^l (x^2 - 1)^l \,. \tag{2.95}$$

The orthonormality relation is given by

$$\int_{-1}^{1} P_l(x) P_{l'}(x) \, dx = \int_{0}^{\pi} P_l(\cos\theta) P_{l'}(\cos\theta) \sin\theta \, d\theta = \frac{2}{2l + 1} \delta_{ll'} \,. \tag{2.96}$$

2.8.1 Pair of Hemispherical Shells at Different Potentials

We could, in principle, solve a generic boundary value problem with the potential $\Phi = \Phi_0(\theta)$ specified on a spherical surface. However, we shall be a little more specific with the boundary conditions. Let us solve the problem discussed in Sect. 2.6.1 again using the method of separation of variables. We need to solve the Laplace equation inside a sphere of radius R consisting of two hemispherical shells with the potential specified on the surface as follows: $\Phi(R, \theta) = +V$ for $0 \leq \theta \leq \frac{\pi}{2}$ and $\Phi(R, \theta) = -V$ for $\frac{\pi}{2} \leq \theta \leq \pi$.

The solution must take the following form:

$$\Phi(r, \theta) = V \sum_{l=0}^{\infty} A_l \left(\frac{r}{R} \right)^l P_l(\cos\theta). \tag{2.97}$$

Here, we have put $B_l = 0$ to avoid the singularity at the origin $r = 0$. Moreover, the terms V and R have been introduced for reasons related to dimensionality. In addition to the Legendre polynomials $P_l(\cos\theta)$, all other terms inside the summation such as the factor $\frac{r}{R}$ and the coefficients A_l are now dimensionless. The introduction of the potential term V outside the summation sets the scale of the potential and makes the entire expansion dimensionally correct.

On the spherical boundary surface, $r = R$ and the potential is given by

$$\Phi(R, \theta) = V \sum_{l=0}^{\infty} A_l \, P_l(\cos \theta). \tag{2.98}$$

Now we utilize the orthonormality relation in Eq. (2.96) to rewrite Eq. (2.98) as follows:

$$\int_0^{\pi} \Phi(R, \theta) P_{l'}(\cos \theta) \sin \theta \, d\theta = V \sum_{l=0}^{\infty} A_l \int_0^{\pi} P_l(\cos \theta) P_{l'}(\cos \theta) \sin \theta \, d\theta$$

$$= V \sum_{l=0}^{\infty} A_l \frac{2}{2l+1} \delta_{ll'} = V \, A_{l'} \frac{2}{2l'+1}. \tag{2.99}$$

Therefore,

$$A_l = \frac{2l+1}{2V} \int_0^{\pi} \Phi(R, \theta) P_l(\cos \theta) \sin \theta \, d\theta$$

$$= \frac{2l+1}{2V} \left[\int_0^{\frac{\pi}{2}} (+V) \, P_l(\cos \theta) \sin \theta \, d\theta + \int_{\frac{\pi}{2}}^{\pi} (-V) \, P_l(\cos \theta) \sin \theta \, d\theta \right]$$

$$= \frac{2l+1}{2} \left[\int_0^{\frac{\pi}{2}} P_l(\cos \theta) \sin \theta \, d\theta - \int_{\frac{\pi}{2}}^{\pi} P_l(\cos \theta) \sin \theta \, d\theta \right]. \tag{2.100}$$

Using the result $P_l(x) = (-1)^l \, P_l(-x)$, we can write

$$A_l = \frac{2l+1}{2} \left[\int_0^1 P_l(x) \, dx - (-1)^l \int_0^1 P_l(x) \, dx \right]. \tag{2.101}$$

When l is even,

$$A_l = 0, \tag{2.102}$$

and when l is odd,

$$A_l = (2l+1) \int_0^1 P_l(x) \, dx. \tag{2.103}$$

We now evaluate the integral (for odd l) using the following recurrence relation:

$$\frac{d P_{l+1}}{dx} = (2l+1) P_l + \frac{d P_{l-1}}{dx}. \tag{2.104}$$

Substituting P_l from Eq. (2.104) in Eq. (2.103) and noting that $P_l(1) = 1$ for any l, we get

$$A_l = \int_0^1 \left[\frac{d P_{l+1}}{dx} - \frac{d P_{l-1}}{dx} \right] dx = P_{l-1}(0) - P_{l+1}(0). \qquad (2.105)$$

Since the coefficients with odd l only survives, we can write down the solution for potential inside the sphere as

$$\Phi(r, \theta) = V \sum_{m=0}^{\infty} A_{2m+1} \left(\frac{r}{R} \right)^{2m+1} P_{2m+1}(\cos\theta), \qquad (2.106)$$

where $m = 0, 1, 2, ...,$ and

$$A_{2m+1} = P_{2m}(0) - P_{2m+2}(0). \qquad (2.107)$$

We compute the non-zero coefficients here using the Legendre polynomials[5] as follows: $A_1 = \frac{3}{2}$, $A_3 = -\frac{7}{8}$, $A_5 = \frac{11}{16}$,....,etc. We can now expand the series solution (up to some lowest order terms) as follows:

$$\Phi(r, \theta) = V \left[A_1 \left(\frac{r}{R} \right) P_1(\cos\theta) + A_3 \left(\frac{r}{R} \right)^3 P_3(\cos\theta) + A_5 \left(\frac{r}{R} \right)^5 P_5(\cos\theta) + \cdots \right]$$

$$= \frac{3Vr}{2R} \left[P_1(\cos\theta) - \frac{7}{12} \left(\frac{r}{R} \right)^2 P_3(\cos\theta) + \frac{11}{24} \left(\frac{r}{R} \right)^4 P_5(\cos\theta) + \cdots \right]. \qquad (2.108)$$

This is the solution for potential inside the sphere. We have already obtained the solution for potential outside the sphere using the Green function method (Eq. 2.71). It is easy to obtain the solution outside the sphere from Eq. (2.94) now. Note that now we have to put $A_l = 0$ as otherwise the solution outside will blow up. However, the coefficients B_l are the same as given by Eq. (2.107), and we merely replace $\left(\frac{r}{R} \right)^l$

---

[5] The first few polynomials are listed below.

$$P_0(x) = 1$$
$$P_1(x) = x$$
$$P_2(x) = \frac{(3x^2 - 1)}{2}$$
$$P_3(x) = \frac{(5x^3 - 3x)}{2}$$
$$P_4(x) = \frac{(35x^4 - 30x^2 + 3)}{8}$$
$$P_5(x) = \frac{(63x^5 - 70x^3 + 15x)}{8}$$
$$P_6(x) = \frac{(231x^6 - 315x^4 + 105x^2 - 5)}{16}$$

and so on.

by $\left(\frac{R}{r}\right)^{l+1}$ in Eq. (2.97) with everything else remaining exactly the same as that for the solution inside and get

$$\Phi(r, \theta) = \frac{3VR^2}{2r^2} \left[P_1(\cos\theta) - \frac{7}{12}\left(\frac{R}{r}\right)^2 P_3(\cos\theta) + \frac{11}{24}\left(\frac{R}{r}\right)^4 P_5(\cos\theta) + \cdots \right].$$

(2.109)

This is exactly the same expression we derived using the method of the Green function in Eq. (2.85).

2.9   Laplace Equation in Cylindrical Coordinates

We now intend to solve the Laplace equation and find out the potential inside a cylindrical region with the potential specified on the surface of the cylinder.

The symmetry of the problem suggests that we use cylindrical coordinates to write down the Laplace equation

$$\frac{\partial^2 \Phi}{\partial \rho^2} + \frac{1}{\rho}\frac{\partial \Phi}{\partial \rho} + \frac{1}{\rho^2}\frac{\partial^2 \Phi}{\partial \varphi^2} + \frac{\partial^2 \Phi}{\partial z^2} = 0.$$

(2.110)

We now look for solutions of the form

$$\Phi(\rho, \varphi, z) = \mathcal{R}(\rho)Q(\varphi)Z(z).$$

(2.111)

On substituting Eq. (2.111) in Eq. (2.110) and rearranging the terms such that the left-hand side is a function of ρ and φ while the right-hand side is a function of z alone, we have

$$\frac{1}{\mathcal{R}}\frac{d^2\mathcal{R}}{d\rho^2} + \frac{1}{\mathcal{R}\rho}\frac{d\mathcal{R}}{d\rho} + \frac{1}{Q\rho^2}\frac{d^2Q}{d\varphi^2} = -\frac{1}{Z}\frac{d^2Z}{dz^2} = -k^2,$$

(2.112)

where k is a constant. Then,

$$\frac{d^2Z}{dz^2} - k^2 Z = 0.$$

(2.113)

Using the same argument, it follows that

$$\frac{\rho^2}{\mathcal{R}}\frac{d^2\mathcal{R}}{d\rho^2} + \frac{\rho}{\mathcal{R}}\frac{d\mathcal{R}}{d\rho} + k^2\rho^2 = -\frac{1}{Q}\frac{d^2Q}{d\varphi^2} = \nu^2,$$

(2.114)

where ν is a constant. Therefore,

$$\frac{d^2Q}{d\varphi^2} + \nu^2 Q = 0.$$

(2.115)

Finally, we are left with

$$\frac{d^2 \mathcal{R}}{d\rho^2} + \frac{1}{\rho}\frac{d\mathcal{R}}{d\rho} + \left(k^2 - \frac{\nu^2}{\rho^2}\right)\mathcal{R} = 0. \tag{2.116}$$

Thus, we have separated the Laplace equation in cylindrical coordinates into three ordinary differential equations Eqs. (2.113), (2.115) and (2.116).

As before, we opt for sine hyperbolic function as solution of Eq. (2.113) and write

$$Z(z) \sim \sinh kz, \tag{2.117}$$

where k depends on the boundary conditions. On the other hand, the solution for Eq. (2.115) is given by

$$Q(\varphi) = A \sin m\varphi + B \cos m\varphi, \tag{2.118}$$

where $\nu = m$ is an integer. Note that ν is not an integer, in general. However, for the kind of boundary value problems we shall be dealing with, the potential should be single valued under a full rotation in the azimuthal direction. Therefore, ν must be an integer, i.e., $\nu = m = 0, \pm 1, \pm 2, ...$, etc. Let us now define a dimensionless variable $x = k\rho$. The radial equation Eq. (2.116) then reduces to the following standard form called the Bessel equation

$$\frac{d^2 \mathcal{R}}{dx^2} + \frac{1}{x}\frac{d\mathcal{R}}{dx} + \left(1 - \frac{\nu^2}{x^2}\right)\mathcal{R} = 0. \tag{2.119}$$

The solutions of the equation are called the Bessel functions of the first kind of order $\pm \nu$ and are given by

$$J_\nu(x) = \left(\frac{x}{2}\right)^\nu \sum_{j=0}^\infty \frac{(-1)^j}{j!\Gamma(j+\nu+1)}\left(\frac{x}{2}\right)^{2j}, \tag{2.120}$$

$$J_{-\nu}(x) = \left(\frac{x}{2}\right)^{-\nu} \sum_{j=0}^\infty \frac{(-1)^j}{j!\Gamma(j-\nu+1)}\left(\frac{x}{2}\right)^{2j}. \tag{2.121}$$

The 'gamma function'[6] is defined as $\Gamma(n) = (n-1)!$ when n is a positive integer. For $n = 0$ or negative, Γ is singular.

Since $\nu = m$ is an integer here, it can be shown

$$J_{-m}(x) = (-1)^m J_m(x). \tag{2.122}$$

---

[6] In general, $\Gamma(z) = \int_0^\infty t^{z-1} e^{-t} \, dt$.

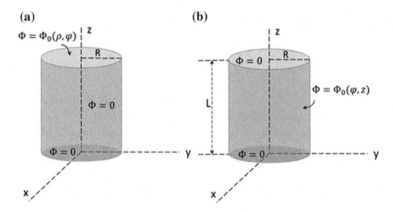

Fig. 2.11 Two basic boundary value problems in cylindrical geometry

Thus, the two solutions $J_{\pm v}(x)$ in Eq. (2.120) and Eq. (2.121), respectively, are not linearly independent when v is an integer. It is anyway customary, and in this case necessary, to introduce another solution which is linearly independent to $J_{\pm v}(x)$. The solution is called the Bessel function of second kind of order v or simply, the Neumann function, and it is given by

$$N_v(x) = \frac{J_v(x)\cos v\pi - J_{-v}(x)}{\sin v\pi}. \tag{2.123}$$

For $v = m$, where m is a positive integer, one can also define $N_m(x)$ using a series representation as

$$N_m(x) = -\frac{1}{\pi}\left(\frac{x}{2}\right)^{-m}\sum_{j=0}^{m-1}\frac{(n-j-1)!}{j!}\left(\frac{x}{2}\right)^{2j} + \frac{2}{\pi}\log\left(\frac{x}{2}\right)J_m(x)$$

$$-\frac{1}{\pi}\left(\frac{x}{2}\right)^{m}\sum_{j=0}^{\infty}[\psi(j+1)+\psi(m+j+1)]\frac{(-1)^j\left(\frac{x}{2}\right)^{2j}}{j!(m+j)!}, \tag{2.124}$$

where ψ is known as the digamma function.[7]

The solution of the radial equation can be finally written as

$$\mathcal{R}(\rho) = CJ_m(k\rho) + DN_m(k\rho). \tag{2.125}$$

Let us now state the boundary conditions more explicitly. We consider a cylinder free of charges with radius R and height L as shown in Fig. 2.11. If we want to solve the Laplace equation inside the cylinder with the potential specified everywhere on the

---

[7] $\psi(z) = \frac{d}{dz}\log\Gamma(z).$

cylindrical surface, it is necessary to break up the problem into independent parts with different boundary conditions. We can assume that the potential is zero everywhere except at the top flat surface (Fig. 2.11A) or we assume that the potential is zero everywhere except on the curved surface (Fig. 2.11B). Once we get the solutions to these problems, it is easy to construct a solution for the more general case of non-zero potential specified everywhere on the surface by taking appropriate superpositions of the solutions. Here, we shall discuss only one such problem.

2.9.1 Non-zero Potential only at the Top Surface

The boundary conditions are: (1) $\Phi = \Phi_0(\rho, \varphi)$ at $z = L$ and (2) $\Phi = 0$ everywhere else on the surfaces $\rho = R$ and $z = 0$ (Fig. 2.11A). The Neumann function $N_m(x)$ diverges at $\rho = 0$. Therefore, we put $D = 0$ in Eq. (2.125) so that Φ is finite at $\rho = 0$. The solution should thus be of the form

$$\Phi = J_m(k\rho)(A \sin m\varphi + B \cos m\varphi) \sinh kz . \tag{2.126}$$

Since the potential Φ is zero at $\rho = R$, $J_m(k_m R) = 0$. It follows that $k_{mn} = \frac{x_{mn}}{R}$, where $n = 1, 2, 3, \ldots$ and x_{mn} is the n-th root of the equation $J_m(x_{mn}) = 0$. Therefore,

$$\Phi(\rho, \varphi, z) = \sum_{m=0}^{\infty} \sum_{n=1}^{\infty} J_m(k_{mn}\rho)(A_{mn} \sin m\varphi + B_{mn} \cos m\varphi) \sinh k_{mn} z. \tag{2.127}$$

Again, at $z = L$, $\Phi = \Phi_0(\rho, \varphi)$. Then,

$$\Phi_0(\rho, \varphi) = \sum_{m=0}^{\infty} \sum_{n=1}^{\infty} J_m(k_{mn}\rho)(A_{mn} \sin m\varphi + B_{mn} \cos m\varphi) \sinh k_{mn} L . \tag{2.128}$$

We can determine the coefficients A_{mn} and B_{mn} separately by making use of orthogonality relations of the basis functions of both φ and ρ,

$$\int_0^{2\pi} \sin m\varphi \sin l\varphi d\varphi = \pi \delta_{lm}, \tag{2.129}$$

$$\int_0^R \rho \, d\rho J_\nu(k_{\nu m}\rho) J_\nu(k_{\nu n}\rho) = \frac{1}{2}R^2 \left[J_{\nu+1}(k_{\nu m}R) \right]^2 \delta_{mn}. \tag{2.130}$$

As an illustration, let us calculate A_{mn} first. Multiplying both sides of Eq. (2.128) by $\rho J_l(k_{lr}\rho) \sin(l\varphi)$ and integrating over ρ and φ,

$$\int_0^R \int_0^{2\pi} \Phi_0(\rho,\varphi)\rho J_l(k_{lr}\rho)\sin(l\varphi)d\rho\,d\varphi$$

$$= \sum_{m=0}^{\infty}\sum_{n=1}^{\infty}\int_0^R d\rho\rho J_l(k_{lr}\rho)J_m(k_{mn}\rho)A_{mn}\left(\int_0^{2\pi}\sin(m\varphi)\sin(l\varphi)d\varphi\right)\sinh(k_{mn}L)$$

$$= \sum_{m=0}^{\infty}\sum_{n=1}^{\infty}\int_0^R d\rho\rho J_l(k_{lr}\rho)J_m(k_{mn}\rho)A_{mn}\pi\delta_{lm}\sinh(k_{mn}L)$$

$$= \sum_{n=1}^{\infty} A_{ln}\pi\sinh(k_{ln}L)\int_0^R d\rho\rho J_l(k_{lr}\rho)J_l(k_{ln}\rho)$$

$$= \frac{\pi}{2}R^2\left[J_{l+1}(k_{lr}R)\right]^2\sinh(k_{rl}L)A_{rl}. \tag{2.131}$$

After rearranging the terms and relabeling the indices, we finally obtain

$$A_{mn} = \frac{2\operatorname{cosech}(k_{mn}L)}{\pi R^2\left[J_{m+1}(k_{mn}R)\right]^2}\int_0^{2\pi}d\varphi\sin(m\varphi)\int_0^R \Phi_0(\rho,\varphi)\rho J_m(k_{mn}\rho)d\rho. \tag{2.132}$$

Similarly, we can calculate B_{mn}.

$$B_{mn} = \frac{2\operatorname{cosech}(k_{nm}L)}{\pi R^2\left[J_{m+1}(k_{mn}R)\right]^2}\int_0^{2\pi}d\varphi\cos(m\varphi)\int_0^R \Phi_0(\rho,\varphi)\rho J_m(k_{mn}\rho)d\rho. \tag{2.133}$$

Thus, the potential anywhere within the cylindrical region for the given boundary condition is described by Eq. (2.127). The coefficients can be calculated using Eqs. (2.132) and (2.133) once the concrete functional form of $\Phi_0(\rho,\varphi)$ on the top surface is given.

2.10 Problems

1. We have derived the integral statement in Eq. (2.17) for a region of volume V bounded by a surface ∂V. Utilize the statement to prove that for a charge-free space, the electrostatic potential at a point $\mathbf{x}$ is its average over the surface of any sphere centered at that point. Show that there is no minimum and maximum for Φ for a charge-free region and that a charged particle cannot be held in stable equilibrium by electrostatic forces alone.

2. A point charge q is at a distance d away from a grounded infinite plane conductor. Using the method of images, find
   (a) the surface-charge density induced on the plane;
   (b) the force between the plane and the charge;
   (c) the total force acting on the plane.
   Calculate the work done to bring the charge q from infinity and the potential

energy due to the charge q and the image charge. Which one is the correct expression for the electrostatic energy stored in the system?
3. Using the method of images, discuss the problem of a point charge q inside a hollow, grounded, conducting sphere of inner radius R. Find
(a) the potential inside the sphere;
(b) the induced surface-charge density;
(c) the magnitude and direction of the force acting on q.
What will be the solution if the sphere is not grounded and instead kept at a fixed potential V, with everything else remaining the same?
4. A charge q is placed inside an isolated, uncharged, hollow conducting sphere of inner radius R_1 and outer radius R_2. Find the electrostatic force on the charge using the method of images.
5. Assume that the Dirichlet boundary conditions are given for the 'half-space' defined by $z \geq 0$, with the boundaries at the plane $z = 0$ and at infinity.
(a) Write down the appropriate Green function $G(x, x')$.
(b) Now suppose that the potential on the plane $z = 0$ is $\Phi = V$ inside a circle of radius R on the plane centered at the origin, and $\Phi = 0$ outside the circle on the plane. Find an integral expression for the potential at any point in the half-space in cylindrical coordinates (ρ, φ, z).
(c) Find the potential Φ along the axis of the circle ($\rho = 0$).
(d) Find the potential Φ at large distances ($\rho^2 + z^2 \gg R^2$) as a power series expansion in $(\rho^2 + z^2)^{-1}$. Check the validity of the expression by comparing the potential at $\rho = 0$ with that obtained in (c).
6. A grounded conducting sphere of radius a is placed inside a ring of radius R ($R > a$) and total charge Q. Find the potential outside the sphere using the Green function method.
7. A conducting sphere of radius R carries charge q uniformly distributed over the surface. The sphere is placed in a uniform electric field $\mathbf{E}_0$. Find the potential everywhere inside and outside the sphere. Calculate the induced dipole moment on the sphere.
8. An uncharged conducting sphere of radius R is placed in a uniform electric field $\mathbf{E}$ along the z-direction. We have already solved the problem using the method of images. Now find the potential in the region outside the sphere by solving the Laplace equation subject to the boundary conditions.
9. Suppose the charge distribution over the surface of a spherical shell is $\sigma(\theta) = k \cos \theta$. Find the potential everywhere (inside and outside the sphere).
10. Consider a conducting cubic box with sides of length L. Five of the six faces are grounded. The remaining face is electrically isolated from the five faces and held at a constant potential V_0. Find the potential inside the box. Compute the numerical value of the potential at the center of the cube. Find the charge density on the surface held at a constant potential.
11. The charge distribution over the curved surface of an infinite cylinder is given by $\sigma(\varphi) = k \sin \beta \varphi$, where k and β are constants. Find the potential inside and outside the cylinder.

12. Consider a hollow cylinder of radius b with symmetry axis along z-direction and with planar end faces at $z = 0$ and $z = L$. The potential on the end faces is zero, while the given potential on the curved surface is $V(\varphi, z)$. Find an expression for the potential anywhere inside the cylinder by solving the Laplace equation.

13. A long hollow conducting cylinder (with symmetry axis along z-direction) of inner radius R is separated into two halves. Assuming that the two halves are electrically isolated from each other by a small gap along the z-direction, and that both halves are kept at different potentials V_1 and V_2, find the potential $\Phi(\rho, \varphi)$ inside the cylinder by solving the Laplace equation. Calculate the surface-charge density on the cylinder.

Chapter 3
Electrodynamics of Material Media

In contrast to the microscopic equations of electrodynamics in absence of material media discussed in Chap. 1, the macroscopic equations of electrodynamics in material media are constructed by averaging the electromagnetic fields over macroscopically infinitesimal volumes of space in the matter, thus ignoring microscopic variations or fluctuations of fields resulting from local inhomogeneities at the molecular length scale. Although we shall use the same notations for $\mathbf{E}$, $\mathbf{B}$ fields and introduce new vector fields such as $\mathbf{H}$ and $\mathbf{D}$, what we essentially mean by these fields in the matter is their space-averaged (and not time-averaged) quantities.

All macroscopic matter can be classified into two broad categories so far as their electrical properties are concerned: conductors and dielectrics. These two categories fundamentally differ from each other in the way they respond to the externally applied electric field.

Conductors allow charge flow or current in presence of the external electric field. It can be easily shown that any free charges reside on the surface of the conductor under the static condition in absence of the external electric field and the electric field is zero in the bulk of the conductor. If the electric field was non-zero inside the conductor that would lead to energy dissipative charge flow, which is not possible without pumping energy from an external agency such as a battery. Thus, in the absence of the externally applied electric field, the charges distribute themselves on the surface in such a way as to cancel out the internal electric field caused by them. We have already applied the microscopic equations of electrostatics on a conductor in Chap. 2. This is because of the fact that charges reside on the surface of the conductor. The resulting electric field is only just outside the conductor.

On the other hand, dielectrics are poor conductors. There are no free charges in an ideal dielectric. However, the charges can move over microscopic length scales and their collective behavior under the external electric field determines the dielectric properties. The basic building blocks of dielectric materials are electric dipoles. In this chapter, starting with electric multipole expansion, we shall go on to discuss various boundary value problems in dielectric media.

© Springer Nature Singapore Pte Ltd. 2021
K. Bhattacharya and S. Mukhopadhyay, *Introduction to Advanced Electrodynamics*,
https://doi.org/10.1007/978-981-16-7802-8_3

Subsequently, we discuss magnetic multipole expansion and boundary value problems in magnetized media. We end the chapter with a discussion on generalized Ampere's law in material media and the conservation laws related to electromagnetic field energy and momenta.

3.1 Electric Multipole Expansion

The scalar potential $\Phi(\mathbf{x})$ due to a charge distribution $\rho(\mathbf{x}')$ is given by

$$\Phi(\mathbf{x}) = \int \frac{\rho(\mathbf{x}')}{|\mathbf{x} - \mathbf{x}'|} \, d^3x' . \tag{3.1}$$

Far away from the localized charge distribution, the observer will see the total charge Q of the distribution, effectively as a point charge, and the potential will fall off as inverse of the distance between the charge and the observation point. However, even if the total charge Q is zero, the scalar potential at $\mathbf{x}$ outside the charge distribution may not be zero. It turns out that $\Phi(\mathbf{x})$ can be expressed in terms of 'multipole moments' dictated by the geometry of the charge distribution. We note that $|\mathbf{x}| = r$ and $|\mathbf{x}'| = r'$. Then,

$$\frac{1}{|\mathbf{x} - \mathbf{x}'|} = (x^2 + x'^2 - 2\mathbf{x} \cdot \mathbf{x}')^{-1/2}$$

$$= \frac{1}{r}\left[1 + \left(\frac{r'}{r}\right)^2 - \frac{2}{r^2}\sum_i x_i x_i'\right]^{-\frac{1}{2}} .$$

For points outside the charge distribution, $r' \ll r$, we can use the binomial expansion

$$(1 + \varepsilon)^{-\frac{1}{2}} = \sum_{n=0}^{\infty} (-1)^n \frac{(2n-1)!!}{n!\,2^n} \varepsilon^n ,$$

where $\varepsilon = \left(\frac{r'}{r}\right)^2 - \frac{2}{r^2}\sum_i x_i x_i'$. Therefore,

$$\frac{1}{|\mathbf{x} - \mathbf{x}'|} = \frac{1}{r}\sum_{n=0}^{\infty} (-1)^n \frac{(2n-1)!!}{n!\,2^n} \left[\left(\frac{r'}{r}\right)^2 - \frac{2}{r^2}\sum_i x_i x_i'\right]^n$$

$$= \sum_{n=0}^{\infty} \frac{(-1)^n}{r^{2n+1}} \frac{(2n-1)!!}{n!\,2^n} \left[r'^2 - 2\sum_i x_i x_i'\right]^n .$$

Let us first write down some of the lowest order terms in the expansion separately. For $n = 0$: $\frac{1}{r}$; for $n = 1$: $-\frac{1}{2}\frac{r'^2 - 2\sum_i x_i x_i'}{r^3}$; for $n = 2$: $\frac{3}{8}\frac{(r'^2 - 2\sum_i x_i x_i')^2}{r^5}$; etc. We collect the coefficients of $\frac{1}{r^n}$ and write down the expansion in the following manner:

$$
\frac{1}{|\mathbf{x} - \mathbf{x}'|} = \frac{1}{r} + \frac{\sum_i x_i x_i'}{r^3} + \frac{1}{2r^5} \left[3 \left(\sum_i x_i x_i' \right)^2 - r^2 r'^2 \right] + \cdots
$$

$$
= \frac{1}{r} + \frac{\mathbf{x} \cdot \mathbf{x}'}{r^3} + \frac{1}{2r^5} \left[3 \left(\sum_{ij} x_i x_j x_i' x_j' \right) - \left(\sum_{ij} x_i x_j \delta_{ij} \right) r'^2 \right] + \cdots
$$

$$
= \frac{1}{r} + \frac{\mathbf{x} \cdot \mathbf{x}'}{r^3} + \frac{1}{2r^5} \sum_{ij} \left(3x_i' x_j' - r'^2 \delta_{ij} \right) x_i x_j + \cdots . \tag{3.2}
$$

The potential Φ in Eq. (3.1) can now be written as

$$
\Phi(\mathbf{x}) = \frac{1}{r} \int \rho(\mathbf{x}') \, d^3 x' + \frac{\mathbf{x}}{r^3} \cdot \int \mathbf{x}' \rho(\mathbf{x}') \, d^3 x'
$$

$$
+ \frac{x_i x_j}{2r^5} \int \sum_{ij} \left(3x_i' x_j' - r'^2 \delta_{ij} \right) \rho(\mathbf{x}') \, d^3 x' + \cdots . \tag{3.3}
$$

The first term in the expansion is easily identified as the monopole term. The total charge q (or the monopole moment) is given by

$$
q = \int \rho(\mathbf{x}') \, d^3 x' . \tag{3.4}
$$

For the second term, we define the electric dipole moment $\mathbf{p}$ as

$$
\mathbf{p} = \int \mathbf{x}' \rho(\mathbf{x}') \, d^3 x' . \tag{3.5}
$$

For the third term in the expansion, we define the quadrupole moment Q_{ij}, which is a traceless tensor, as follows:

$$
Q_{ij} = \int \left(3x_i' x_j' - r'^2 \delta_{ij} \right) \rho(\mathbf{x}') \, d^3 x' . \tag{3.6}
$$

Note that the quadrupole moment tensor Q_{ij} is real, symmetric ($Q_{ij} = Q_{ji}$) and traceless ($\text{Tr}(Q) = 0$) so that there can be at most five independent components. The trace vanishes as

$$
\text{Tr}(Q) = \sum_i Q_{ii} = \sum_i \int \left(3x_i' x_i' - r'^2 \delta_{ii} \right) \rho(\mathbf{x}') \, d^3 x'
$$

$$
= \int \rho(\mathbf{x}') \, d^3 x' \left(3 \sum_i x_i'^2 - 3r'^2 \right) = 0 . \tag{3.7}
$$

The potential can be finally expressed in Cartesian coordinates as follows:

$$\Phi(x) = \frac{q}{r} + \frac{\mathbf{p} \cdot \mathbf{x}}{r^3} + \frac{\frac{1}{2}\sum_{ij} Q_{ij} x_i x_j}{r^5} + \cdots. \tag{3.8}$$

The potential can also be expressed in terms of the orthogonal Legendre polynomials. Assuming that the angle between $\mathbf{x}$ and $\mathbf{x}'$ is θ, we start from

$$\frac{1}{|\mathbf{x} - \mathbf{x}'|} = (r^2 + r'^2 - 2rr' \cos\theta)^{-1/2} \tag{3.9}$$

and carry out the expansion as before[1]:

$$\frac{1}{|\mathbf{x} - \mathbf{x}'|} = \frac{1}{r} \sum_{n=0}^{\infty} (-1)^n \frac{(2n-1)!!}{n!\,2^n} \left[\left(\frac{r'}{r}\right)^2 - \frac{2r'}{r} \cos\theta \right]^n$$

$$= \frac{1}{r} \sum_{n=0}^{\infty} (-1)^n \frac{(2n-1)!!}{n!\,2^n} \left[\frac{r'}{r} \left(\frac{r'}{r} - 2\cos\theta \right) \right]^n.$$

We now write down the lower order terms (up to $n = 3$) explicitly and collect the coefficients of $\left(\frac{r'}{r}\right)^n$ as

$$\frac{1}{|\mathbf{x} - \mathbf{x}'|} = \frac{1}{r} \left[1 + \left(\frac{r'}{r}\right) \cos\theta + \left(\frac{r'}{r}\right)^2 \left(\frac{3}{2} \cos^2\theta - \frac{1}{2} \right) \right.$$

$$\left. + \left(\frac{r'}{r}\right)^3 \left(\frac{5}{2} \cos^3\theta - \frac{3}{2} \cos\theta \right) + \cdots \right].$$

It turns out that the coefficients are the Legendre polynomials. We finally have

$$\frac{1}{|\mathbf{x} - \mathbf{x}'|} = \frac{1}{r} \sum_{n=0}^{\infty} \left(\frac{r'}{r}\right)^n P_n(\cos\theta). \tag{3.10}$$

Substituting the above expression in Eq. (3.1), we obtain the multipole expansion in terms of the Legendre polynomials. The use of spherical coordinates is manifestly more convenient compared to Cartesian coordinates so far as computing the higher order multipoles (octupole and beyond) is concerned. The expansion looks like

$$\Phi(x) = \sum_{n=0}^{\infty} \frac{1}{r^{n+1}} \int r'^n P_n(\cos\theta) \rho(\mathbf{x}') d^3 x'. \tag{3.11}$$

---

[1] For multipole expansion using spherical harmonics, see Jackson, Classical Electrodynamics, Third edition.

3.1.1 The Electric Dipole Moment

The monopole moment q is the net charge of the distribution. Since the monopole term in the potential falls off as $\frac{1}{r}$, it will dominate the potential at large r, whenever q is non-zero. The electric dipole moment $\mathbf{p}$ is non-zero for any spatially extended charge distribution where the 'centers' of the positively charged and negatively charged components are displaced with respect to each other. A given charge distribution in a macroscopic medium possesses 'permanent electric dipole moment' when $\mathbf{p}$ is non-zero in the absence of the electric field. We could also have a scenario where the electric dipole moments in a medium are produced by the external electric field. These are called 'induced dipole moments'. Regardless of the physical origin of the electric dipole moment, $\Phi_{\mathrm{dip}} = \frac{\mathbf{p} \cdot \mathbf{x}}{r^3}$ describes the true potential far from the distribution.

In general, the multipole moments are dependent on the choice of origin of the coordinate system. This can be easily understood using a simple example. Take a point charge q and calculate the potential anywhere by shifting the origin of the coordinate-ordinate system away from the location of the charge. You will see all sorts of multipole potentials appearing in the expansion! The monopole moment q itself is a scalar and as such invariant under coordinate transformation. The dipole moment, on the other hand, is dependent on the choice of the origin so long as the net charge is non-zero. This can be proven with a simple argument. Let $\mathbf{p}'$ be the 'new' dipole moment when the origin is shifted by δ so that in the 'new' coordinate system, $\mathbf{x}'' = \mathbf{x}' + \delta$ and $\rho(\mathbf{x}'') = \rho(\mathbf{x}')$. Then, $\mathbf{p}' = \int \mathbf{x}'' \rho(\mathbf{x}'') \, d^3 x'' = \int (\mathbf{x}' + \delta) \rho(\mathbf{x}') \, d^3 x' = \mathbf{p} + \delta \int \rho(\mathbf{x}') \, d^3 x' = \mathbf{p} + q\delta$. Clearly, $\mathbf{p} = \mathbf{p}'$ only when $q = 0$.

As an illustration, let us calculate the dipole moment of two point charges $\pm q$ placed at $\mathbf{x}' = \pm \frac{1}{2}\mathbf{d}$, respectively. The charge density can be expressed as $\rho(\mathbf{x}') = q\delta(\mathbf{x}' - \frac{1}{2}\mathbf{d}) - q\delta(\mathbf{x}' + \frac{1}{2}\mathbf{d})$. Using Eq. (3.5), dipole moment is given by

$$\mathbf{p} = \int \left[q\,\delta\left(\mathbf{x}' - \frac{1}{2}\mathbf{d}\right) - q\,\delta\left(\mathbf{x}' + \frac{1}{2}\mathbf{d}\right) \right] \mathbf{x}' \, d^3 x'$$

$$= q\left(\frac{1}{2}\mathbf{d}\right) - q\left(-\frac{1}{2}\mathbf{d}\right) = q\mathbf{d}. \tag{3.12}$$

In this case, the net charge is zero, and the dipole moment $\mathbf{p} = q\mathbf{d}$ is invariant under coordinate transformation.

The electric field corresponding to each multipole potential can be obtained by using $\mathbf{E} = -\nabla\Phi$. For example, for the dipolar potential, the corresponding electric field is

$$\mathbf{E}_{\mathrm{dip}}(\mathbf{x}) = -\nabla\left(\frac{\mathbf{p} \cdot \mathbf{x}}{r^3}\right) = \frac{\left[3\mathbf{x}(\mathbf{p} \cdot \mathbf{x}) - r^2\mathbf{p}\right]}{r^5}. \tag{3.13}$$

Noting that $\frac{\mathbf{x}}{r} = \hat{\mathbf{r}}$ is the unit vector along $\mathbf{x}$, we express the dipolar electric field as follows:

$$\mathbf{E}_{\text{dip}} = \frac{1}{r^3}\left[3(\mathbf{p}\cdot\hat{\mathbf{r}})\hat{\mathbf{r}} - \mathbf{p}\right] . \tag{3.14}$$

For an idealized point dipole at the origin, the electric field given by the above expression is exact for $r \neq 0$. In order to take into account the singular behavior at $\mathbf{r} = 0$ and calculate the electric field for an idealized point dipole anywhere, we make use of the following identity[2]:

$$\partial_i \partial_j \frac{1}{r} = \frac{3x_i x_j - r^2 \delta_{ij}}{r^5} - \frac{4\pi}{3}\delta_{ij}\delta(\mathbf{r}) . \tag{3.15}$$

Now we apply the identity in Eq. (3.15) to find the electric field for a point dipole as follows:

$$
\begin{aligned}
E_i &= -\partial_i \left(\sum_j \frac{p_j x_j}{r^3}\right) = \sum_j p_j \partial_i \left(\frac{x_j}{r^3}\right) \\
&= -\sum_j p_j \partial_i \partial_j \left(\frac{1}{r}\right) \\
&= \sum_j p_j \left[\frac{3x_i x_j - r^2 \delta_{ij}}{r^5} - \frac{4\pi}{3}\delta_{ij}\delta(\mathbf{r})\right] \\
&= \frac{\sum_j 3x_i\left(p_j x_j\right) - r^2 p_j \delta_{ij}}{r^5} - \frac{4\pi}{3}\sum_j p_j \delta_{ij}\delta(\mathbf{r}) \\
&= \frac{3x_i\left(\mathbf{p}\cdot\mathbf{x}\right) - r^2 p_i}{r^5} - \frac{4\pi}{3}p_i\delta(\mathbf{r}) .
\end{aligned}
\tag{3.16}
$$

Therefore, the exact electric field anywhere due to an idealized point dipole placed at the origin is given by[3]

$$
\begin{aligned}
\mathbf{E}_{\text{dip}} &= \frac{3\mathbf{x}\left(\mathbf{p}\cdot\mathbf{x}\right) - r^2\mathbf{p}}{r^5} - \frac{4\pi}{3}\mathbf{p}\delta(\mathbf{r}) \\
&= \frac{1}{r^3}\left[3(\mathbf{p}\cdot\hat{\mathbf{r}})\hat{\mathbf{r}} - \mathbf{p}\right] - \frac{4\pi}{3}\mathbf{p}\delta(\mathbf{r}) .
\end{aligned}
\tag{3.17}
$$

Let us check the consistency of the above expression with the Maxwell equations. We first evaluate the volume integral of the electrostatic field due to an arbitrary charge distribution $\rho(\mathbf{r}')$, for a spherical volume V of radius R, centered at the origin. The charge distribution is assumed to be completely within the spherical region. The volume integral of the field is

---

[2] For a proof of the identity, see C. P. Frahm, American Journal of Physics **51**, 826 (1983).

[3] The importance of the additional delta-function term is more fundamental, as will be clearer when we calculate the corresponding expression for magnetic field.

$$\int_V \mathbf{E}(\mathbf{x})\, d^3x = \int_V \left[\int_V \frac{\rho(\mathbf{x}')(\mathbf{x} - \mathbf{x}')}{|\mathbf{x} - \mathbf{x}'|^3} d^3x' \right] d^3x . \tag{3.18}$$

We now rearrange the terms to invert the order of integration as follows:

$$\int_V \mathbf{E}(\mathbf{x})\, d^3x = -\int_V \rho(\mathbf{x}') \left[\int_V \frac{(\mathbf{x}' - \mathbf{x})}{|\mathbf{x}' - \mathbf{x}|^3} d^3x \right] d^3x' . \tag{3.19}$$

The expression within bracket is easily identified as the electric field produced at $\mathbf{x}'$ (anywhere within the sphere) by an uniformly charged sphere of radius R and unit charge density. The electric field inside a sphere with uniform charge density is easily obtained using Gauss's law[4] and is given by

$$\mathbf{E}(\mathbf{x}) = \frac{4\pi}{3}\rho\mathbf{x} . \tag{3.20}$$

Therefore,

$$\int_V \mathbf{E}(\mathbf{x})\, d^3x = -\int \rho(\mathbf{x}') \left(\frac{4\pi\mathbf{x}'}{3} \right) d^3x' = -\frac{4\pi}{3}\mathbf{p} . \tag{3.21}$$

Here $\mathbf{p}$ is the electric dipole moment of the charge distribution inside the sphere of radius R. The result is correct as it follows directly from the Maxwell equations. Let us now integrate the electric field given by Eq. (3.14) over the same sphere where the dipole moment is of the spherical charge distribution. Without any loss of generality, we can assume that the dipole moment is directed along z-axis. The Cartesian components of $\mathbf{E}_{\text{dip}}$ due to the dipole moment $p\hat{\mathbf{z}}$ in spherical coordinates are given by

$$E_x = \frac{3p}{2} \frac{\sin 2\theta \cos\varphi}{r^3} ,$$
$$E_y = \frac{3p}{2} \frac{\sin 2\theta \cos\varphi}{r^3} ,$$
$$E_z = \frac{p(3\cos^2\theta - 1)}{r^3} .$$

Clearly, due to the azimuthal symmetry around z-axis, the volume integrals of E_x and E_y vanish on angular integration. We have to be a little careful with evaluating the volume integral for E_z. Keeping in mind the singularity at the origin, we exclude a small spherical region of radius R_0 centered at the origin and integrate over the angular coordinates-ordinates first. Finally, we take the limit $R_0 \to 0$. The volume integral is

---

[4] Assume a sphere of uniform charge density ρ with center at the origin. Now imagine a Gaussian spherical surface of radius R centered at the origin and inside the sphere. Then the application of integral form of Gauss's law gives $E 4\pi R^2 = 4\pi\rho\frac{4\pi}{3}R^3$. Therefore, $E = \frac{4\pi}{3}\rho R$. In general, the electric field at any point inside a sphere with uniform charge density ρ is $E(\mathbf{x}) = \frac{4\pi}{3}\rho\mathbf{x}$.

$$\int E_z d^3x = \int p \frac{(3\cos^2\theta - 1)}{r^3} r^2 \sin\theta \, d\theta \, d\varphi \, dr$$

$$= 2\pi p \int_{R_0}^{R} \frac{dr}{r} \int_{0}^{\pi} (3\cos^2\theta - 1)\sin\theta \, d\theta \, . \qquad (3.22)$$

The integral vanishes again because of the angular integration. The result is, therefore, in contradiction with the same volume integral evaluated using the Maxwell equations. If we use the electric field given by Eq. (1.24), the volume integral of the first term vanishes, as we have already witnessed. We are finally left with the following integral:

$$\int_V \mathbf{E}(\mathbf{x}) \, d^3x = -\frac{4\pi}{3} \int_V \mathbf{p}\delta(\mathbf{r}) \, d^3r = -\frac{4\pi}{3}\mathbf{p} \, . \qquad (3.23)$$

The result is consistent with Eq. (3.21).

3.2 Electrostatic Interaction of an Extended Charge Distribution with External Electric Field

The electrostatic energy W of a given charge distribution $\rho(\mathbf{x}')$ is the work done to assemble the charge distribution (not creating it!). However, what we are concerned about here is the interaction energy of the charge distribution placed in an external electric field $E(\mathbf{x}') = -\nabla'\Phi(\mathbf{x}')$. The electrostatic energy is given by

$$W = \int \rho(\mathbf{x}')\Phi(\mathbf{x}') \, d^3x' \, . \qquad (3.24)$$

Note that there is a subtle difference between Eqs. (3.24) and (1.32) in Chap. 1. A factor of $\frac{1}{2}$ is missing in Eq. (3.24). The reason is simple: unlike before, we are dealing here with the interaction energy of a charge distribution with an external electric field. The electric field in the present case is not produced by the charge distribution $\rho(\mathbf{x}')$. Hence, there is no question of double counting arising here.

The potential Φ can be expanded in a Taylor series about an origin chosen to be within the charge distribution and written as

$$\Phi(\mathbf{x}') = \left[\Phi(\mathbf{x}')\right]_{\mathbf{x}'=0} + \mathbf{x}' \cdot \left[\nabla'\Phi\right]_{\mathbf{x}'=0} + \frac{1}{2!}(\mathbf{x}' \cdot \nabla')^2\Phi\big|_{\mathbf{x}'=0} + \cdots$$

$$= \Phi(0) - \mathbf{x}' \cdot \mathbf{E}(0) + \frac{1}{2!}\sum_{i,j} x'_i x'_j \left[\partial'_i \partial'_j \Phi\right]_{\mathbf{x}'=0} + \cdots$$

$$\Phi(0) - \mathbf{x}' \cdot \mathbf{E}(0) - \frac{1}{2!}\sum_{i,j} x'_i x'_j \left[\partial'_i E_j\right]_{\mathbf{x}'=0} + \cdots \, . \qquad (3.25)$$

Note that $\mathbf{E}$ is an external field such that the sources producing the field are outside the region occupied by the charge distribution ρ. Therefore,

$$\nabla' \cdot \mathbf{E} = \sum_{ij} \partial_i' E_j \delta_{ij} = 0. \tag{3.26}$$

The third term in the expansion in Eq. (3.25) is accordingly adjusted using Eq. (3.26), so as to include a quadrupole moment term. Now we can write

$$\Phi(\mathbf{x}') = \Phi(0) - \mathbf{x}' \cdot \mathbf{E}(0) - \frac{1}{6} \sum_{i,j} \left(3x_i'x_j' - r'^2\delta_{ij}\right) \left[\frac{\partial E_j}{\partial x_i'}\right]_{\mathbf{x}'=0} + \cdots . \tag{3.27}$$

Therefore, the interaction energy of the charge distribution with an external electric field is given by

$$W = \int \rho(\mathbf{x}') \left[\Phi(0) - \mathbf{x}' \cdot \mathbf{E}(0) - \frac{1}{6} \sum_{i,j} \left(3x_i'x_j' - r'^2\delta_{ij}\right) \frac{\partial E_j}{\partial x_i'}\bigg|_{\mathbf{x}'=0} + \cdots \right] d^3x'$$

$$= q\Phi(0) - \mathbf{p} \cdot \mathbf{E}(0) - \frac{1}{6} \sum_{ij} Q_{ij} \frac{\partial E_j}{\partial x_i'}\bigg|_{\mathbf{x}'=0} + \cdots . \tag{3.28}$$

Here, q, $\mathbf{p}$ and Q_{ij} are given by Eq. (3.4), Eq. (3.5) and Eq. (3.6), respectively. The interaction energy of the charge distribution with the external electric field separates out into terms describing how different multipoles interact with the field. For example, the first term describes the interaction of the monopole moment q (charge) with the potential $\Phi(0)$ corresponding to the external electric field; the second, the interaction of net dipole moment $\mathbf{p}$ with the electric field $\mathbf{E}$; the third, interaction between the quadrupole moment and the spatial gradient of the field and so on.

3.3 Macroscopic Field Equations of Electrostatics

In contrast to the conducting media, no constant current can flow in a dielectric, and hence the constant electric field inside a dielectric is not necessarily zero. Hence, the equations describing the electrostatic field inside a dielectric must differ from that in a vacuum. It turns out that the macroscopic electric field is still derivable as gradient of scalar potential $\Phi(\mathbf{x})$ and $\nabla \times \mathbf{E} = 0$. On the other hand, $\nabla \cdot \mathbf{E} = 4\pi \bar{\rho}$, where $\bar{\rho}$ is the localized or 'bound' charge density averaged over a macroscopically small yet microscopically large volume, centered at $\mathbf{x}$. In general, the total charge inside a dielectric as well as the average macroscopic multipole moments are zero in the absence of the external electric field. When an electric field is established in the dielectric, the local charge distribution or the shape of the molecular orbital undergo distortion, leading to multipole moments while the net charge still remains zero

Fig. 3.1 An electrically
polarized body of volume V
and bounding surface ∂V is
shown, along with
polarization $\mathbf{P}(\mathbf{x}')$ and the
corresponding volume
element d^3x'. The
observation point $\mathcal{P}$ is at $\mathbf{x}$

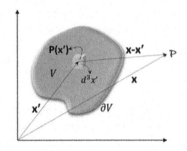

in absence of external electric fields. If the dominant contribution to the multipole moments in presence of the electric field is due to electric dipoles, we say that the medium is electrically polarized.

Since $\int \bar{\rho}(\mathbf{x}) \, d^3x = 0$ for a macroscopic body of any shape, $\bar{\rho}$ can be written as divergence of a vector quantity $\mathbf{P}$ as follows:

$$\bar{\rho} = -\nabla \cdot \mathbf{P}. \tag{3.29}$$

The minus sign seems arbitrary at the moment. We will justify it a little later. Outside the body, $\mathbf{P} = 0$. We can easily verify that $\int \bar{\rho} \, d^3x = 0$, over a volume V enclosed by a surface ∂V which is everywhere outside the body: $\int_V \bar{\rho} \, d^3x = -\int_V \nabla \cdot \mathbf{P} \, d^3x = -\oint_{\partial V} \mathbf{P} \cdot \hat{\mathbf{n}} \, da = 0$. The closed surface integral is zero since $\mathbf{P} = 0$ everywhere on the surface ∂V. The vector quantity $\mathbf{P}$ is called the electric polarization and is defined as the net dipole moment per unit volume.

Let us consider a macroscopic body of electrically polarized matter of volume V, bounded by a surface ∂V (Fig. 3.1). The dipole moment of the macroscopically infinitesimal volume element d^3x' centered at $\mathbf{x}'$ is $\mathbf{P}d^3x'$. The total scalar potential at $\mathbf{x}$ due the macroscopic body is obtained by integrating over the potential due to all such volume elements as

$$\Phi(\mathbf{x}) = \int_V \frac{\mathbf{P}(\mathbf{x}') \cdot (\mathbf{x} - \mathbf{x}')}{|\mathbf{x} - \mathbf{x}'|^3} \, d^3x'. \tag{3.30}$$

Noting that $\nabla' \frac{1}{|\mathbf{x}-\mathbf{x}'|} = \frac{(\mathbf{x}-\mathbf{x}')}{|\mathbf{x}-\mathbf{x}'|^3}$ and utilizing standard vector product rule, Eq. (3.30) can be rewritten as follows:

$$\Phi(\mathbf{x}) = \int_V \mathbf{P}(\mathbf{x}') \cdot \nabla' \frac{1}{|\mathbf{x} - \mathbf{x}'|} \, d^3x'$$
$$= \int_V \left[\nabla' \cdot \frac{\mathbf{P}(\mathbf{x}')}{|\mathbf{x} - \mathbf{x}'|} - \frac{1}{|\mathbf{x} - \mathbf{x}'|} \nabla' \cdot \mathbf{P} \right] d^3x'. \tag{3.31}$$

The first term in the integrand can be converted into surface integral using divergence theorem:

$$\Phi(\mathbf{x}) = \oint_{\partial V} \frac{\mathbf{P}(\mathbf{x}')}{|\mathbf{x} - \mathbf{x}'|} \cdot \hat{\mathbf{n}} \, da - \int_V \frac{\nabla' \cdot \mathbf{P}}{|\mathbf{x} - \mathbf{x}'|} \, d^3 x'. \tag{3.32}$$

While the first integral of Eq. (3.32) is similar in appearance to scalar potential due to surface-charge distribution, the second integral looks like the same due to volume charge distribution. Moreover, we can readily identify the average 'bound' volume charge density (over a macroscopically infinitesimal volume) $\bar{\rho} = -\nabla \cdot \mathbf{P}$ of Eq. (3.29) in the second integral. Equation (3.32) can be interpreted as follows: the scalar potential Φ and consequently, electric field $\mathbf{E}$ at a point outside the electrically polarized macroscopic body of volume V bounded by surface ∂V is equal to the potential produced by the surface charge on ∂V with density $\bar{\sigma} = \mathbf{P} \cdot \hat{\mathbf{n}}$ and volume charge in V with density $\bar{\rho} = -\nabla \cdot \mathbf{P}$. We reiterate that $\bar{\sigma}$ and $\bar{\rho}$ are averaged quantities over a macroscopically infinitesimal area and volume, respectively. By 'averaging over macroscopically infinitesimal' volume or area, we mean averaging over region large enough so that the microscopic fluctuations of the relevant quantity (in this case, charge density) are smoothed out and yet the region should be small enough (compared to the dimension of the object) so that variations of the same quantity over larger length scales can be described. The potential Φ in the present case is

$$\Phi(\mathbf{x}) = \oint_{\partial V} \frac{\bar{\sigma}(\mathbf{x}')}{|\mathbf{x} - \mathbf{x}'|} \, da - \int_V \frac{\bar{\rho}(\mathbf{x}')}{|\mathbf{x} - \mathbf{x}'|} \, d^3 x'. \tag{3.33}$$

Till now, we have assumed that there are no free charges in the object, or, in other words, there are no delocalized conduction electrons. If we take any free charge (for whatever reason it may appear) into account, at a macroscopic level, Eq. (3.30) can be rewritten as

$$\Phi(\mathbf{x}) = \int_V \frac{\rho(\mathbf{x}')}{|\mathbf{x} - \mathbf{x}'|} \, d^3 x' + \int_V \frac{\mathbf{P}(\mathbf{x}') \cdot (\mathbf{x} - \mathbf{x}')}{|\mathbf{x} - \mathbf{x}'|^3} \, d^3 x'. \tag{3.34}$$

The second term can be modified as before leading to

$$\Phi(\mathbf{x}) = \int_V \frac{\rho(\mathbf{x}')}{|\mathbf{x} - \mathbf{x}'|} \, d^3 x' + \oint_{\partial V} \frac{\mathbf{P}(\mathbf{x}')}{|\mathbf{x} - \mathbf{x}'|} \cdot \hat{\mathbf{n}} \, da - \int_V \frac{\nabla' \cdot \mathbf{P}}{|\mathbf{x} - \mathbf{x}'|} \, d^3 x'. \tag{3.35}$$

Since $\mathbf{P} = 0$ outside the object, we can always take a larger volume V, so that the combined volume of the localized and the delocalized charge distribution remains completely inside V, without changing the result of the integral. The corresponding surface ∂V is now completely outside the volume of distribution where $\mathbf{P} = 0$. As a result, the surface integral vanishes and we are left with

$$\Phi(\mathbf{x}) = \int_V \frac{[\rho(\mathbf{x}') - \nabla' \cdot \mathbf{P}(\mathbf{x}')]}{|\mathbf{x} - \mathbf{x}'|} \, d^3 x'. \tag{3.36}$$

Equation (3.36) is readily identified with the standard expression for potential due to a continuous charge distribution $\rho \rightarrow \rho - \nabla \cdot \mathbf{P}$. Gauss's law now reads

$$\nabla \cdot \mathbf{E}(\mathbf{x}) = 4\pi \left[\rho(\mathbf{x}) - \nabla \cdot \mathbf{P}(\mathbf{x}) \right] . \tag{3.37}$$

We define the electric displacement vector

$$\mathbf{D} = \mathbf{E} + 4\pi \mathbf{P} . \tag{3.38}$$

Equation (3.37) then reduces to

$$\nabla \cdot \mathbf{D}(\mathbf{x}) = 4\pi \rho(\mathbf{x}) . \tag{3.39}$$

Equation (3.39) is the macroscopic counterpart of microscopic Gauss's law. The macroscopic version of other microscopic Maxwell equation of electrostatics in vacuum

$$\nabla \times \mathbf{E}(\mathbf{x}) = 0 , \tag{3.40}$$

of course, remains unchanged in form. We can readily write down Eq. (3.39) in SI system, using the recipe discussed in the first chapter. The transformations are: $[\rho]_G \longrightarrow \frac{1}{4\pi\epsilon_0} [\rho]_{SI}, [\mathbf{P}]_G \longrightarrow \frac{1}{4\pi\epsilon_0} [\mathbf{P}]_{SI}$ and $[\mathbf{E}]_G \longrightarrow [\mathbf{E}]_{SI}$ applied to Eq. (3.38) combined with Eq. (3.39). Then we obtain $\nabla \cdot [\mathbf{D}]_{SI} = [\rho]_{SI}$.

What is the physical origin of $\bar{\sigma}$ and $\bar{\rho}$? First, these are real accumulations of charges which originate from the spatial variation of electric polarization. However, unlike 'free' charges, these cannot be moved around and are, therefore, called bound charges. The volume charge density $\bar{\rho}$ is expressed as the negative divergence of $\mathbf{P}$ and arises due to non-uniform polarization inside the dielectric. The surface charge $\bar{\sigma}$ arises out of electric charges piling up at the surface where the physical dipole moments get terminated. The surface-bound charge will be non-zero irrespective of whether the bulk electric polarization is uniform or not.

The macroscopic equations in media, viz., Eqs. (3.40) and (3.39) can be solved provided we can write down a constitutive relation between $\mathbf{D}$ and $\mathbf{E}$. In most of the cases, except for ferroelectric materials, the relation between $\mathbf{D}$ and $\mathbf{E}$ is linear. The higher order terms in the expansion of $\mathbf{P}$ in powers of $\mathbf{E}$ can be neglected so long as the externally applied electric field is small compared to the internal molecular electric field. In an isotropic dielectric, $\mathbf{P}$ and $\mathbf{E}$ are in the same direction and we have

$$\mathbf{P} = \chi_e \mathbf{E} , \tag{3.41}$$

where the constant χ_e is called the electrical susceptibility. Therefore,

$$\mathbf{D} = (1 + 4\pi \chi_e) \mathbf{E} = \epsilon \mathbf{E} . \tag{3.42}$$

Here, ϵ is called the dielectric permittivity of the medium. Note that ϵ is a dimensionless quantity in Gaussian system and is also called the dielectric constant in this system. (In SI system, the dielectric permittivity $\epsilon = \epsilon_0 [1 + 4\pi \chi_e]$ is not dimensionless and the dimensionless quantity ϵ/ϵ_0 is called the dielectric constant.) If the isotropic medium (space rotation symmetry) is also homogeneous (space translation symme-

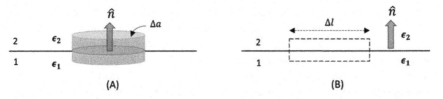

Fig. 3.2 **A** Gaussian pillbox and **B** Rectangular loop across the interface of two linear dielectrics

try), then ϵ is independent of position. Taking the divergence (spatial derivative) on both sides of Eq. (3.42) gives $\nabla \cdot \mathbf{D} = \nabla \cdot (\epsilon \mathbf{E}) = \epsilon \nabla \cdot \mathbf{E} = 4\pi \rho$. Consequently,

$$\nabla \cdot \mathbf{E} = \frac{4\pi \rho}{\epsilon} . \tag{3.43}$$

This suggests that the electric field generated by the charge distribution inside a macroscopic body is reduced by a factor of $\frac{1}{\epsilon}$ compared to that in a vacuum. The polarization induced by the electric field inside the body acts in opposition to it, thereby reducing the overall electric field. There is a great practical use of this phenomenon. Take the example of a parallel plate capacitor. The overall electric field inside two capacitor plates of given charge and geometry separated by a dielectric as opposed to air or vacuum will be reduced by a factor of $\frac{1}{\epsilon}$, where ϵ is the dielectric constant of the medium. Consequently, the capacitance which is inversely proportional to the potential difference (or electric field) between the plates will be increased by a factor of ϵ.

3.4 Boundary Conditions at the Interface of Two Linear Media

Let us now work out the boundary condition of $\mathbf{E}$ and $\mathbf{D}$ fields at the interface between two linear media with dielectric constants ϵ_1 and ϵ_2, respectively. To find out the behavior of $\mathbf{D}$ across the interface, we apply the integral form of Eq. (3.39), i.e., $\oint_{\partial V} \mathbf{D} \cdot \hat{\mathbf{n}} \, da = 4\pi \int_V \rho \, d^3x$ over a cylindrical Gaussian pillbox of vanishingly small thickness, half of which is above and the other half below the interface (Fig. 3.2A). Suppose, the area of each of the two flat surfaces parallel to the interface plane is Δa. Assuming that there is free or excess charge density σ at the interface, $\int_V \rho \, d^3x = \sigma \, \Delta a$. While for the curved surface, $\int \mathbf{D} \cdot \hat{\mathbf{n}} \, da = 0$. Consequently, $(\mathbf{D}_1 - \mathbf{D}_2) \cdot \hat{\mathbf{n}} \, \Delta a = 4\pi \sigma \, \Delta a$, where $\mathbf{D}_1$ and $\mathbf{D}_2$ are the displacement vectors near the interface, corresponding to the media with dielectric constants ϵ_1 (medium 1) and ϵ_2 (medium 2), respectively. It follows that

$$(\mathbf{D}_1 - \mathbf{D}_2) \cdot \hat{\mathbf{n}} = 4\pi \sigma . \tag{3.44}$$

Here, $\hat{\mathbf{n}}$ is the unit vector normal to the interface pointing from medium 1 toward medium 2. Thus, the normal component of $\mathbf{D}$ is discontinuous at the interface with free charge. Otherwise, if the surface charge is only due to the polarization, the normal component of $\mathbf{D}$ is continuous across the interface. To find out the boundary condition for $\mathbf{E}$ field, we apply the integral form of Eq. (3.40), i.e., $\oint \mathbf{E} \cdot \mathbf{dl} = 0$ over a rectangular loop straddling the interface (Fig. 3.2B). The contribution from the vanishingly small arms vertical to the interface is zero. If the length of the parallel arms is Δl each, then $(\mathbf{E}_1 - \mathbf{E}_2).\Delta \mathbf{l} = 0$. We formally write down the boundary condition using the unit vector $\hat{\mathbf{n}}$ normal to the interface as

$$(\mathbf{E}_1 - \mathbf{E}_2) \times \hat{\mathbf{n}} = 0 . \tag{3.45}$$

Thus, the parallel component of the electric field $\mathbf{E}$ is continuous across the interface, regardless of whether there is free charge at the interface or not.

Note that the parallel component of $\mathbf{D}$ is not continuous across the interface as $\nabla \times \mathbf{D} \neq 0$, in general. Only when $\nabla \times \mathbf{P}$ vanishes, parallel component $\mathbf{D}$ becomes continuous.

3.5 Boundary Value Problems in Presence of Dielectrics

3.5.1 Image Problem Involving Dielectrics

Consider a boundary between two dielectric media labeled as 1 and 2 with dielectric constants ϵ_1 and ϵ_2, respectively, at $z = 0$. The medium 1 occupies the half-space $z > 0$, while medium 2, the half-space $z < 0$ (Fig. 3.3). Let there be a point charge q in medium 1 at $\mathbf{x}_0 = (0, 0, z_0)$. We want to find the solutions of the following equations:

$$\nabla^2 \Phi = 4\pi q \delta(\mathbf{x} - \mathbf{x}_0), \quad z > 0, \tag{3.46}$$
$$\nabla^2 \Phi = 0, \qquad\qquad z < 0. \tag{3.47}$$

To find the potential for $z > 0$, we place an image charge q' at $z = -z_0$. The potential anywhere at $z > 0$ due to the two charges q and q' at $z = \pm z_0$ in a medium of dielectric constant ϵ_1 is given by

$$\Phi_{z>0}(\mathbf{x}) = \frac{1}{\epsilon_1} \left(\frac{q}{\sqrt{x^2 + y^2 + (z - z_0)^2}} + \frac{q'}{\sqrt{x^2 + y^2 + (z + z_0)^2}} \right) . \tag{3.48}$$

There is no charge anywhere for $z < 0$ in the original problem. The potential for $z < 0$ is then given by adding an image charge q'' exactly at the location of the charge q and finding the potential due to the combined charges $q + q''$ in a medium of dielectric constant ϵ_2. The potential is

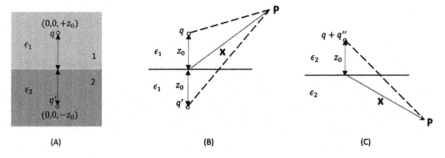

Fig. 3.3 **A** Image problem for a charge near the boundary between two semi-infinite dielectric media. **B** Image charge configuration for calculating potential anywhere in medium 1. **C** Image charge configuration for calculating potential anywhere in medium 2

$$\Phi_{z<0}(\mathbf{x}) = \frac{1}{\epsilon_2} \frac{q + q''}{\sqrt{x^2 + y^2 + (z - z_0)^2}}. \tag{3.49}$$

Noting that there is no free charge at the boundary between the two dielectrics, we now apply the boundary conditions Eqs. (3.44) and (3.45) at $z = 0$ to arrive at the following equations:

$$\epsilon_1 E_z|_{z=0+} = \epsilon_2 E_z|_{z=0-}, \tag{3.50}$$

$$E_{x,y}|_{z=0+} = E_{x,y}|_{z=0-}. \tag{3.51}$$

We now express the Cartesian components of the electric field $\mathbf{E}$ in terms of spatial derivatives of the potential Φ and obtain

$$-\epsilon_1 \frac{\partial \Phi_{z>0}}{\partial z}\bigg|_{z=0} = -\epsilon_2 \frac{\partial \Phi_{z<0}}{\partial z}\bigg|_{z=0}, \tag{3.52}$$

$$-\epsilon_1 \frac{\partial \Phi_{z>0}}{\partial x, y}\bigg|_{z=0} = -\epsilon_2 \frac{\partial \Phi_{z<0}}{\partial x, y}\bigg|_{z=0}, \tag{3.53}$$

which give

$$q' = -q'', \tag{3.54}$$

$$\frac{1}{\epsilon_1}(q + q') = \frac{1}{\epsilon_2}(q + q''). \tag{3.55}$$

Eliminating q'' from Eqs. (3.54) and (3.55), we obtain

$$q' = \left(\frac{\epsilon_1 - \epsilon_2}{\epsilon_1 + \epsilon_2}\right) q. \tag{3.56}$$

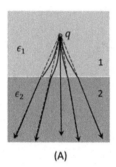
(A)

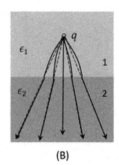
(B)

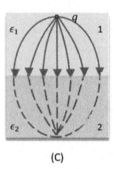
(C)

Fig. 3.4 Electric field lines for a point charge q near the boundary between two dielectrics. **A** For $\epsilon_1 > \epsilon_2$, **B** For $\epsilon_1 < \epsilon_2$. The dashed lines in **A** and **B** correspond to the case $\epsilon_1 = \epsilon_2$. **C** For $\epsilon_2 = \infty$

Therefore,

$$q + q'' = q - \left(\frac{\epsilon_1 - \epsilon_2}{\epsilon_1 + \epsilon_2}\right)q = \left(\frac{2\epsilon_2}{\epsilon_1 + \epsilon_2}\right)q. \tag{3.57}$$

The potential at $z > 0$ is given by Eq. (3.48) with q' from Eq. (3.56). On the other hand, Eq. (3.57) is used to calculate the potential at $z < 0$, given by Eq. (3.49). Since $\nabla \cdot \mathbf{D} = 4\pi q\delta(\mathbf{x} - \mathbf{x_0})$, $\nabla \cdot \mathbf{E}$ is zero everywhere except at $\mathbf{x_0}$, the location of the point charge q. It follows that $\nabla \cdot \mathbf{P} = \nabla \cdot (\chi_e \mathbf{E}) = \nabla \chi_e \cdot \mathbf{E} + \chi_e \nabla \cdot \mathbf{E} = \nabla \chi_e \cdot \mathbf{E} + \chi_e 4\pi q\delta(\mathbf{x} - \mathbf{x_0})$. At $z = 0$, the boundary between the two dielectrics $\chi_e = \frac{\epsilon - 1}{4\pi}$ takes a discontinuous jump, leading to non-zero $\nabla \chi_e$. Therefore, $\nabla \cdot \mathbf{P}$ is non-zero at the boundary between medium 1 and 2. The surface-charge density at the boundary is then given by $\bar{\sigma} = -\int_{0^-}^{0^+} \nabla \cdot \mathbf{P}\, dz = -(\mathbf{P_2} - \mathbf{P_1}) \cdot \hat{\mathbf{n}}$ (consistent with the expression for polarization surface charge $\bar{\sigma} = \mathbf{P} \cdot \hat{\mathbf{n}}$ discussed before). Here, $\hat{\mathbf{n}} = -\hat{\mathbf{z}}$ is the unit normal at the boundary directed from medium 1 to 2. Therefore,

$$\bar{\sigma} = \left(\frac{\epsilon_2 - 1}{4\pi}\mathbf{E_2} - \frac{\epsilon_1 - 1}{4\pi}\mathbf{E_1}\right) \cdot \hat{z} \tag{3.58}$$

$$= \frac{1}{4\pi}\left[-(\epsilon_2 - 1)\,\nabla\Phi_{z<0}|_{z=0} + (\epsilon_1 - 1)\,\nabla\Phi_{z>0}|_{z=0}\right] \cdot \hat{z}. \tag{3.59}$$

Note that the direction of $\nabla\Phi_{z<0}$ is exactly opposite to $\hat{z}$ at $z = 0$. Replacing $\Phi_{z<0}$ and $\Phi_{z>0}$ from Eq. (3.49) and Eq. (3.48), respectively, in Eq. (3.59),

$$\bar{\sigma} = \frac{1}{4\pi}\left[\frac{(\epsilon_2 - 1)}{\epsilon_2}\frac{(q + q'')(-z_0)}{(x^2 + y^2 + z_0^2)^{\frac{3}{2}}} + \frac{(\epsilon_1 - 1)}{\epsilon_1}\frac{(q - q')(z_0)}{(x^2 + y^2 + z_0^2)^{\frac{3}{2}}}\right]. \tag{3.60}$$

Again, using the expressions given by Eqs. (3.54) and (3.57) in Eq. (3.60), we finally obtain

$$\bar{\sigma} = \frac{1}{4\pi} \frac{z_0(q + q'')}{(x^2 + y^2 + z_0^2)^{\frac{3}{2}}} \left[-\frac{(\epsilon_2 - 1)}{\epsilon_2} + \frac{(\epsilon_1 - 1)}{\epsilon_1} \right]$$

$$= \frac{q}{2\pi\epsilon_1} \frac{(\epsilon_1 - \epsilon_2)}{(\epsilon_1 + \epsilon_2)} \cdot \frac{z_0}{(x^2 + y^2 + z_0^2)^{\frac{3}{2}}} \cdot \tag{3.61}$$

For $\epsilon_1 = \epsilon_2 = \epsilon$, the induced charge $\bar{\sigma}$ at the interface $z = 0$ becomes zero. This effectively means that the electric field varies as $\frac{q}{\epsilon r^2}$ (r being the distance from charge q) over the entire region (shown by dashed lines Fig. 3.4A, B). When $\epsilon_1 < \epsilon_2$, the induced charge at the interface is negative and the field lines are attractive, whereas for $\epsilon_1 > \epsilon_2$, field lines are repulsive. For $\epsilon_1 \ll \epsilon_2$, medium 2 increasingly behaves like a conductor. In the limit $\epsilon_2 = \infty$, the potential inside the conductor vanishes and the field lines at $z = 0$ become perpendicular to the interface (Fig. 3.4C).

3.5.2 Dielectric Sphere in Uniform External Field

Consider a sphere of radius R and dielectric constant ϵ, placed in an originally uniform external electric field E_0 along z-axis. There are a few points to be noted at the outset. Since $\nabla \times \mathbf{E} = 0$ everywhere, the electric field $\mathbf{E}$ at any point in space can be expressed as the gradient of the scalar potential Φ at that point. For a homogeneous, isotropic medium, $\mathbf{D}(\mathbf{x}) = \epsilon \mathbf{E}(\mathbf{x})$, where ϵ is the dielectric constant of the medium. Since $\nabla \cdot \mathbf{D} = 0$, with the macroscopic charge density ρ being zero inside the sphere, it follows that $\nabla \cdot \mathbf{E} = 0$. The electric field outside the sphere has zero divergence anyway. Hence, the scalar potential Φ satisfies the Laplace equation both inside and outside the sphere, and we want to find out the potential in these two regions.

The azimuthal symmetry in the problem suggests that the solution must be an expansion in terms of the Legendre polynomials.[5] For $r < R$, the solution must not have any singularity at $r = 0$. Therefore,

$$\Phi_{in}(r, \theta) = \sum_{l=0}^{\infty} A_l \, r^l P_l(\cos\theta). \tag{3.62}$$

For $r > R$, potential must not blow up at infinity. Plus, at large distances, the potential must be $-E_0 \, r \cos\theta$ such that it leads to uniform electric field E_0 along z-axis. It follows that the solution outside the sphere should be

$$\Phi_{out}(r, \theta) = -E_0 \, r \, \cos\theta + \sum_{l=0}^{\infty} B_l \, r^{-(l+1)} P_l(\cos\theta). \tag{3.63}$$

---

[5] See boundary value problems in spherical polar coordinates discussed in detail in Chap. 2. The general solution is given by Eq. (2.94).

The second term in the expansion for $r > R$ is associated with the polarization charge induced on the sphere. The boundary conditions at the surface of the sphere are

$$- \frac{1}{r} \frac{\partial \Phi_{in}}{\partial \theta} \bigg|_{r=R} = - \frac{1}{r} \frac{\partial \Phi_{out}}{\partial \theta} \bigg|_{r=R}, \tag{3.64}$$

$$- \epsilon \frac{\partial \Phi_{in}}{\partial r} \bigg|_{r=R} = - \frac{\partial \Phi_{out}}{\partial r} \bigg|_{r=R}. \tag{3.65}$$

The boundary conditions given by Eqs. (3.64) and (3.65) describe the continuity of the tangential component of $\mathbf{E}$ and the normal component of $\mathbf{D}$, respectively, at the surface of the sphere. From Eq. (3.64),

$$\sum_{l=0}^{\infty} A_l R^{l-1} P_l' = -E_0 P_1' + \sum_{l=0}^{\infty} B_l R^{-(l+1)} P_l'. \tag{3.66}$$

Comparing the coefficients on both sides for $l = 1$ and $l > 1$, respectively, we get

$$A_1 = -E_0 + B_1 R^{-2}, \tag{3.67}$$

$$A_l = B_l R^{-(2l+1)}. \tag{3.68}$$

From Eq. (3.65),

$$\epsilon \sum_{l=0}^{\infty} A_l l R^{l-1} P_l = -E_0 \cos \theta - \sum_{l=0}^{\infty} B_l (l+1) R^{-(l+2)} P_l. \tag{3.69}$$

Again, comparing the coefficients on both sides for $l = 1$ and $l > 1$, respectively, we have

$$\epsilon A_1 = -E_0 - 2B_1 R^{-3}, \tag{3.70}$$

$$\epsilon A_l l = B_l (l+1) R^{-(2l+1)}. \tag{3.71}$$

Clearly, for $l > 1$, the two boundary conditions cannot be simultaneously satisfied unless $A_l = B_l = 0$. On the other hand, for $l = 1$, we have

$$A_1 = \frac{-3}{\epsilon + 2} E_0, \tag{3.72}$$

$$B_1 = \frac{\epsilon - 1}{\epsilon + 2} R^3 E_0. \tag{3.73}$$

Plugging in the values of the coefficients in Eq. (3.62) and Eq. (3.63), respectively, we obtain the solution.

$$\Phi_{in}(r, \theta) = \left(\frac{-3}{\epsilon + 2}\right) E_0 r \cos \theta. \tag{3.74}$$

$$\Phi_{out}(r, \theta) = -E_0 r \cos \theta + \left(\frac{\epsilon - 1}{\epsilon + 2}\right) \frac{R^3}{r^2} E_0 \cos \theta. \tag{3.75}$$

Note that when $\epsilon \to \infty$, we recover the result for the conducting sphere discussed in Chap. 2.

The electric field inside the sphere is given by

$$\mathbf{E}_{in} = -\nabla \Phi_{in} = \left(\frac{3}{\epsilon + 2}\right) E_0 \hat{z}. \tag{3.76}$$

The electric field is constant inside the sphere and along the direction of the applied electric field E_0 (for $\epsilon > 1$, $E < E_0$). The displacement vector inside the sphere is

$$\mathbf{D}_{in} = \epsilon \mathbf{E}_{in} = \left(\frac{3\epsilon}{\epsilon + 2}\right) E_0 \hat{z}. \tag{3.77}$$

Therefore, the polarization $\mathbf{P}$ is

$$\mathbf{P} = \frac{1}{4\pi}(\mathbf{D}_{in} - \mathbf{E}_{in}) = \frac{3}{4\pi}\left(\frac{\epsilon - 1}{\epsilon + 2}\right) E_0 \hat{z}. \tag{3.78}$$

The polarization is uniform inside the sphere. The dipole moment $\mathbf{p}$ of the sphere is then

$$\mathbf{p} = \frac{4\pi}{3} R^3 \mathbf{P} = \left(\frac{\epsilon - 1}{\epsilon + 2}\right) R^3 E_0 \hat{z}. \tag{3.79}$$

The corresponding potential outside the sphere is

$$\Phi_{dip} = \frac{\mathbf{p} \cdot \mathbf{x}}{r^3} = \left(\frac{\epsilon - 1}{\epsilon + 2}\right) R^3 E_0 \frac{\hat{z} \cdot \mathbf{x}}{r^3} = \left(\frac{\epsilon - 1}{\epsilon + 2}\right) \frac{R^3}{r^2} E_0 \cos \theta. \tag{3.80}$$

This is precisely the additional term in the expression for the potential outside the sphere in Eq. (3.75). It is now clear that the potential outside the sphere is due to the external electric field plus the induced dipole moment $\mathbf{p}$ of the sphere given by Eq. (3.79). The macroscopic charge density inside the sphere is zero. The volume polarization-charge density is zero as well because the polarization everywhere inside the sphere is uniform or $\nabla \cdot \mathbf{P} = 0$. However, the polarization surface-charge density is non-zero and is given by

$$\bar{\sigma} = \mathbf{P} \cdot \hat{\mathbf{r}} = P \cos \theta = \frac{3}{4\pi}\left(\frac{\epsilon - 1}{\epsilon + 2}\right) E_0 \cos \theta. \tag{3.81}$$

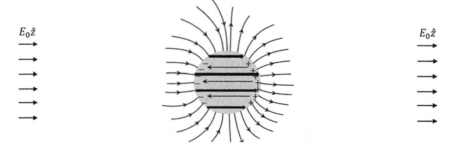

$E_0\hat{z}$ $E_0\hat{z}$

Fig. 3.5 Dielectric sphere in a uniform external electric field: the thick arrows inside the sphere are the induced dipole moment fields; the thin arrows inside the sphere represent field due to polarization charges

The surface charges produce an internal field that is directed opposite to the applied electric field, thus reducing the overall field inside the sphere (Fig. 3.5).

3.5.3 Dielectric with Spherical Cavity in Presence of Uniform Electric Field

Let us consider a dielectric medium with dielectric constant ϵ having a spherical cavity of radius R. The uniform external electric field is E_0 parallel to z-axis. This is an inversion of problem of the dielectric sphere in uniform electric field. The problem is easy to solve once we recognize that the relative dielectric constant of the cavity to that of the surrounding dielectric medium is now $\frac{1}{\epsilon}$. Then we can immediately write down the potential inside and outside the cavity as

$$\Phi_{\text{in}}(r, \theta) = \left(\frac{-3\epsilon}{1+2\epsilon}\right) E_0 r \cos\theta. \tag{3.82}$$

$$\Phi_{\text{out}}(r, \theta) = -E_0 r \cos\theta + \left(\frac{1-\epsilon}{1+2\epsilon}\right) \frac{R^3}{r^2} E_0 \cos\theta. \tag{3.83}$$

The electric field inside the cavity is uniform, parallel to external electric field and is given by

$$\mathbf{E}_{\text{in}} = -\nabla\Phi_{\text{in}} = \left(\frac{3\epsilon}{1+2\epsilon}\right) E_0\hat{z}. \tag{3.84}$$

The electric field outside the cavity is the external electric field plus the electric field due to the dipole moment (oriented opposite to the external field) at the center of the cavity. The dipole moment inside the cavity is

$$\mathbf{p} = \frac{4\pi}{3} R^3 \mathbf{P} = \left(\frac{1-\epsilon}{1+2\epsilon}\right) R^3 E_0 \hat{z} . \tag{3.85}$$

3.6 Multipole Magnetic Moments

The magnetic vector potential due to a localized current distribution $\mathbf{J}(\mathbf{x}')$ at any point $\mathbf{x}$ is given by

$$\mathbf{A}(\mathbf{x}) = \frac{1}{c} \int \frac{\mathbf{J}(\mathbf{x}')}{|\mathbf{x} - \mathbf{x}'|} d^3 x' . \tag{3.86}$$

We expand the vector potential to the lowest order approximation using Eq. (3.2) as follows:

$$
\begin{aligned}
\mathbf{A}(\mathbf{x}) &= \frac{1}{c} \int \mathbf{J}(\mathbf{x}') \left[\frac{1}{r} + \frac{\mathbf{x} \cdot \mathbf{x}'}{r^3} + \frac{1}{2r^5} \sum_{ij} \left(3x_i' x_j' - r'^2 \delta_{ij} \right) x_i x_j + \cdots \right] d^3 x' \\
&= \frac{1}{cr} \int \mathbf{J}(\mathbf{x}') d^3 x' + \frac{1}{cr^3} \int \mathbf{J}(\mathbf{x}') \mathbf{x} \cdot \mathbf{x}' d^3 x' + \cdots ,
\end{aligned}
\tag{3.87}
$$

where $r = |\mathbf{x}|$. Noting that $\nabla' \cdot \mathbf{J} = 0$ for a steady current distribution and using standard vector identity, we have $\nabla' \cdot (x_i' \mathbf{J}) = \nabla' x_i' \cdot \mathbf{J} + x_i' \nabla' \cdot \mathbf{J} = \nabla' x_i' \cdot \mathbf{J}$. Taking the volume integral on both sides and using divergence theorem, we get $\oint_{\partial V} x_i' \mathbf{J} \cdot \hat{\mathbf{n}} \, da = \int_V \nabla' x_i' \cdot \mathbf{J} \, d^3 x$. Since the surface enclosing the volume is outside the current distribution, $\mathbf{J} = 0$ on ∂V and the surface integral vanishes. We are now left with $\int_V \nabla' x_i' \cdot \mathbf{J} \, d^3 x = \int_V J_i \, d^3 x = 0$. Since the relation is true for any component J_i,

$$\int_V \mathbf{J} \, d^3 x = 0 . \tag{3.88}$$

The first term of the expansion in Eq. (3.87) is called the monopole term, similar to the case of electrostatic expansion. The monopole term for a localized steady current distribution thus vanishes unlike the case of electrostatics.

The second term in Eq. (3.87) can be simplified similarly. In this case, we note $\nabla' \cdot (x_i' x_j' \mathbf{J}) = \nabla' x_i' \cdot x_j' \mathbf{J} + \nabla' x_j' \cdot x_i' \mathbf{J} + x_i' x_j' \nabla' \cdot \mathbf{J} = x_j' \nabla' x_i' \cdot \mathbf{J} + x_i' \nabla' x_j' \cdot \mathbf{J} = x_j' J_i + x_i' J_j$. Again, using divergence theorem and pushing the surface outside the current distribution, we have $\oint_{\partial V} (x_i' x_j' \mathbf{J}) \cdot \hat{\mathbf{n}} \, da = \int_V (x_j' J_i + x_i' J_j) d^3 x'$. The surface integral vanishes as $\mathbf{J} = 0$ outside the current distribution. Therefore,

$$\int_V x_j' J_i \, d^3 x' = - \int_V x_i' J_j \, d^3 x' . \tag{3.89}$$

We turn our attention to the dipole term in Eq. (3.87). We note that

$$\int \mathbf{x} \cdot \mathbf{x}' J_i \, d^3x' = \sum_{j=1}^{3} x_j \int x_j' J_i \, d^3x' = -\sum_{j=1}^{3} x_j \int_V x_i' J_j \, d^3x' \,. \tag{3.90}$$

Therefore,

$$\int \mathbf{x} \cdot \mathbf{x}' J_i \, d^3x' = -\frac{1}{2} \sum_{j=1}^{3} x_j \int (x_i' J_j - x_j' J_i) \, d^3x'$$

$$= -\frac{1}{2} \sum_{j=1}^{3} \sum_{k=1}^{3} \epsilon_{ijk} \int (\mathbf{x}' \times \mathbf{J})_k \, d^3x' \tag{3.91}$$

$$= -\frac{1}{2} \left[\mathbf{x} \times \int (\mathbf{x}' \times \mathbf{J}) \, d^3x' \right]_i \,. \tag{3.92}$$

The relation is true for any component J_i. Consequently,

$$\int (\mathbf{x} \cdot \mathbf{x}') \mathbf{J} \, d^3x' = -\frac{1}{2} \left[\mathbf{x} \times \int (\mathbf{x}' \times \mathbf{J}) \, d^3x' \right] \,. \tag{3.93}$$

We define the magnetic dipole moment of the current distribution as

$$\mathbf{m} = \frac{1}{2c} \int (\mathbf{x}' \times \mathbf{J}) \, d^3x' \,, \tag{3.94}$$

such that the dipolar vector potentialis given by

$$\mathbf{A}_{\text{dip}}(\mathbf{x}) = \frac{\mathbf{m} \times \mathbf{x}}{r^3} \,. \tag{3.95}$$

The magnetization or the magnetic dipole moment per unit volume is given by

$$\mathbf{M}(\mathbf{x}') = \frac{1}{2c} \mathbf{x}' \times \mathbf{J}(\mathbf{x}') \,. \tag{3.96}$$

The magnetic field corresponding to each dipole vector potential $\mathbf{A}_{\text{dip}}$ can be obtained by using $\mathbf{B}_{\text{dip}} = \nabla \times \mathbf{A}_{\text{dip}}$. It is straightforward to prove using standard vector identity

$$\mathbf{B}_{\text{dip}} = \frac{1}{r^3} \left[3(\mathbf{m} \cdot \hat{\mathbf{r}})\hat{\mathbf{r}} - \mathbf{m} \right] \,. \tag{3.97}$$

Similar to the case for the electric dipole, Eq. (3.97) is correct for any point in space outside the current distribution. To calculate the magnetic field anywhere, we consider an idealized point magnetic dipole placed at the origin. We again utilize the identity in Eq. (3.15) to take care of the singularity at the origin in the following way:

$$B_i = \nabla \times \left(\frac{\mathbf{m} \times \mathbf{x}}{r^3} \right)_i$$

$$= -\left[\nabla \times \mathbf{m} \times \nabla \left(\frac{1}{r} \right) \right]_i$$

$$= -\epsilon_{ijk} \partial_j \left(\mathbf{m} \times \nabla \frac{1}{r} \right)_k$$

$$= -\epsilon_{ijk} \partial_j \epsilon_{klm} m_l \partial_m \frac{1}{r}$$

$$= -m_l \epsilon_{kij} \epsilon_{klm} \partial_j \partial_m \frac{1}{r}$$

$$= -m_l \left(\delta_{il} \delta_{jm} - \delta_{im} \delta_{jl} \right) \left(-\frac{4\pi}{3} \delta_{jm} \delta(\mathbf{r}) + \frac{3x_j x_m - r^2 \delta_{jm}}{r^5} \right) .$$

Here, we have introduced the so-called Levi-Civita symbol ϵ_{ijk}. It is sufficient at this point to note that ϵ_{ijk} is antisymmetric, i.e., it gives a minus sign under interchange of any pair of indices, with $\epsilon_{123} = 1$. It reduces to zero if any index is repeated. We shall discuss about the Levi-Civita in detail in Chap. 7. Here, we have also used the identity $\epsilon_{kij} \epsilon_{klm} = \delta_{il} \delta_{jm} - \delta_{im} \delta_{jl}$.[6] Keeping in mind that the repeated indices are summed over, it is now straightforward to show

$$B_i = \frac{8\pi}{3} m_i \delta(\mathbf{r}) + \frac{3x_i (\mathbf{m} \cdot \mathbf{x}) - r^2 m_i}{r^5} . \tag{3.98}$$

Therefore, the exact magnetic field anywhere due to an idealized point magnetic dipole placed at the origin is given by

$$\mathbf{B}_{\mathrm{dip}} = \frac{3\mathbf{x}(\mathbf{m} \cdot \mathbf{x}) - r^2 \mathbf{m}}{r^5} + \frac{8\pi}{3} \mathbf{m} \delta(\mathbf{r})$$

$$= \frac{1}{r^3} \left[3(\mathbf{m} \cdot \hat{\mathbf{r}}) \hat{\mathbf{r}} - \mathbf{m} \right] + \frac{8\pi}{3} \mathbf{m} \delta(\mathbf{r}) . \tag{3.99}$$

Note that the two delta-function terms appearing in Eqs. (3.17) and (3.99) are different from each other in sign as well in the magnitude of the coefficients. It arises mainly from the fact that the source for magnetic moments is not any magnetic equivalent of electric charges. Isolated magnetic charges do not exist in nature or in other words, the magnetic field is divergence-free everywhere.

We demonstrate this explicitly with the help of a point electric dipole of moment $\mathbf{p}$ located at $\mathbf{x}_0$. It is easy to show that the dipole can be described by an equivalent charge distribution $\rho_{\mathrm{eff}}(\mathbf{x}) = -\mathbf{p} \cdot \nabla \delta(\mathbf{x} - \mathbf{x}_0)$. We equate the potential at $\mathbf{x}$ due to the point dipole moment $\mathbf{p}$ at $\mathbf{x}_0$ and the equivalent charge distribution ρ_{eff} and get

---

[6] The proof of the identity follows from the more general one developed for four-dimensional metric in Chap. 7.

$$\int \frac{\rho_{\text{eff}}(\mathbf{x}')}{\mathbf{x} - \mathbf{x}'} d^3 x' = \frac{\mathbf{p} \cdot (\mathbf{x} - \mathbf{x}_0)}{|\mathbf{x} - \mathbf{x}_0|^3} . \tag{3.100}$$

We rewrite the right-hand side in integral form. It follows that

$$\int \frac{\rho(\mathbf{x}')}{\mathbf{x} - \mathbf{x}'} d^3 x' = \int \mathbf{p} \cdot \nabla' \frac{1}{|\mathbf{x} - \mathbf{x}'|} \delta(\mathbf{x}' - \mathbf{x}_0) d^3 x' . \tag{3.101}$$

Note that the minus sign is absorbed since the differentiation is with respect to the primed variable. Integration by parts will simply shift the derivative to the delta function, accompanied by a minus sign, and as a result, we get

$$\int \frac{\rho(\mathbf{x}')}{\mathbf{x} - \mathbf{x}'} d^3 x' = -\int \frac{1}{|\mathbf{x} - \mathbf{x}'|} \mathbf{p} \cdot \nabla' \delta(\mathbf{x}' - \mathbf{x}_0) d^3 x' . \tag{3.102}$$

Assuming that the integrals are true at every point in space, we have the following expression for the equivalent charge distribution of a point electric dipole located at $\mathbf{x}_0$:

$$\rho_{\text{eff}}(\mathbf{x}) = -\mathbf{p} \cdot \nabla \delta(\mathbf{x} - \mathbf{x}_0) . \tag{3.103}$$

Here, we have replaced the dummy primed variable by unprimed variable. Further, if the point dipole is located at the origin, we can simply write

$$\rho_{\text{eff}} = -\mathbf{p} \cdot \nabla \delta(\mathbf{r}) . \tag{3.104}$$

Hence, the divergence of the corresponding dipolar electric field using Gauss's law can be written as

$$\nabla \cdot \mathbf{E}_{\text{dip}} = -4\pi \mathbf{p} \cdot \nabla \delta(\mathbf{r}) . \tag{3.105}$$

Again from Eq. (1.24), we can also write

$$\nabla \cdot \mathbf{E}_{\text{dip}} = \nabla \cdot \frac{1}{r^3} \left[3(\mathbf{p} \cdot \hat{\mathbf{r}}) \hat{\mathbf{r}} - \mathbf{p} \right] - \frac{4\pi}{3} \mathbf{p} \cdot \nabla \delta \mathbf{r} . \tag{3.106}$$

It follows that

$$\nabla \cdot \frac{1}{r^3} \left[3(\mathbf{p} \cdot \hat{\mathbf{r}}) \hat{\mathbf{r}} - \mathbf{p} \right] = -\frac{8\pi}{3} \mathbf{p} \cdot \nabla \delta(\mathbf{r}) . \tag{3.107}$$

Replacing $\mathbf{p}$ by $\mathbf{m}$ and utilizing the resulting expression in Eq. (3.99), we find that the magnetic field is indeed divergence-free, as expected.[7]

$$\nabla \cdot \mathbf{B}_{\text{dip}} = 0 . \tag{3.108}$$

---

[7] We could have presented arguments similar to that in Sect. 3.1.1 and checked the consistency of the expression for the point dipole field with Maxwell's equations. This is left as an exercise.

3.7 Magnetized Matter

What is a 'magnetized matter'? How do we 'magnetize' an object? In this section, we shall briefly dwell on these questions. A 'magnetized' object produces a magnetic field of its own and the object is 'magnetized' in presence of the external magnetic field. Depending on the type of material, the response to the external magnetic field could be qualitatively and quantitatively different and accordingly, we can classify them. Although the origin of magnetism in the matter is essentially quantum mechanical, we shall restrict ourselves to a classical description that works fine for the problems to be discussed in this chapter.

A macroscopic object is made of specific molecules or atoms. These basic building blocks are associated with microscopic charge distribution which in turn could be responsible for permanent or induced dipole moments. For example, paramagnets and ferromagnets have permanent magnetic dipole moments whereas in diamagnets, the dipole moments are induced, which means that the microscopic dipole moment vanishes once the external field is withdrawn. However, for our purpose, the nature of the magnetic material is irrelevant. All we need to know is how the dipole moments are distributed in material media. We define a macroscopic variable called 'magnetization' $\mathbf{M}$ at a point $\mathbf{x}'$ as the average magnetic moment per unit volume such that the total magnetic dipole moment $\mathbf{m}$ of the macroscopically infinitesimal volume element d^3x' centered at $\mathbf{x}'$ is $\mathbf{M}\, d^3x'$ (Fig. 3.6). If $\mathbf{M}$ is the same everywhere, the macroscopic object is said to be uniformly magnetized.

The central question that we ask is the following. Given a piece of magnetized material of volume V, bounded by closed surface ∂V and with magnetization $\mathbf{M}(\mathbf{x}')$ where $\mathbf{x}'$ is any point in the bulk (inside) and on the surface of the material, what is the magnetic field produced by the material everywhere? We also assume the existence of a macroscopic volume current density $\mathbf{J}(\mathbf{x}')$ due to the flow of free charges inside the material and a macroscopic surface current density $\mathbf{K}(\mathbf{x}')$ due to the flow of free charges on the surface of the materials. We repeat that these are macroscopic currents averaged over microscopic variations. The vector potential at any point $\mathbf{x}$ due to the contributions of free volume current $\mathbf{J}$, surface current $\mathbf{K}$ and the magnetization distribution $\mathbf{M}$ is given by

Fig. 3.6 A magnetized object with free volume and surface current

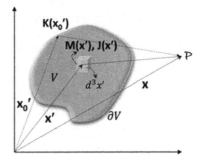

$$\mathbf{A}(\mathbf{x}) = \frac{1}{c} \int_V \frac{\mathbf{J}(\mathbf{x}')}{|\mathbf{x} - \mathbf{x}'|} d^3 x' + \int_V \frac{\mathbf{M}(\mathbf{x}') \times \mathbf{x}}{|\mathbf{x} - \mathbf{x}'|^3} d^3 x' + \frac{1}{c} \int_{\partial V} \frac{\mathbf{K}(\mathbf{x}')}{|\mathbf{x} - \mathbf{x}'|} da' . \quad (3.109)$$

The third term on the right-hand side of Eq. (3.109) is a surface integral over ∂V. Let us, for now, focus our attention on the second term on the right-hand side. Recalling that $\nabla' \frac{1}{\mathbf{x} - \mathbf{x}'} = \frac{\mathbf{x} - \mathbf{x}'}{|\mathbf{x} - \mathbf{x}'|^3}$ and using standard vector identity[8], the second term can be written as

$$\int_V \frac{\mathbf{M}(\mathbf{x}') \times \mathbf{x}}{|\mathbf{x} - \mathbf{x}'|^3} d^3 x' = \int_V \mathbf{M}(\mathbf{x}') \times \nabla' \frac{1}{|\mathbf{x} - \mathbf{x}'|}$$

$$= -\int_V \nabla' \times \frac{\mathbf{M}(\mathbf{x}')}{|\mathbf{x} - \mathbf{x}'|} d^3 x'$$

$$+ \int_V \frac{\nabla' \times \mathbf{M}(\mathbf{x}')}{|\mathbf{x} - \mathbf{x}'|} d^3 x' . \quad (3.110)$$

There is another useful identity,[9] a look-alike for Gauss's divergence theorem, now involving volume integral of curl of any vector $\mathbf{A}$, given as

$$\int_V (\nabla \times \mathbf{A}) \, d^3 x = -\int_{\partial V} \mathbf{A} \times \hat{\mathbf{n}} \, da = \int_{\partial V} (\hat{\mathbf{n}} \times \mathbf{A}) \, da . \quad (3.111)$$

Using the identity, we further convert the first term on the right-hand side of Eq. (3.110) to a surface integral as

$$\int_V \nabla' \times \frac{\mathbf{M}(\mathbf{x}')}{|\mathbf{x} - \mathbf{x}'|} d^3 x' = \int_{\partial V} \frac{\hat{\mathbf{n}}' \times \mathbf{M}(\mathbf{x}')}{|\mathbf{x} - \mathbf{x}'|} da' . \quad (3.112)$$

Collecting all the integrals, we have from Eq. (3.109)

$$\mathbf{A}(\mathbf{x}) = \frac{1}{c} \int_V \frac{\mathbf{J}(\mathbf{x}')}{|\mathbf{x} - \mathbf{x}'|} d^3 x' + \frac{1}{c} \int_{\partial V} \frac{\mathbf{K}(\mathbf{x}')}{|\mathbf{x} - \mathbf{x}'|} da'$$

$$- \int_{\partial V} \frac{\hat{\mathbf{n}}' \times \mathbf{M}(\mathbf{x}')}{|\mathbf{x} - \mathbf{x}'|} da' + \int_V \frac{\nabla' \times \mathbf{M}(\mathbf{x}')}{|\mathbf{x} - \mathbf{x}'|} d^3 x' .$$

Finally, we have an expression for the vector potential $\mathbf{A}$ due to the combined effect of magnetization and free current densities:

$$\mathbf{A}(\mathbf{x}) = \frac{1}{c} \int_V \frac{\mathbf{J}(\mathbf{x}') + c \nabla' \times \mathbf{M}(\mathbf{x}')}{|\mathbf{x} - \mathbf{x}'|} d^3 x' + \frac{1}{c} \int_{\partial V} \frac{\mathbf{K}(\mathbf{x}') + c \mathbf{M}(\mathbf{x}') \times \hat{\mathbf{n}}'}{|\mathbf{x} - \mathbf{x}'|} da' . \quad (3.113)$$

We conclude that the magnetization distribution $\mathbf{M}(\mathbf{x})$ in a magnetized object generates an effective volume current density,

---

[8] $\nabla \times (\phi \mathbf{V}) = \nabla \phi \times \mathbf{V} + \phi \nabla \times \mathbf{V}$.

[9] The proof of the identity is straightforward and is left as an exercise.

$$\mathbf{J}_M(\mathbf{x}) = c\nabla \times \mathbf{M}(\mathbf{x}), \tag{3.114}$$

and an effective surface current density,

$$\mathbf{K}_M(\mathbf{x}) = c\mathbf{M}(\mathbf{x}) \times \hat{\mathbf{n}}'. \tag{3.115}$$

Henceforth, we shall call them 'bound' magnetization currents. The volume and surface magnetization currents have the same origin and they are real, physical currents producing magnetic fields in the same way as free currents. The magnetization or, equivalently, the distribution of magnetic dipole moments can be represented by tiny localized current loops. If the magnetization is non-uniform, the area vector as well as the current associated with these loops might vary accordingly, giving rise to a net macroscopic volume as well as surface current distribution. If an object is uniformly magnetized, the magnetization volume current is zero although the corresponding surface current may not necessarily be zero.

Ampere's law given by Eq. (1.64) can now be written in terms of the 'free charge current' $\mathbf{J}$ and the 'bound magnetization current' $\mathbf{J}_M$ as follows:

$$\begin{aligned}
\nabla \times \mathbf{B} &= \frac{4\pi}{c}(\mathbf{J} + \mathbf{J}_M) \\
&= \frac{4\pi}{c}\mathbf{J} + 4\pi\nabla \times \mathbf{M}.
\end{aligned} \tag{3.116}$$

We now define the so-called 'auxiliary' magnetic field $\mathbf{H}$ which is a part of the 'total' magnetic field and is generated only by free charge currents.

$$\mathbf{H} = \mathbf{B} - 4\pi\mathbf{M}. \tag{3.117}$$

Equation (3.116) then reduces to the following expression of Ampere's law in terms of free charge current alone:

$$\nabla \times \mathbf{H} = \frac{4\pi}{c}\mathbf{J}. \tag{3.118}$$

In general, for weakly magnetized medium (usually non-ferromagnetic), the magnetization $\mathbf{M}$ is linearly proportional to the external magnetic field $\mathbf{H}$ and is given as

$$\mathbf{M} = \chi_m\mathbf{H}. \tag{3.119}$$

The constant of proportionality χ_m is called the magnetic susceptibility of the medium. If the medium is isotropic, the field $\mathbf{B}$ is linearly proportional to $\mathbf{H}$ as follows:

$$\mathbf{B} = \mu\mathbf{H}. \tag{3.120}$$

Here, μ is called the magnetic permeability of the medium. Combining Eqs. (3.119) and (3.120), it is straightforward to show that

$$\mu = 1 + 4\pi \chi_m \,. \tag{3.121}$$

Clearly, when $\mu > 1$, χ_m is positive (paramagnets), whereas for $\mu < 1$, χ_m is negative (diamagnets).

3.8 Boundary Conditions on B and H at the Interface Between Two Linear Isotropic Media

Let us now work out the boundary condition of **B** and **H** fields at the interface between two linear isotropic magnetic media with permeabilities μ_1 and μ_2, respectively. To find out the behavior of **B** across the interface, we apply the integral form of $\nabla \cdot \mathbf{B} = 0$, i.e., $\oint_{\partial V} \mathbf{B} \cdot \hat{\mathbf{n}} \, da = 0$ over a cylindrical Gaussian pillbox (Fig. 3.7A). Suppose, the area of each of the two flat surfaces parallel to the interface plane is Δa. In the limit of vanishingly small thickness of the pillbox, the magnetic flux through the curved surface vanishes. On the other hand, the flux through the flat surface is $(\mathbf{B_1} - \mathbf{B_2}) \cdot \hat{\mathbf{n}} \, \Delta a$, where $\mathbf{B_1}$ and $\mathbf{B_2}$ are the magnetic fields near the interface, corresponding to the media with permeabilities μ_1 (medium 1) and μ_2 (medium 2), respectively. It follows that

$$(\mathbf{B_1} - \mathbf{B_2}) \cdot \hat{\mathbf{n}} = 0. \tag{3.122}$$

Here, $\hat{\mathbf{n}}$ is the unit vector normal to the interface pointing from medium 1 toward medium 2. Thus, the normal component of **B** is continuous across the interface. The boundary condition for the normal component of **H** field follows automatically from Eq. (3.122), by using Eq. (3.117). The magnetization of the two media close to interface is given by $\mathbf{M}_1$ and $\mathbf{M}_2$, respectively. We have

$$(\mathbf{H_1} - \mathbf{H_2}) \cdot \hat{\mathbf{n}} = -4\pi \Delta M^{\perp} \,. \tag{3.123}$$

Here, $\Delta M^{\perp} = (\mathbf{M_1} - \mathbf{M_2}) \cdot \hat{\mathbf{n}}$. Thus, the normal component of **H** field is discontinuous across the interface by an amount proportional to the change in magnetization at the interface.

To find out the boundary condition for parallel component **H** field, we apply the integral form of Eq. (3.118), i.e.,

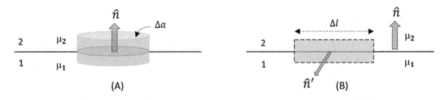

Fig. 3.7 **A** Gaussian pillbox and **B** Rectangular loop across the interface of two linear media

$$\oint \mathbf{H} \cdot \mathbf{dl} = \frac{4\pi}{c} \int_{\partial V} \mathbf{J} \cdot \mathbf{da} \tag{3.124}$$

over a rectangular loop straddling the interface (Fig. 3.7B). The contributions to the line integral from the vanishingly small arms vertical to the interface are zero. If the length of the parallel arms is Δl each, and if there is no free current $\mathbf{K}$ at the interface, then $(\mathbf{H}_1 - \mathbf{H}_2) \cdot \Delta l = 0$. It follows that $(\mathbf{H}_1 - \mathbf{H}_2) \times \hat{\mathbf{n}} = 0$, where $\hat{\mathbf{n}}$ is the unit normal to the interface as shown in Fig. 3.7A. Thus, the parallel component of $\mathbf{H}$ field is continuous across the interface if there is no free charge current at the interface.

Now suppose that the interface carries free surface current $\mathbf{K}$. Then the right-hand side of Eq. (3.124) describing the total flux of current density through the surface attached to the loop is zero only if the plane of the loop, or more precisely, the unit normal $\hat{\mathbf{n}}'$ to the flat surface attached to the loop (Fig. 3.7B), is perpendicular to the direction of free surface current $\mathbf{K}$. Otherwise, the enclosed current will be non-zero due to the existence of surface current $\mathbf{K}$. When the area vector of the loop is parallel to the direction of free surface current $\mathbf{K}$, Eq. (3.124) reduces to $\mathbf{H}_1 - \mathbf{H}_2 \cdot \delta l = \frac{4\pi}{c} K \Delta l$. We formally write down the boundary condition for the parallel components of $\mathbf{H}$ field using the unit vector $\hat{\mathbf{n}}$ normal to the interface as

$$(\mathbf{H}_1 - \mathbf{H}_2) \times \hat{\mathbf{n}} = \frac{4\pi}{c} \mathbf{K} . \tag{3.125}$$

The boundary condition for the parallel components of $\mathbf{B}$ field follows similarly from the integral form of Ampere's law discussed in Chap. 1 and is given by

$$(\mathbf{B}_1 - \mathbf{B}_2) \times \hat{\mathbf{n}} = \frac{4\pi}{c} \mathbf{K} . \tag{3.126}$$

In Eq. (3.126), $\mathbf{K}$ is the total surface current at the interface, including the magnetization current.

3.9 Boundary Value Problems in Magnetostatics

The equations that govern the boundary value problems in presence of magnetized macroscopic linear media are as follows:

$$\nabla \cdot \mathbf{B} = 0 , \tag{3.127}$$

$$\nabla \times \mathbf{H} = \frac{4\pi}{c} \mathbf{J} . \tag{3.128}$$

The vector fields $\mathbf{B}$, $\mathbf{H}$ and $\mathbf{M}$ are related to each other by the following set of equations:

$$\mathbf{B} = \mathbf{H} + 4\pi\mathbf{M}, \tag{3.129}$$

$$\mathbf{M} = \chi_m\mathbf{H}, \tag{3.130}$$

$$\mathbf{B} = \mu\mathbf{H}. \tag{3.131}$$

Typically, the spatial distribution of $\mathbf{M}$ inside the macroscopic object of volume V bounded by surface ∂V is given, which is 'frozen' in time. We are particularly interested in situations where the free current $\mathbf{J}_f = 0$, which means $\nabla \times \mathbf{H} = 0$ everywhere. In such cases, $\mathbf{H}$ can be expressed as a gradient of 'magnetic' scalar potential Φ_M defined as

$$\mathbf{H} = -\nabla\Phi_M. \tag{3.132}$$

It then follows from Eqs. (3.127) and (3.129)

$$\nabla \cdot \mathbf{H} = -4\pi\nabla \cdot \mathbf{M} \tag{3.133}$$

Combining Eqs. (3.132) and (3.133), we finally arrive at the following equation:

$$\nabla^2\Phi_M = 4\pi\nabla \cdot \mathbf{M} = -4\pi\rho_M. \tag{3.134}$$

This is strikingly similar to the Poisson equation we encountered in electrostatics, albeit with an effective or fictitious 'magnetic charge' $\rho_M = -\nabla \cdot \mathbf{M}$. Since we are already familiar with the formal solution of the Poisson equation for electrostatics,[10] we can readily write down the solution in absence of boundaries as

$$\Phi_M(\mathbf{x}) = \int \frac{-\nabla' \cdot \mathbf{M}(\mathbf{x}')}{|\mathbf{x} - \mathbf{x}'|}\, d^3x'. \tag{3.135}$$

Further, if $\mathbf{M}$ is well behaved and localized,

$$\Phi_M(\mathbf{x}) = -\int \nabla' \cdot \frac{\mathbf{M}(\mathbf{x}')}{|\mathbf{x} - \mathbf{x}'|}\, d^3x' + \int \nabla' \frac{1}{|\mathbf{x} - \mathbf{x}'|} \cdot \mathbf{M}(\mathbf{x}')\, d^3x'. \tag{3.136}$$

Interchanging primed and unprimed variables in the gradient term, we can write

$$\nabla' \frac{1}{|\mathbf{x} - \mathbf{x}'|} = -\nabla \frac{1}{|\mathbf{x} - \mathbf{x}'|}. \tag{3.137}$$

Since in the second term, the differentiation is with respect to unprimed variable, we can take it outside the integral which is over primed variables and write

$$\Phi_M(\mathbf{x}) = -\int \nabla' \cdot \frac{\mathbf{M}(\mathbf{x}')}{|\mathbf{x} - \mathbf{x}'|}\, d^3x' - \nabla \cdot \int \frac{\mathbf{M}(\mathbf{x}')}{|\mathbf{x} - \mathbf{x}'|}\, d^3x'. \tag{3.138}$$

---

[10] Refer to Eq. (1.22).

The first volume integral on the right-hand side can be converted into surface integral using divergence theorem. The surface term vanishes when integrated over the surface where $\mathbf{M}$ becomes zero. We are finally left with

$$\Phi_M(\mathbf{x}) = -\nabla \cdot \int \frac{\mathbf{M}(\mathbf{x}')}{|\mathbf{x} - \mathbf{x}'|} \, d^3x'. \tag{3.139}$$

Far from the region of non-vanishing magnetization where $|\mathbf{x}| \gg |\mathbf{x}'|$, the expression of Φ_M in Eq. (3.139) resembles the electric dipolar potential or in other words the $\mathbf{H}$ field is approximately similar to dipolar field far from a localized magnetization distribution.

$$\Phi_M(\mathbf{x}) \simeq -\nabla \left(\frac{1}{|\mathbf{x}|} \right) \cdot \int \mathbf{M}(\mathbf{x}') \, d^3x' = \frac{\mathbf{m} \cdot \mathbf{x}}{r^3}. \tag{3.140}$$

Now suppose, instead of having a localized smoothly varying magnetization distribution, we have a magnetized matter of volume V and bounded surface ∂V where $\mathbf{M}$ falls abruptly from a finite value inside to zero just outside the boundary. It turns out that if we integrate the charge density $\rho_M = -\nabla \cdot \mathbf{M}$ over the volume of a Gaussian pillbox (as in Fig. 3.7) of cross section A and vanishing thickness, a non-zero surface charge σ_M is obtained. The charge density is given by

$$\sigma_M = \frac{1}{A} \int \rho_M \, d^3x = -\frac{1}{A} \int \nabla \cdot \mathbf{M}(\mathbf{x}') \, d^3x' = \frac{1}{A} \int \mathbf{M} \cdot \hat{\mathbf{n}}' \, da'. \tag{3.141}$$

In the infinitesimal limit of cross section A,

$$\sigma_M = \mathbf{M} \cdot \hat{\mathbf{n}}. \tag{3.142}$$

Here, $\hat{\mathbf{n}}$ is the outward drawn unit normal to the surface. The potential Φ_M is then due to the volume charge ρ_M and the surface charge σ_M:

$$\Phi_M(\mathbf{x}) = \int_V \frac{-\nabla' \cdot \mathbf{M}(\mathbf{x}')}{|\mathbf{x} - \mathbf{x}'|} \, d^3x' + \int_{\partial V} \frac{\mathbf{M} \cdot \hat{\mathbf{n}}'}{|\mathbf{x} - \mathbf{x}'|} \, da'. \tag{3.143}$$

We have already discussed how to calculate the vector potential for such a system using magnetization currents. Assuming $J = 0$ and, $K = 0$ and using Eqs. (3.113), (3.109) and (3.115), we have the following expression for the vector potential of a magnetized matter with sharp boundaries.

$$\mathbf{A}(\mathbf{x}) = \int_V \frac{\nabla' \times \mathbf{M}(\mathbf{x}')}{|\mathbf{x} - \mathbf{x}'|} \, d^3x' + \int_{\partial V} \frac{\mathbf{M}(\mathbf{x}') \times \hat{\mathbf{n}}'}{|\mathbf{x} - \mathbf{x}'|} \, da'. \tag{3.144}$$

Note that just as only the surface magnetization current contributes to the vector potential of a uniformly magnetized matter with sharp boundaries, given by Eq. (3.144), in terms of 'magnetic charge' description, the volume integral of

Fig. 3.8 Uniformly
magnetized sphere

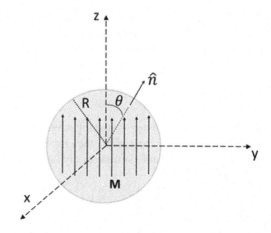

Eq. (3.143) vanishes and only the surface integral has a non-zero contribution to
the magnetic scalar potential.

We now apply the method of solving boundary value problems in magnetostatics
using the magnetic scalar potential for a specific scenario.

3.9.1 A Uniformly Magnetized Sphere

We consider a uniformly magnetized sphere of radius R (Fig. 3.8) without any free
current. Outside the sphere, the magnetization vanishes abruptly. Inside the sphere,

$$\mathbf{M} = M\hat{\mathbf{z}}. \tag{3.145}$$

In this case, the volume magnetic charge density $\rho_M = -\nabla \cdot \mathbf{M} = 0$ and the surface
magnetic charge density $\sigma_M = \mathbf{M} \cdot \hat{\mathbf{n}} = M \cos\theta$.

Since $\rho_M = 0$, the magnetic scalar potential Φ_M satisfies the Laplace equation not
just outside but inside the sphere as well. We want to find out the potential in these
two regions (as we have already done in electrostatics), albeit using the boundary
conditions of magnetostatics. Again, the azimuthal symmetry in the problem suggests
that the solution must be an expansion in terms of the Legendre polynomials. Inside
the sphere, for $r < R$, the solution is

$$\Phi_M^{\text{in}}(r, \theta) = \sum_{l=0}^{\infty} A_l\, r^l P_l(\cos\theta). \tag{3.146}$$

Outside the sphere, for $r > R$,

$$\Phi_M^{\text{out}}(r, \theta) = \sum_{l=0}^{\infty} B_l \, r^{-(l+1)} P_l(\cos \theta) \,. \tag{3.147}$$

Again, inside the sphere,

$$\mathbf{H}_{\text{in}} = -\nabla \Phi_M^{\text{in}} \,, \tag{3.148}$$

$$\mathbf{B}_{\text{in}} = \mathbf{H}_{\text{in}} + 4\pi M \hat{\mathbf{z}} \,. \tag{3.149}$$

Outside the sphere,

$$\mathbf{H}_{\text{out}} = -\nabla \Phi_M^{\text{out}} \,, \tag{3.150}$$

$$\mathbf{B}_{\text{out}} = \mathbf{H}_{\text{out}} \,. \tag{3.151}$$

We now write down the boundary conditions. From Eq. (3.123), we have

$$\left. \frac{\partial \Phi_M^{\text{in}}}{\partial r} \right|_{r=R} - \left. \frac{\partial \Phi_M^{\text{out}}}{\partial r} \right|_{r=R} = 4\pi M \cos \theta \,. \tag{3.152}$$

Moreover, since there is no free charge current, Eq. (3.125) implies that the tangential component of **H** is continuous at the surface of the sphere

$$\left. \frac{\partial \Phi_M^{\text{in}}}{\partial \theta} \right|_{r=R} = \left. \frac{\partial \Phi_M^{\text{out}}}{\partial \theta} \right|_{r=R} \,. \tag{3.153}$$

From Eq. (3.152), we have

$$\sum_{l=0}^{\infty} A_l \, l \, R^{l-1} \, P_l = 4\pi M \cos \theta - \sum_{l=0}^{\infty} B_l \, (l+1) R^{-(l+2)} P_l \,. \tag{3.154}$$

Comparing the coefficients on both sides for $l = 1$ and $l > 1$, respectively,

$$A_1 = 4\pi M - 2B_1 R^{-3} \,, \tag{3.155}$$

$$A_l \, l = B_l (l+1) \, R^{-(2l+1)} \,. \tag{3.156}$$

Again, from Eq. (3.153), we obtain

$$\sum_{l=0}^{\infty} A_l \, R^l \, P_l' = \sum_{l=0}^{\infty} B_l \, R^{-(l+1)} P_l' \,. \tag{3.157}$$

Comparing the coefficients on both sides, we get, for $l = 1$ and $l > 0$,

$$A_1 = B_1 R^{-3}, \tag{3.158}$$

$$A_l = B_l R^{-(2l+1)}. \tag{3.159}$$

Clearly, for $l > 1$, the two boundary conditions cannot be simultaneously satisfied unless $A_l = B_l = 0$. For $l = 1$, we have

$$A_1 = B_1 R^3, \tag{3.160}$$

$$B_1 = \frac{4\pi}{3} R^3 M. \tag{3.161}$$

Plugging in the values of the coefficients in Eq. (3.146) and Eq. (3.147), respectively, we obtain the solution

$$\Phi_M^{\text{in}}(r, \theta) = \frac{4\pi}{3} M r \cos\theta. \tag{3.162}$$

$$\Phi_M^{\text{out}}(r, \theta) = \frac{4\pi R^3 M}{3r^2} \cos\theta = \frac{\mathbf{m} \cdot \mathbf{r}}{r^3}. \tag{3.163}$$

Here, $\mathbf{m} = \frac{4\pi R^3 M}{3}$ is the total dipole moment of the sphere. Therefore, outside the sphere,

$$\mathbf{B}_{\text{out}} = \mathbf{H}_{\text{out}} = \frac{1}{r^3} \left[3(\mathbf{m} \cdot \hat{\mathbf{r}})\hat{\mathbf{r}} - \mathbf{m} \right]. \tag{3.164}$$

Inside the sphere,

$$\mathbf{H}_{\text{in}} = -\frac{4\pi}{3} \mathbf{M}, \tag{3.165}$$

$$\mathbf{B}_{\text{in}} = -\frac{4\pi}{3} \mathbf{M} + 4\pi \mathbf{M} = \frac{8\pi}{3} \mathbf{M}. \tag{3.166}$$

There are a few points to be noted here. First, in general, the magnetic field is dipolar far away from the magnetization distribution. However, in this case, the magnetic field is strictly dipolar even just outside the sphere. Second, inside the sphere, the **B** field and **H** field are constant and oppositely directed. Third, **B** field forms closed loops. However, **H** field lines either originate from or terminate on the surface of the sphere (Fig. 3.9).

We now apply a uniform magnetic field $\mathbf{B}_a$ over all space. Using the principle of linear superposition, for a permanently magnetized sphere (**M** independent of external magnetic field), we can write

$$\mathbf{B}_{\text{in}} = \mathbf{B}_a + \frac{8\pi}{3} \mathbf{M}, \tag{3.167}$$

$$\mathbf{H}_{\text{in}} = \mathbf{B}_a - \frac{4\pi}{3} \mathbf{M}. \tag{3.168}$$

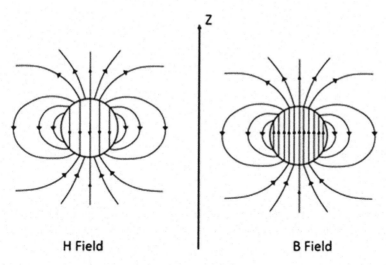

H Field   B Field

Fig. 3.9 'B' and 'H' field lines of a uniformly magnetized sphere

If the sphere is not permanently magnetized or, rather, if **M** itself is caused by the external magnetic field, then the following relation also holds:

$$\mathbf{B}_{\text{in}} = \mu \mathbf{H}_{\text{in}} . \tag{3.169}$$

Therefore,

$$\mathbf{B}_a + \frac{8\pi}{3}\mathbf{M} = \mu(\mathbf{B}_a - \frac{4\pi}{3}\mathbf{M}) . \tag{3.170}$$

The magnetization can now be expressed as follows:

$$\mathbf{M} = \frac{3}{4\pi}\left(\frac{\mu - 1}{\mu + 2}\right)\mathbf{B}_a . \tag{3.171}$$

The above result is valid for a paramagnetic ($\mu > 1$) or a diamagnetic material ($\mu < 1$) and is strikingly analogous to the case of electric polarization **P** of a dielectric sphere placed in external electric field.[11]

3.10   Generalized Ampere's Law in Material Media

We now revisit generalized Ampere's law already discussed in Chap. 1 in the light of the new vector field **H** developed in the present chapter. We are not entirely surprised to note that Eq. (3.128) is only valid for free charge current in steady state, i.e., when

---

[11] See Eq. (3.78).

$\nabla \cdot \mathbf{J} = 0$. For non-steady current,

$$\nabla \cdot \mathbf{J} = -\frac{\partial \rho}{\partial t} . \tag{3.172}$$

Again, from Eq. (3.39),

$$\rho = \frac{1}{4\pi} \nabla \cdot \mathbf{D} . \tag{3.173}$$

It follows that

$$\nabla \cdot \left(\mathbf{J} + \frac{1}{4\pi} \frac{\partial \mathbf{D}}{\partial t} \right) = 0 . \tag{3.174}$$

Clearly, Eq. (3.128) will be valid for non-steady current if we replace $\mathbf{J}$ by $\mathbf{J} + \frac{1}{4\pi} \frac{\partial \mathbf{D}}{\partial t}$. We can now write down generalized Ampere's law in the following form:

$$\nabla \times \mathbf{H} = \frac{4\pi}{c} \left(\mathbf{J} + \frac{1}{4\pi} \frac{\partial \mathbf{D}}{\partial t} \right) = \frac{4\pi}{c} \mathbf{J} + \frac{1}{c} \frac{\partial \mathbf{D}}{\partial t} . \tag{3.175}$$

We can approach the problem from a different direction. We begin with Ampere's law discussed in the first chapter and break up the total current into free charge current $\mathbf{J}$ and magnetization current $\mathbf{J}_M$. Additionally, we introduce a new term $\mathbf{J}_P = \frac{\partial \mathbf{P}}{\partial t}$ and call it polarization current and write

$$\nabla \times \mathbf{B} = \frac{4\pi}{c} \left(\mathbf{J} + c\nabla \times \mathbf{M} + \frac{\partial \mathbf{P}}{\partial t} \right) + \frac{1}{c} \frac{\partial \mathbf{E}}{\partial t} . \tag{3.176}$$

We rearrange the terms as follows:

$$\nabla \times (\mathbf{B} - 4\pi \mathbf{M}) = \frac{4\pi}{c} \mathbf{J} + \frac{1}{c} \frac{\partial}{\partial t} (\mathbf{E} + 4\pi \mathbf{P}) . \tag{3.177}$$

Finally, we arrive at the same equation written down earlier using the argument of local conservation of free charges and get

$$\nabla \times \mathbf{H} = \frac{4\pi}{c} \mathbf{J} + \frac{1}{c} \frac{\partial \mathbf{D}}{\partial t} . \tag{3.178}$$

However, we need to justify the somewhat arbitrary inclusion of the polarization current term $\mathbf{J}_P$. We find that inclusion of $\mathbf{J}_P$ ensures the conservation of bound charges as follows:

$$\nabla \cdot \mathbf{J}_P = \nabla \cdot \left(\frac{\partial \mathbf{P}}{\partial t} \right) = \frac{\partial}{\partial t} (\nabla \cdot \mathbf{P}) . \tag{3.179}$$

Identifying the polarization charge $\bar{\rho} = -\nabla \cdot \mathbf{P}$, we write down the corresponding equation of continuity as

$$\nabla \cdot \mathbf{J}_P = -\frac{\partial \bar{\rho}}{\partial t} . \tag{3.180}$$

Thus, the polarization current $\mathbf{J}_P$ is associated with the movement of polarization charges.

3.11 Energy and Momentum Conservation: Poynting Theorem

Suppose we are given a time-varying, continuous distribution of charges and currents $\rho(\mathbf{x}', t), \mathbf{J}(\mathbf{x}', t)$, which produces electric and magnetic fields $\mathbf{E}(\mathbf{x}, t), \mathbf{B}(\mathbf{x}, t)$. The work done dW by the electromagnetic fields on charges $\rho\, d^3x'$ in time dt is given by

$$dW = \rho\, d^3x \left(\mathbf{E} + \frac{1}{c} \mathbf{v} \times \mathbf{B} \right) \cdot \mathbf{v}\, dt = \rho\, d^3x\, \mathbf{E} \cdot \mathbf{v}\, dt , \tag{3.181}$$

where $\mathbf{v}$ is the velocity. The rate at which work is done by the electromagnetic fields[12] on the source charges and currents contained within volume V is then given by

$$\frac{dW}{dt} = \int_V (\mathbf{J} \cdot \mathbf{E})\, d^3x' , \tag{3.182}$$

where $\mathbf{J} = \rho \mathbf{v}$ is the current density. The term $\mathbf{J} \cdot \mathbf{E}$ can, thus, be interpreted as the power delivered per unit volume.

From generalized Ampere's law or the Ampere-Maxwell law,

$$\mathbf{J} = \frac{c}{4\pi} \left(\nabla \times \mathbf{H} - \frac{1}{c} \frac{\partial \mathbf{D}}{\partial t} \right) . \tag{3.183}$$

Therefore,

$$\frac{dW}{dt} = \frac{1}{4\pi} \int_V \left(c\mathbf{E} \cdot \nabla \times \mathbf{H} - \mathbf{E} \cdot \frac{\partial \mathbf{D}}{\partial t} \right) d^3x' . \tag{3.184}$$

Again,

$$\nabla \cdot (\mathbf{E} \times \mathbf{H}) = \mathbf{H} \cdot (\nabla \times \mathbf{E}) - \mathbf{E} \cdot (\nabla \times \mathbf{H}) . \tag{3.185}$$

We replace $\mathbf{E} \cdot \nabla \times \mathbf{H}$ in Eq. (3.184) by using Eq. (3.185) as follows:

$$\frac{dW}{dt} = \frac{1}{4\pi} \int_V \left[-c\nabla \cdot (\mathbf{E} \times \mathbf{H}) + c\mathbf{H} \cdot (\nabla \times \mathbf{E}) - \mathbf{E} \cdot \frac{\partial \mathbf{D}}{\partial t} \right] d^3x' . \tag{3.186}$$

We replace $\nabla \times \mathbf{E}$ term in the integrand using Faraday's law and get

---

[12] Note that magnetic field does no work.

$$\frac{dW}{dt} = -\frac{1}{4\pi} \int_V \left[c\nabla \cdot (\mathbf{E} \times \mathbf{H}) + \mathbf{H} \cdot \frac{\partial \mathbf{B}}{\partial t} + \mathbf{E} \cdot \frac{\partial \mathbf{D}}{\partial t} \right] d^3x'. \tag{3.187}$$

The negative sign in the above equation is interpreted as representing the loss of energy stored in the fields which is transferred to the charges as kinetic energy. The equation is true in general, even in absence of material media.[13] However, instead of going ahead with the exercise in absence of material media, we concentrate on the special case of a linear dielectric and magnetic medium. For linear media, we can write

$$\mathbf{E} \cdot \frac{\partial \mathbf{D}}{\partial t} = \frac{1}{2} \frac{\partial}{\partial t} (\mathbf{E} \cdot \mathbf{D}), \tag{3.188}$$

$$\mathbf{H} \cdot \frac{\partial \mathbf{B}}{\partial t} = \frac{1}{2} \frac{\partial}{\partial t} (\mathbf{H} \cdot \mathbf{B}). \tag{3.189}$$

Therefore, the expression for the rate of work done by the fields on the charges can be written as

$$\frac{dW}{dt} = -\frac{c}{4\pi} \int_V \nabla \cdot (\mathbf{E} \times \mathbf{H}) \, d^3x' - \frac{1}{8\pi} \int_V \frac{\partial}{\partial t} (\mathbf{E} \cdot \mathbf{D} + \mathbf{H} \cdot \mathbf{B}) \, d^3x'. \tag{3.190}$$

Next, we convert the integral with divergence term into a surface integral and write

$$\frac{dW}{dt} = \int_V (\mathbf{J} \cdot \mathbf{E}) \, d^3x' \tag{3.191}$$

$$= -\oint_{\partial V} \frac{c}{4\pi} (\mathbf{E} \times \mathbf{H}) \cdot \hat{\mathbf{n}}' \, da' - \int_V \frac{\partial}{\partial t} \left[\frac{1}{8\pi} (\mathbf{E} \cdot \mathbf{D} + \mathbf{H} \cdot \mathbf{B}) \right] d^3x'. \tag{3.192}$$

We now identify the electromagnetic energy density as

$$u = \frac{1}{8\pi} (\mathbf{E} \cdot \mathbf{D} + \mathbf{H} \cdot \mathbf{B}), \tag{3.193}$$

such that the total electromagnetic energy is given by $U = \int_V u \, d^3x'$. The surface integral represents a flux of the vector field $\mathbf{S}$, called the 'Poynting vector', defined as

$$\mathbf{S} = \frac{c}{4\pi} (\mathbf{E} \times \mathbf{H}). \tag{3.194}$$

The Poynting vector $\mathbf{S}$ has the dimension of energy per unit area per unit time and represents the flow of energy in space. We can write Eq. (3.192) as

---

[13] Replace $\mathbf{H}$ by $\mathbf{B}$ and $\mathbf{D}$ by $\mathbf{E}$ at this point to derive the corresponding statement for energy conservation in absence of material media.

$$\frac{dW}{dt} = -\frac{d}{dt} \int_V u \, d^3x' - \oint_{\partial V} \mathbf{S} \cdot \hat{\mathbf{n}}' \, da' . \qquad (3.195)$$

Thus, the rate of decrease of electromagnetic energy within any volume V is equal to the rate of work done on the charges by the electromagnetic force (or the increase in mechanical energy of the charges) plus the rate of energy flux out of the surface ∂V enclosing the volume. This is the statement of the Poynting theorem. The above equation can be expressed in differential form as well. We convert the surface integral in Eq. (3.195) into volume integral using divergence theorem. Then using Eq. (3.191), we can express the Poynting theorem in the following form:

$$-\int_V (\mathbf{J} \cdot \mathbf{E}) \, d^3x' = \int_V \frac{\partial u}{\partial t} d^3x' + \int_V \nabla \cdot \mathbf{S} \, d^3x' . \qquad (3.196)$$

The volume V being arbitrary, we write down Poynting's theorem in differential form similar to a continuity equation as

$$\frac{\partial u}{\partial t} + \nabla \cdot \mathbf{S} = -\mathbf{J} \cdot \mathbf{E} . \qquad (3.197)$$

The equation can also be written by expressing W as a volume integral over a u_{mech}, the mechanical energy density as

$$\frac{dW}{dt} = \frac{d}{dt} \int_V u_{\text{mech}} \, d^3x' . \qquad (3.198)$$

Then, we have a simpler differential form of the equation of continuity for energy flow:

$$-\nabla \cdot \mathbf{S} = \frac{\partial}{\partial t} (u_{\text{mech}} + u) . \qquad (3.199)$$

The crucial point about the Poynting theorem is that it is a statement on conservation of energy taking into account the transfer of energy between the charges and the fields. Now suppose we have a region of space without any charges. In that case, $\frac{dW}{dt} = 0$ and the Poynting theorem reduces to the statement on the local conservation of electromagnetic field energy, exclusively.

3.11.1 Conservation of Linear Momentum

As in case of energy conservation, we want to derive an equation of continuity which describes the flow of 'field momentum' across the surface enclosing a volume in space and the transfer of momentum between the fields and the charges within that volume. Total electromagnetic force on the charges in volume V is given by

$$\mathbf{F} = \frac{d\mathbf{P}}{dt} = \int_V \left(\mathbf{E} + \frac{1}{c}\mathbf{v} \times \mathbf{B} \right) \rho \, d^3x' ,$$

$$= \int_V \left(\rho\mathbf{E} + \frac{1}{c}\mathbf{J} \times \mathbf{B} \right) d^3x' . \tag{3.200}$$

Here, $\mathbf{P}$ is the total mechanical momenta associated with the charges in volume V. We replace the source terms with field terms using the Maxwell equations as follows:

$$\frac{d\mathbf{P}}{dt} = \frac{1}{4\pi} \int \left[\mathbf{E}(\nabla \cdot \mathbf{D}) - \mathbf{B} \times (\nabla \times \mathbf{H}) + \frac{1}{c}\mathbf{B} \times \frac{\partial\mathbf{D}}{\partial t} \right] . \tag{3.201}$$

Using Faraday's law, we can write

$$\mathbf{B} \times \frac{\partial\mathbf{D}}{\partial t} = \frac{\partial}{\partial t}(\mathbf{B} \times \mathbf{D}) - \frac{\partial B}{\partial t} \times \mathbf{D} = \frac{\partial}{\partial t}(\mathbf{B} \times \mathbf{D}) + c(\nabla \times \mathbf{E}) \times \mathbf{D} . \tag{3.202}$$

Using the above expression in Eq. (3.201), we get

$$\frac{d\mathbf{P}}{dt} = \frac{1}{4\pi} \int \left[\mathbf{E}(\nabla \cdot \mathbf{D}) - \mathbf{B} \times (\nabla \times \mathbf{H}) - \mathbf{D} \times (\nabla \times \mathbf{E}) - \frac{1}{c}\frac{\partial}{\partial t}(\mathbf{D} \times \mathbf{B}) \right] d^3x' . \tag{3.203}$$

In the absence of material media, we can simplify the calculation considerably by assuming $\mathbf{D} = \mathbf{E}$ and $\mathbf{H} = \mathbf{B}$. Since $\nabla \cdot \mathbf{B} = 0$, we can add a term $\mathbf{B}(\nabla \cdot \mathbf{B})$ in the integrand on the right-hand side. Then Eq. (3.203) can be written as

$$\frac{d\mathbf{P}}{dt} + \frac{1}{4\pi c} \int \frac{\partial}{\partial t}(\mathbf{E} \times \mathbf{B}) \, d^3x' =$$
$$\frac{1}{4\pi} \int [\mathbf{E}(\nabla \cdot \mathbf{E}) - \mathbf{E} \times (\nabla \times \mathbf{E}) + \mathbf{B}(\nabla \cdot \mathbf{B})$$
$$-\mathbf{B} \times (\nabla \times \mathbf{B})] \, d^3x' . \tag{3.204}$$

Since the terms involving $\mathbf{E}$ and $\mathbf{B}$ are similar, it will suffice to evaluate any one of the Cartesian components ($i = 1, 2, 3$) and extend the result to other terms. The objective is to convert them into divergence terms, going by our previous experience of dealing with energy conservation. We pick up the $i = 1$ term involving $\mathbf{E}$ and write

$$[\mathbf{E}(\nabla \cdot \mathbf{E}) - \mathbf{E} \times (\nabla \times \mathbf{E})]_1$$
$$= E_1 \left(\frac{\partial E_1}{\partial x_1} + \frac{\partial E_2}{\partial x_2} + \frac{\partial E_3}{\partial x_3} \right) - E_2 \left(\frac{\partial E_2}{\partial x_1} - \frac{\partial E_1}{\partial x_2} \right) + E_3 \left(\frac{\partial E_1}{\partial x_3} - \frac{\partial E_3}{\partial x_1} \right) .$$

We now rearrange the terms on the right-hand side as follows:

$$[\mathbf{E}(\nabla \cdot \mathbf{E}) - \mathbf{E} \times (\nabla \times \mathbf{E})]_1$$

$$= \frac{1}{2}\left(\frac{\partial E_1^2}{\partial x_1}\right) - \frac{1}{2}\left(\frac{\partial E_2^2}{\partial x_1}\right) - \frac{1}{2}\left(\frac{\partial E_3^2}{\partial x_1}\right) + \frac{\partial}{\partial x_2}(E_1 E_2) + \frac{\partial}{\partial x_3}(E_1 E_3)$$

$$= \frac{\partial E_1^2}{\partial x_1} - \frac{1}{2}\frac{\partial E^2}{\partial x_1} + \frac{\partial}{\partial x_2}(E_1 E_2) + \frac{\partial}{\partial x_3}(E_1 E_3)$$

$$= \frac{\partial}{\partial x_1}(E_1 E_1) + \frac{\partial}{\partial x_2}(E_1 E_2) + \frac{\partial}{\partial x_3}(E_1 E_3) - \frac{1}{2}\frac{\partial E^2}{\partial x_1}. \qquad (3.205)$$

Therefore, in general, we can write

$$[\mathbf{E}(\nabla \cdot \mathbf{E}) - \mathbf{E} \times (\nabla \times \mathbf{E})]_i = \sum_{j=1}^{3} \frac{\partial}{\partial x_j}\left(E_i E_j - \frac{1}{2}\delta_{ij}\mathbf{E} \cdot \mathbf{E}\right). \qquad (3.206)$$

And similarly,

$$[\mathbf{B}(\nabla \cdot \mathbf{B}) - \mathbf{B} \times (\nabla \times \mathbf{B})]_i = \sum_{j=1}^{3} \frac{\partial}{\partial x_j}\left(B_i B_j - \frac{1}{2}\delta_{ij}\mathbf{B} \cdot \mathbf{B}\right). \qquad (3.207)$$

We introduce a second-rank tensor called the Maxwell stress tensor T_{ij}, which is defined as follows:

$$T_{ij} = \frac{1}{4\pi}\left[E_i E_j + B_i B_j - \frac{1}{2}\delta_{ij}(\mathbf{E} \cdot \mathbf{E} + \mathbf{B} \cdot \mathbf{B})\right]. \qquad (3.208)$$

We rewrite Eq. (3.204) in component form using T_{ij} on the right-hand side as

$$\frac{dP_i}{dt} = -\frac{d}{dt}\int \frac{1}{4\pi c}(\mathbf{E} \times \mathbf{B})_i \, d^3x' + \int \left(\sum_{j=1}^{3}\frac{\partial T_{ij}}{\partial x_j}\right) d^3x'. \qquad (3.209)$$

The second integrand on the right-hand side is a divergence term. Hence, the corresponding volume integral can be converted into a surface integral. As in case of energy conservation, we identify this term as representing momentum flux across the surface ∂V enclosing volume V. As a result, we have

$$\frac{dP_i}{dt} = -\frac{d}{dt}\int \frac{1}{4\pi c}(\mathbf{E} \times \mathbf{B})_i \, d^3x' + \oint_{\partial V}\sum_{j=1}^{3} T_{ij}\, n_j \, da. \qquad (3.210)$$

Here, $\mathbf{n}$ is the outward unit normal to the surface ∂V. The first term on the right-hand side is the total time derivative of a volume integral. Comparing with the left-hand side, we define the corresponding integrand as the ith component of momentum density $\mathbf{g}_i$. Therefore, the total electromagnetic field momentum density[14] is given by

$$\mathbf{g} = \frac{1}{4\pi c} \, (\mathbf{E} \times \mathbf{B}) \, . \tag{3.211}$$

The total electromagnetic field momenta is then given by

$$\mathbf{P}_{\text{field}} = \frac{1}{4\pi c} \int_V \mathbf{E} \times \mathbf{B} \, d^3 x' \, . \tag{3.212}$$

Finally, we write down the equation describing conservation of linear momenta of the combined system of electromagnetic fields and charged particles in compact form as

$$\frac{d}{dt} \, (\mathbf{P}_{\text{field}} + \mathbf{P})_i = \oint_{\partial V} \sum_{j=1}^{3} T_{ij} \, n_j \, da \, . \tag{3.213}$$

The term $\sum_{j=1}^{3} T_{ij} \, n_j$ describes flux of field momentum $\mathbf{g}$ across the surface ∂V along ith direction. Alternatively, it represents the force per unit area or stress transmitted across the surface ∂V enclosing volume V.

3.12 Problems

1. Consider a collection of three point charges located on the z-axis: a charge $+2q$ at the origin and two charges $-q$ each at $z = \pm a$. Find the lowest non-vanishing electric multipole moment.
2. Two pairs of point charges $\pm q$ are placed on the vertices of a square of length l on the xy plane such that one arm of the square is along x-direction and another along y-direction. The like charges are placed diagonally, or in other words, the nearest neighbors are oppositely charged. Find the electric quadrupole tensor. Diagonalize the tensor using coordinate-ordinate transformation.
3. A point charge q is placed at a distance d from the center of a dielectric sphere of radius R and dielectric constant ϵ. Solve the Laplace equation using appropriate boundary conditions to find the force between the dielectric sphere and the point charge.
4. Two concentric conducting spherical shells of radii $r = a$ and $r = b$ $(a < b)$, respectively, carry charges $\pm Q$. The volume of space between the two shells is 'half-filled' by a dielectric material (of dielectric constant ϵ), such that one

---

[14] It turns out that in material media, the field momentum density is usually defined as

$$\mathbf{g} = \frac{1}{4\pi c} \, (\mathbf{E} \times \mathbf{H}) \, .$$

.

half of the space between the two shells is completely empty and the other half completely filled by the dielectric. Find the electric field everywhere between the two shells. Calculate the surface-charge distribution on the inner shell and the polarization-charge density on the inner surface of the dielectric at $r = a$.

5. Consider a long thick cylindrical shell of dielectric constant ϵ and inner and outer radii a and b, respectively, placed in an otherwise uniform electric field $\mathbf{E}_0$. The electric field is directed perpendicular to the symmetry axis of the cylinder. The dielectric constant of the medium inside and outside the cylinder is unity. Find the potential and electric field everywhere. The cylinders are long enough such that end effects can be neglected and the problem effectively reduces to solving the Laplace equation in polar coordinates-ordinates subject to appropriate boundary conditions.

6. The magnetic dipole moment of a volume current distribution, as discussed in the text, is given by

$$\mathbf{m} = \frac{1}{2} \int \mathbf{x}' \times \mathbf{J}(\mathbf{x}') \, d^3 x'.$$

Using the above expression, find the magnetic dipole moment of a spherical shell of radius R, carrying a surface-charge density σ and rotating with angular velocity $\omega \hat{\mathbf{z}}$.

7. Consider a sphere of radius R, having frozen-in uniform magnetization M pointing toward the north pole. Find the 'auxiliary H' field inside the sphere using an electrostatic analogy. Find the 'B' field inside the sphere.

8. Estimate the magnetic field $\mathbf{B}$ for the following objects having frozen-in uniform magnetization by using bound current densities (and the boundary conditions, wherever applicable).

   (a) Disk of radius R, thickness t and magnetization M is parallel to the axis of the disk. Find the field at a point on the axis just inside and just outside.

   (b) An infinitely long cylinder carries a uniform magnetization M parallel to its axis. Find the magnetic field inside and outside the cylinder.

9. Find the magnetic dipole moment of a solid sphere carrying uniform volume charge density ρ and rotating with angular velocity $\omega \hat{\mathbf{z}}$.

10. A solid sphere is uniformly magnetized along the $\hat{\mathbf{z}}$-direction (magnetization $\mathbf{M} = M\hat{\mathbf{z}}$).

    (a) Find all the bound currents, and using them, find the magnetic field $\mathbf{B}$ inside the sphere.

    (b) Find the field at the north pole, outside the sphere.

    (c) Find the magnetic field $\mathbf{B}$ just outside the sphere at the equator using appropriate boundary conditions.

    (d) Neatly sketch the magnetic field lines $\mathbf{B}$ everywhere (inside and outside the sphere).

11. A finite solid cylinder of radius R and length $L \gg R$ has a frozen-in constant magnetization M parallel to the axis.

    (a) Calculate the bound volume current and bound surface currents everywhere.

    (b) Write down the boundary conditions for the parallel and normal components of **B** and **H** fields at the flat surfaces (top and bottom) and curved surface of the cylinder.

(*c*) Sketch the **B** field lines everywhere, keeping in mind the boundary conditions.

12. Consider a solid cylinder of length L and radius R ($R < L$) with uniform, 'frozen-in' magnetization $\mathbf{M}_0$ along the symmetry axis. Calculate the magnetic scalar potential Φ_M and thereby the **H** and **B** fields at all points on the symmetry axis, both inside and outside the cylinder.

13. A capacitor with two circular plates of radius R is being charged by a constant current. Calculate the Poynting vector at radial distance r inside the capacitor, and verify that its flux equals the rate of change of the energy stored in the region bounded by radius r.

14. A long solenoid of radius R carries a weakly time-dependent current $I(t)$. The solenoid is encircled by a symmetrically placed conducting loop of radius $4R$. Calculate the energy inflow to the external loop and show that it replenishes the energy dissipated by the Joule heating.

15. A setup consists of three very long coaxial parts: (i) a nonconducting cylinder with radius a with total charge Q, (ii) another nonconducting cylinder with radius $b > a$ having total charge $-Q$ and (iii) a solenoid with radius $R > b$, having n turns per unit length and carrying current I. The two cylinders are free to rotate about the common axis and are initially at rest. The current in the solenoid is then switched off.

(*a*) Find the total mechanical angular momenta imparted to the charged cylinders by the time the current in the solenoid is decreased to zero. Ignore the **B** fields generated by rotating charged cylinders.

(*b*) We have already defined in the text the linear momentum stored in the electromagnetic field. Now calculate explicitly the total initial angular momentum stored in the field.

Chapter 4
Initiation to Electromagnetic Radiation

Till now we have written down the Maxwell's equations. Those equations can be solved in the electrostatic limit or the magnetostatic limit. In this chapter, we will give the general time-dependent solution of the Maxwell's equations. The general time-dependent solution turns out to be the electromagnetic wave solution. Here, we will first discuss about the plane electromagnetic wave solutions and then present a particular case where radiation can be produced from charges moving non-relativistically. Later, we will introduce radiation from relativistically moving point charges.

4.1 Plane-Wave Solutions

Presently we will study plane electromagnetic waves and first show that this kind of waves satisfy Maxwell's equations. The plane-wave solution is obtained in the absence of any sources, consequently the relevant Maxwell equations are

$$\nabla \cdot \mathbf{E} = 0, \qquad \nabla \times \mathbf{E} + \frac{1}{c}\frac{\partial \mathbf{B}}{\partial t} = 0,$$

$$\nabla \cdot \mathbf{B} = 0, \qquad \nabla \times \mathbf{B} - \frac{\mu \epsilon}{c}\frac{\partial \mathbf{E}}{\partial t} = 0. \tag{4.1}$$

Here we have assumed

$$\mathbf{D} = \epsilon \mathbf{E}, \quad \text{and} \quad \mathbf{B} = \mu \mathbf{H}, \tag{4.2}$$

where the dielectric constant and magnetic permeability are supposed to be independent of wave frequency. This last assumption is not absolutely essential and we will soon consider medium dependent effects due to which the dielectric constant

© Springer Nature Singapore Pte Ltd. 2021
K. Bhattacharya and S. Mukhopadhyay, *Introduction to Advanced Electrodynamics*,
https://doi.org/10.1007/978-981-16-7802-8_4

becomes a function of the frequency of the waves. If we apply the curl operator on the Faraday's law, we get

$$\nabla \times (\nabla \times \mathbf{E}) + \frac{1}{c} \frac{\partial}{\partial t} (\nabla \times \mathbf{B}) = 0. \tag{4.3}$$

Now $\nabla \times (\nabla \times \mathbf{E}) = \nabla(\nabla \cdot \mathbf{E}) - \nabla^2 \mathbf{E} = -\nabla^2 \mathbf{E}$ as divergence of the electric field is zero in the absence of charges. Using the corrected Ampere's law in the above equation, we get

$$\nabla^2 \mathbf{E} - \left(\frac{\mu\epsilon}{c^2}\right) \frac{\partial^2 \mathbf{E}}{\partial t^2} = 0. \tag{4.4}$$

This is the wave equation showing that time varying electric field components will travel as a plane wave. We will come to the explicit solution later. If we apply the curl operator on the corrected Ampere's law, we will have

$$\nabla^2 \mathbf{B} - \left(\frac{\mu\epsilon}{c^2}\right) \frac{\partial^2 \mathbf{B}}{\partial t^2} = 0, \tag{4.5}$$

showing that time varying magnetic fields can also travel as plane waves. The components of the electric as well as the magnetic fields satisfy the equation

$$\nabla^2 u - \left(\frac{1}{v^2}\right) \frac{\partial^2 u}{\partial t^2} = 0, \tag{4.6}$$

where u represents components of electric or magnetic field. Here

$$v \equiv \frac{c}{\sqrt{\mu\epsilon}}, \tag{4.7}$$

is a medium dependent constant with the dimension of velocity. The wave equation, Eq. (4.6), has the well known plane-wave solution as

$$u = ()e^{i(\mathbf{k}\cdot\mathbf{x}\pm\omega t)}, \tag{4.8}$$

where the empty bracket represents a spacetime independent constant. The angular frequency ω and the magnitude of the wave vector $\mathbf{k}$ are related by

$$|\mathbf{k}| = k = \frac{\omega}{v} = \sqrt{\mu\epsilon}\frac{\omega}{c}. \tag{4.9}$$

If the unit vector along $\mathbf{k}$ is called $\hat{\mathbf{n}}$, which implies

$$\mathbf{k} = \frac{\omega}{v}\hat{\mathbf{n}}, \tag{4.10}$$

then in general we can write the solution of the wave equation as

$$u(\mathbf{x}, t) = A\, e^{i(\mathbf{k}\cdot\mathbf{x}-\omega t)} + B\, e^{i(\mathbf{k}\cdot\mathbf{x}+\omega t)}\,. \tag{4.11}$$

The above equation can also be written as

$$u(\mathbf{x}, t) = A\, e^{ik(\hat{\mathbf{n}}\cdot\mathbf{x}-vt)} + B\, e^{ik(\hat{\mathbf{n}}\cdot\mathbf{x}+vt)}\,, \tag{4.12}$$

which illuminates the nature of wave propagation. Here, A and B are constants. Suppose we concentrate on the solution $e^{ik(\hat{\mathbf{n}}\cdot\mathbf{x}-vt)}$. From this solution, we see the phase $\phi = k(\hat{\mathbf{n}}\cdot\mathbf{x} - vt)$ is actually a function of time. Suppose we want to see how the surface on which ϕ is constant moves in time. We set $\phi(\mathbf{x}, t) = $ constant. Suppose at time t_1, the constant phase is obtained at position $\mathbf{x}_1$ and at time t_2, the constant phase is obtained at position $\mathbf{x}_2$. As the phase is constant, on the moving surface, by assumption we must have

$$\hat{\mathbf{n}}\cdot\mathbf{x}_1 - vt_1 = \hat{\mathbf{n}}\cdot\mathbf{x}_2 - vt_2 \implies \hat{\mathbf{n}}\cdot(\mathbf{x}_2 - \mathbf{x}_1) = v(t_2 - t_1)\,, \tag{4.13}$$

for two positions of the surface. For infinitesimal changes in time and position, we have from the above equation

$$v = \hat{\mathbf{n}}\cdot\frac{d\mathbf{x}}{dt}\,, \tag{4.14}$$

which gives the instantaneous velocity of the surface of constant phase. This velocity magnitude is a projection along the $\hat{\mathbf{n}}$ direction and this velocity is called the phase velocity.

4.1.1 Plane Electromagnetic Waves in Dispersive Medium

Till now we were assuming that ϵ and μ in the Maxwell equations

$$\nabla\cdot\mathbf{D} = 0\,, \qquad \nabla\times\mathbf{E} + \frac{1}{c}\frac{\partial\mathbf{B}}{\partial t} = 0\,,$$

$$\nabla\cdot\mathbf{B} = 0\,, \qquad \nabla\times\mathbf{H} - \frac{1}{c}\frac{\partial\mathbf{D}}{\partial t} = 0\,. \tag{4.15}$$

were independent of wave frequency. The above quantities ϵ and μ are always assumed to be independent of spatial coordinates but they, in fact, can depend on time or frequency. Henceforth, we assume

$$\epsilon = \epsilon(\omega)\,, \qquad \mu = \mu(\omega)\,, \tag{4.16}$$

which demands that the constitutive relations change to

$$\mathbf{D}(\mathbf{x}, \omega) = \epsilon(\omega)\mathbf{E}(\mathbf{x}, \omega), \quad \text{and} \quad \mathbf{B}(\mathbf{x}, \omega) = \mu(\omega)\mathbf{H}(\mathbf{x}, \omega), \tag{4.17}$$

where now all the fields are expressed in terms of frequency. These above relations are very interesting, they imply that we directly generalize the constitutive relations to the frequency domain and not in the time domain. The last statement means that in general

$$\mathbf{D}(\mathbf{x}, t) \neq \epsilon(t)\mathbf{E}(\mathbf{x}, t), \quad \text{and} \quad \mathbf{B}(\mathbf{x}, t) \neq \mu(t)\mathbf{H}(\mathbf{x}, t). \tag{4.18}$$

One has to apply the convolution theorem of Fourier transforms to obtain the constitutive relations in the time domain. We will talk about the convolution theorem in a later chapter of this book. When all the fields are expressed in terms of frequency, it is better to work with the Fourier transforms

$$\mathbf{E}(\mathbf{x}, t) = \frac{1}{\sqrt{2\pi}} \int_{-\infty}^{\infty} \mathbf{E}(\mathbf{x}, \omega) \, e^{-i\omega t} \, d\omega \tag{4.19}$$

$$\mathbf{B}(\mathbf{x}, t) = \frac{1}{\sqrt{2\pi}} \int_{-\infty}^{\infty} \mathbf{B}(\mathbf{x}, \omega) \, e^{-i\omega t} \, d\omega, \tag{4.20}$$

and the inverse Fourier transforms:

$$\mathbf{E}(\mathbf{x}, \omega) = \frac{1}{\sqrt{2\pi}} \int_{-\infty}^{\infty} \mathbf{E}(\mathbf{x}, t) \, e^{i\omega t} \, d\omega \tag{4.21}$$

$$\mathbf{B}(\mathbf{x}, \omega) = \frac{1}{\sqrt{2\pi}} \int_{-\infty}^{\infty} \mathbf{B}(\mathbf{x}, t) \, e^{i\omega t} \, d\omega. \tag{4.22}$$

In terms of the Fourier transforms, one can rewrite the Maxwell equation

$$\nabla \times \mathbf{E} + \frac{1}{c}\frac{\partial \mathbf{B}}{\partial t} = 0,$$

as

$$\frac{1}{\sqrt{2\pi}} \int_{-\infty}^{\infty} \left[\nabla \times \mathbf{E}(\mathbf{x}, \omega) - \frac{i\omega}{c}\mathbf{B}(\mathbf{x}, \omega) \right] e^{-i\omega t} \, d\omega = 0. \tag{4.23}$$

For this equation to be true, in general, the integrand in the above integration has to vanish and so we get

$$\nabla \times \mathbf{E}(\mathbf{x}, \omega) - \frac{i\omega}{c}\mathbf{B}(\mathbf{x}, \omega) = 0. \tag{4.24}$$

Now we can evaluate the curl of the above equation and write

$$\nabla \times [\nabla \times \mathbf{E}(\mathbf{x}, \omega)] - \frac{i\omega}{c}\nabla \times \mathbf{B}(\mathbf{x}, \omega) = 0. \tag{4.25}$$

The calculation produces

$$\nabla^2 \mathbf{E}(\mathbf{x}, \omega) + \frac{i\omega}{c} \nabla \times \mathbf{B}(\mathbf{x}, \omega) = 0, \qquad (4.26)$$

where we have used the relation $\nabla \cdot \mathbf{E}(\mathbf{x}, \omega) = 0$. Next, we use the other Maxwell equation

$$\nabla \times \mathbf{H} - \frac{1}{c} \frac{\partial \mathbf{D}}{\partial t} = 0,$$

and write it as

$$\frac{1}{\sqrt{2\pi}} \int_{-\infty}^{\infty} \left[\nabla \times \mathbf{H}(\mathbf{x}, \omega) + \frac{i\omega}{c} \mathbf{D}(\mathbf{x}, \omega) \right] e^{-i\omega t} \, d\omega = 0. \qquad (4.27)$$

The above equation yields

$$\nabla \times \mathbf{H}(\mathbf{x}, \omega) + \frac{i\omega}{c} \mathbf{D}(\mathbf{x}, \omega) = 0 \implies \nabla \times \mathbf{B}(\mathbf{x}, \omega) + \frac{i\mu\epsilon\omega}{c} \mathbf{E}(\mathbf{x}, \omega) = 0 \ (4.28)$$

Feeding the information from the above equation in Eq. (4.26), we get

$$\nabla^2 \mathbf{E}(\mathbf{x}, \omega) + \frac{\mu\epsilon\omega^2}{c^2} \mathbf{E}(\mathbf{x}, \omega) = 0. \qquad (4.29)$$

In a similar way, you will be able to show that the magnetic induction also satisfies a similar equation as

$$\nabla^2 \mathbf{B}(\mathbf{x}, \omega) + \frac{\mu\epsilon\omega^2}{c^2} \mathbf{B}(\mathbf{x}, \omega) = 0. \qquad (4.30)$$

In fact, the components of the electric or the magnetic field satisfies the general equation

$$\nabla^2 u(\mathbf{x}, \omega) + \frac{\mu\epsilon\omega^2}{c^2} u(\mathbf{x}, \omega) = 0. \qquad (4.31)$$

It can easily be seen that this equation has a solution as

$$u(\mathbf{x}, \omega) = ()e^{i\mathbf{k} \cdot \mathbf{x}}. \qquad (4.32)$$

Here, k is a function of ω. The plane-wave solution arises when we Fourier transform the above solution in terms of ω and express the solution as $u(\mathbf{x}, t)$. To see that k is a function of ω, we write the expression of k as

$$|\mathbf{k}| = k = \frac{\omega}{v} = \sqrt{\mu(\omega)\epsilon(\omega)} \frac{\omega}{c}. \qquad (4.33)$$

In the present case, the relation between k and ω is not linear as in the non-dispersive case. In the present case, the phase velocity is

$$v(\omega) = \frac{\omega}{k(\omega)} = \frac{c}{\sqrt{\mu(\omega)\epsilon(\omega)}},$$ (4.34)

showing that waves with different frequencies travel with different phase velocities giving rise to dispersion phenomenon. Generally, the wave vector is a function of the frequency. Its inverse is also true, saying that the frequency is a function of the magnitude of the wave vector as

$$\omega = \omega(k).$$ (4.35)

The relation between the frequency and the wave vector is generally called the dispersion relation. The dispersion relation specifies how the electromagnetic wave travels inside a medium.

4.2 Vectorial Properties of Plane Electromagnetic Waves

Till now we were dealing with the components of the electric and magnetic fields and as a result we were not so much careful about the vectorial nature of the fields. We now discuss the full problem where the vector nature of the fields is taken into account. In both non-dispersive, as well as dispersive cases, the electric and magnetic field for a single frequency can be written as:

$$\mathbf{E}(\mathbf{x}, t) = \boldsymbol{\mathcal{E}}\, e^{ik(\hat{\mathbf{n}} \cdot \mathbf{x} - vt)}, \qquad \mathbf{B}(\mathbf{x}, t) = \boldsymbol{\mathcal{B}}\, e^{ik(\hat{\mathbf{n}} \cdot \mathbf{x} - vt)}.$$ (4.36)

where the vectors $\boldsymbol{\mathcal{E}}$, $\boldsymbol{\mathcal{B}}$ and $\hat{\mathbf{n}}$ are vectors which have components independent of space and time. Putting these expression of the fields back in the wave equation gives

$$k^2 \hat{\mathbf{n}} \cdot \hat{\mathbf{n}} = \mu\epsilon \frac{\omega^2}{c^2}.$$ (4.37)

As $\hat{\mathbf{n}}$ is an unit vector we get back the expression of k^2 as we had before. Next comes the divergence free conditions $\nabla \cdot \mathbf{E} = 0$ and $\nabla \cdot \mathbf{B} = 0$. As we have

$$\mathbf{E}(\mathbf{x}, t) = \boldsymbol{\mathcal{E}}\, e^{ik(\hat{\mathbf{n}} \cdot \mathbf{x} - vt)}, \qquad \mathbf{B}(\mathbf{x}, t) = \boldsymbol{\mathcal{B}}\, e^{ik(\hat{\mathbf{n}} \cdot \mathbf{x} - vt)},$$

the divergence free conditions on the fields easily translates to

$$\hat{\mathbf{n}} \cdot \boldsymbol{\mathcal{E}} = 0, \quad \text{and} \quad \hat{\mathbf{n}} \cdot \boldsymbol{\mathcal{B}} = 0.$$ (4.38)

This shows that both $\mathbf{E}(\mathbf{x}, t)$ and $\mathbf{B}(\mathbf{x}, t)$ are perpendicular to the vector $\hat{\mathbf{n}}$ or the wave vector $\mathbf{k}$. Next, we use the equation

$$\nabla \times \mathbf{E} + \frac{1}{c}\frac{\partial \mathbf{B}}{\partial t} = 0,$$

to see the relation between the vectors $\mathbf{E}(\mathbf{x}, t)$, $\mathbf{B}(\mathbf{x}, t)$ and $\hat{\mathbf{n}}$. First, we see that

$$\nabla \times \mathbf{E} = \left[\nabla e^{ik(\hat{\mathbf{n}}\cdot\mathbf{x}-vt)}\right] \times \boldsymbol{\mathcal{E}} = ik(\hat{\mathbf{n}} \times \boldsymbol{\mathcal{E}})e^{ik(\hat{\mathbf{n}}\cdot\mathbf{x}-vt)}. \tag{4.39}$$

The other factor gives

$$\frac{1}{c}\frac{\partial \mathbf{B}}{\partial t} = -\frac{ikv}{c}\boldsymbol{\mathcal{B}}\,e^{ik(\hat{\mathbf{n}}\cdot\mathbf{x}-vt)}. \tag{4.40}$$

As a consequence, the Faraday's law gives

$$\boldsymbol{\mathcal{B}} = \frac{c}{v}(\hat{\mathbf{n}} \times \boldsymbol{\mathcal{E}}) = \sqrt{\mu\epsilon}(\hat{\mathbf{n}} \times \boldsymbol{\mathcal{E}}). \tag{4.41}$$

Assuming $\hat{\mathbf{n}}$, μ and ϵ all real, we see that $\boldsymbol{\mathcal{E}}$ and $\boldsymbol{\mathcal{B}}$ has the same phase. From this result, we see that we can use three mutually perpendicular unit vectors as ϵ_1, ϵ_2 and $\hat{\mathbf{n}}$ to specify the relevant vectors in our hand. They satisfy the property

$$\epsilon_1 \times \epsilon_2 = \hat{\mathbf{n}}, \quad \epsilon_2 \times \hat{\mathbf{n}} = \epsilon_1, \quad \hat{\mathbf{n}} \times \epsilon_1 = \epsilon_2. \tag{4.42}$$

In terms of these vectors, we can write

$$\boldsymbol{\mathcal{E}} = \epsilon_1 E_0, \qquad \boldsymbol{\mathcal{B}} = \epsilon_2 \sqrt{\mu\epsilon}\, E_0, \tag{4.43}$$

or

$$\boldsymbol{\mathcal{E}} = \epsilon_2 E_0', \qquad \boldsymbol{\mathcal{B}} = -\epsilon_1 \sqrt{\mu\epsilon}\, E_0'. \tag{4.44}$$

Here, E_0 and E_0' are possibly complex constants. In general, we see that the electric field and the magnetic field $\mathbf{E}(\mathbf{x}, t)$ and $\mathbf{B}(\mathbf{x}, t)$ are complex quantities. The electric and magnetic fields are perpendicular to each other and oscillate in space and time. The wave propagates in a direction that is perpendicular to the plane containing the oscillating electric and magnetic fields. To find out the power radiated by the plane electromagnetic wave, we have to generalize the concept of the Poynting vector for complex fields.

We generalize the concept of the Poynting vector for systems where the fields may be given as complex vectors. In this approach to electrodynamics based on complex numbers, we represent the real fields as

$$\text{Re}[\mathbf{E}(\mathbf{x}, t)] \equiv \text{Re}[\mathbf{E}(\mathbf{x})e^{-i\omega t}] = \frac{1}{2}[\mathbf{E}(\mathbf{x})e^{-i\omega t} + \mathbf{E}^*(\mathbf{x})e^{i\omega t}]. \tag{4.45}$$

In this case, one must note that $\mathbf{E}(\mathbf{x})$ can also be complex. With this input, we can now write the basic form of Poynting vector as

$$\mathbf{S}(\mathbf{x}, t) = \frac{c}{4\pi}\text{Re}\,[\mathbf{E}(\mathbf{x}, t) \times \mathbf{H}(\mathbf{x}, t)]\,, \tag{4.46}$$

which is a real vector. As the fields are complex, we have

$$\mathbf{S}(\mathbf{x}, t) = \left(\frac{c}{4\pi}\right) \frac{1}{2}[\mathbf{E}(\mathbf{x})e^{-i\omega t} + \mathbf{E}^*(\mathbf{x})e^{i\omega t}] \times \frac{1}{2}[\mathbf{H}(\mathbf{x})e^{-i\omega t} + \mathbf{H}^*(\mathbf{x})e^{i\omega t}].$$

One can now expand the above cross product and write

$$\mathbf{S}(\mathbf{x}, t) = \left(\frac{c}{4\pi}\right) \frac{1}{4} \left[\mathbf{E}(\mathbf{x}) \times \mathbf{H}(\mathbf{x})\,e^{-2i\omega t} + \mathbf{E}(\mathbf{x}) \times \mathbf{H}^*(\mathbf{x}) + \mathbf{E}^*(\mathbf{x}) \times \mathbf{H}(\mathbf{x}) \right.$$
$$\left. + \mathbf{E}^*(\mathbf{x}) \times \mathbf{H}^*(\mathbf{x})\,e^{2i\omega t} \right].$$

The above equation can also be written in the following form

$$\mathbf{S}(\mathbf{x}, t) = \left(\frac{c}{4\pi}\right) \frac{1}{2}\text{Re}\left[\mathbf{E}(\mathbf{x}) \times \mathbf{H}^*(\mathbf{x})\right] + \left(\frac{c}{4\pi}\right) \frac{1}{2}\text{Re}\left[\mathbf{E}(\mathbf{x}) \times \mathbf{H}(\mathbf{x})\,e^{-2i\omega t}\right]. \tag{4.47}$$

Here, the first term on the right-hand side of the above equation is independent of time, while the second term has sinusoidal variation in time. If we define a time averaged Poynting vector as

$$\langle\mathbf{S}(\mathbf{x})\rangle \equiv \lim_{T \to \infty} \frac{1}{T} \int_0^T \mathbf{S}(\mathbf{x}, t)\,dt\,, \tag{4.48}$$

where the time interval T is sufficiently larger than the period of oscillations, then it can easily be seen that the sinusoidal part of the Poynting vector does not contribute in this average value. As a result of this observation, we omit the sinusoidal part as it does not contribute in the time average and define the complex Poynting vector for the fields (where the fields vary sinusoidally) as

$$\mathbf{S}(\mathbf{x}, t) = \frac{c}{8\pi}[\mathbf{E}(\mathbf{x}, t) \times \mathbf{H}^*(\mathbf{x}, t)]. \tag{4.49}$$

We will have to work with the real part of this vector. We will now use this form of the Poynting vector to calculate the power radiated by plane electromagnetic waves. The relevant Poynting vector for complex fields is

$$S = \frac{c}{8\pi} E \times H^* = \frac{c}{8\pi} (\epsilon_1 \times \epsilon_2) \frac{\sqrt{\mu\epsilon}}{\mu} |E_0|^2 = \frac{c}{8\pi} \sqrt{\frac{\epsilon}{\mu}} |E_0|^2 \, \hat{n}. \qquad (4.50)$$

The time averaged energy density of the radiation fields is given by

$$u \equiv \frac{1}{16\pi} \left(\epsilon E \cdot E^* + \frac{1}{\mu} B \cdot B^* \right). \qquad (4.51)$$

which is a simple generalization of the energy density formulae from electric and magnetic fields in a medium where the constitutive relations are linear. When $\mathcal{E}$ and $\mathcal{B}$ are given as in Eq. (4.43), the energy density for the plane wave turns out to be

$$u = \frac{\epsilon}{8\pi} |E_0|^2. \qquad (4.52)$$

In our whole analysis, we have taken the unit vector $\hat{n}$ to be real. In some situations, it becomes useful to assume this unit vector to be complex. As we will not discuss optics in this book, we will not use this complex unit vector in this course.

4.2.1 Linear and Circular Polarization of Plane Electromagnetic Waves

The vector nature of plane electromagnetic waves brings in new features to the wave propagation mode. These new features are related to the polarization degree of freedom of the plane electromagnetic waves. Polarization of the plane electromagnetic wave is related to the direction of oscillation of the electric field in the electromagnetic wave. Once we know the direction of wave propagation and the polarization of the wave, we practically know everything about the electromagnetic wave as the magnetic field oscillates in a direction that is mutually perpendicular to both the propagation direction and the electric field. Previously we noted two possibilities of the electric and magnetic field directions in the plane wave as

$$\mathcal{E} = \epsilon_1 E_0, \qquad \mathcal{B} = \epsilon_2 \sqrt{\mu\epsilon} \, E_0,$$

or

$$\mathcal{E} = \epsilon_2 E_0', \qquad \mathcal{B} = -\epsilon_1 \sqrt{\mu\epsilon} \, E_0'.$$

Plane waves where the electric field oscillation directions are given in such a fashion, where the direction of oscillation of the electric field is fixed in time, are called linearly polarized electromagnetic waves. In the first case, the wave is linearly polarized in the direction of ϵ_1, whereas in the second case, the wave is linearly polarized in the direction of ϵ_2. The linearly polarized waves discussed before can be specified by the fields

$$\mathbf{E}_1 = \epsilon_1 \, E_1 \, e^{i(\mathbf{k}\cdot\mathbf{x}-\omega t)} \,, \qquad \mathbf{B}_1 = \sqrt{\mu\epsilon} \, \frac{\mathbf{k} \times \mathbf{E}_1}{k} \,, \tag{4.53}$$

or

$$\mathbf{E}_2 = \epsilon_2 \, E_2 \, e^{i(\mathbf{k}\cdot\mathbf{x}-\omega t)} \,, \qquad \mathbf{B}_2 = \sqrt{\mu\epsilon} \, \frac{\mathbf{k} \times \mathbf{E}_2}{k} \,. \tag{4.54}$$

These are examples of the simplest possible linearly polarized waves. We can have more general examples of linearly polarized plane waves propagating in the direction $\mathbf{k} = k\hat{\mathbf{n}}$. The most general linearly polarized plane electromagnetic wave is of the form

$$\mathbf{E}(\mathbf{x}, t) = (\epsilon_1 \, E_1 + \epsilon_2 \, E_2) \, e^{i(\mathbf{k}\cdot\mathbf{x}-\omega t)} \,, \tag{4.55}$$

where the amplitudes E_1 and E_2 can be in general complex numbers.

If both the complex quantities E_1 and E_2 have the same phase, then the above electric field is, in general, linearly polarized. The oscillation direction of the electric field, as shown in Fig. 4.1, will always make an angle

$$\theta = \tan^{-1}\left(\frac{E_2}{E_1}\right) \tag{4.56}$$

with the direction of ϵ_1. Although E_1 and E_2 are complex numbers, their ratio E_2/E_1 is real as both the complex numbers have the same phase. The magnitude of the electric field will be given by $E = \sqrt{|E_1|^2 + |E_2|^2}$. Figure 4.1 shows the most general description of the electric field in a linearly polarized electromagnetic wave.

If in the expression

$$\mathbf{E}(\mathbf{x}, t) = (\epsilon_1 \, E_1 + \epsilon_2 \, E_2) \, e^{i(\mathbf{k}\cdot\mathbf{x}-\omega t)} \,,$$

Fig. 4.1 The electric field in a linearly polarized plane electromagnetic wave

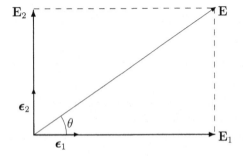

the complex quantities E_1 and E_2 have different phases, then we say that the wave is elliptically polarized. The simplest case of elliptical polarization is circular polarization. Circular polarization occurs when E_1 and E_2 have the same magnitude but differ by a phase of $\pi/2$. In such a case, $E_2 = E_0\, e^{\pm i\pi/2}$ and $E_1 = E_0$, where E_0 is a real quantity. If the phase difference is of such type, then

$$\mathbf{E}(\mathbf{x}, t) = E_0(\epsilon_1 \pm i\epsilon_2)\, e^{i(\mathbf{k}\cdot\mathbf{x}-\omega t)}\,. \tag{4.57}$$

If we suppose that the electromagnetic wave is propagating along the positive z-axis while ϵ_1 and ϵ_2 are along the x and y direction, then the real part of the x-component of the electric field is

$$E_x(\mathbf{x}, t) = E_0\, \mathrm{Re}[e^{i(kz-\omega t)}] = E_0\, \cos(kz - \omega t)\,. \tag{4.58}$$

Similarly we will be able to find out the y-component as

$$E_y(\mathbf{x}, t) = E_0\, \mathrm{Re}[\pm i\, e^{i(kz-\omega t)}] = \mp E_0\, \sin(kz - \omega t)\,. \tag{4.59}$$

The picture in Fig. 4.2 shows the nature of the electric field which is circularly polarized. We see that the x and y components of the electric field oscillates in such a manner that it appears that the electric field in the $x - y$ plane rotates in a circle. In this case, $E_x^2 + E_y^2 = E^2 = E_0^2$. If we fix the plane of rotation at $z = 0$, then we can easily see that as time starts from $t = 0$ the tip of the electric field vector with polarization $(\epsilon_1 + i\epsilon_2)$ rotates in the counterclockwise direction when we look at the plane from the positive direction of z axis. The nature of the electric field is shown in the figure. The electric field magnitude $E = \sqrt{E_x^2 + E_y^2}$ remains constant during this circular motion. Waves polarized in this way are called left circularly polarized. Sometimes waves with this kind of polarization are said to possess positive helicity.

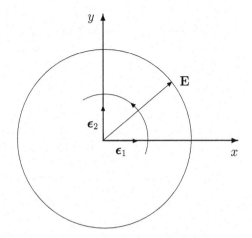

Fig. 4.2 Rotation of the electric field in a circularly polarized plane electromagnetic field. The wave is left circularly polarized

For the same situation, the other polarization degree of freedom corresponding to $(\epsilon_1 - i\epsilon_2)$, rotates in the clockwise direction when observed from the positive direction of z axis. This kind of polarization is called right circular polarization and waves with such polarization are supposed to have negative helicity. Whether be it linear polarization or circular polarization, we see that they can be described by two independent unit vectors. For linear polarization, they are ϵ_1 and ϵ_2, for circular polarization, they are $(\epsilon_1 + i\epsilon_2)$ and $(\epsilon_1 - i\epsilon_2)$. In general, we say that radiation in the form of plane electromagnetic waves has two degrees of freedom. These are the two possible polarization states.

4.2.2   General Polarization Basis

The two vectors, representing two different circular polarizations, can be used to describe the most general polarization state of the plane electromagnetic wave. To verify this statement, we first introduce the two complex orthogonal unit vectors:

$$\epsilon_\pm = \frac{1}{\sqrt{2}}(\epsilon_1 \pm i\epsilon_2) \, . \tag{4.60}$$

We can find out the properties of these polarization basis. First, we see that

$$\epsilon_+^* \cdot \epsilon_+ = \frac{1}{2}(\epsilon_1 - i\epsilon_2) \cdot (\epsilon_1 + i\epsilon_2) = \frac{1}{2}(\epsilon_1 \cdot \epsilon_1 + \epsilon_2 \cdot \epsilon_2 + i\epsilon_1 \cdot \epsilon_2 - i\epsilon_2 \cdot \epsilon_1) \, .$$

As we know that $\epsilon_i \cdot \epsilon_j = \delta_{ij}$ where i, j runs over values 1, 2, one can easily show

$$\epsilon_+^* \cdot \epsilon_+ = 1 \, . \tag{4.61}$$

We can also see that

$$\epsilon_+^* \cdot \epsilon_- = \frac{1}{2}(\epsilon_1 - i\epsilon_2) \cdot (\epsilon_1 - i\epsilon_2) = \frac{1}{2}(\epsilon_1 \cdot \epsilon_1 - \epsilon_2 \cdot \epsilon_2 - i\epsilon_1 \cdot \epsilon_2 - i\epsilon_2 \cdot \epsilon_1) = 0 \, .$$

In general, we can write

$$\epsilon_+^* \cdot \epsilon_+ = \epsilon_-^* \cdot \epsilon_- = 1 \, , \quad \text{and} \quad \epsilon_+^* \cdot \epsilon_- = \epsilon_-^* \cdot \epsilon_+ = 0 \, , \tag{4.62}$$

showing that these vectors form an orthonormal set of basis vectors. Of course, we have another condition which define these basis vectors

$$\epsilon_\pm \cdot \hat{\mathbf{n}} = 0 \, , \tag{4.63}$$

where $\hat{\mathbf{n}}$ specifies the wave propagation direction. From the above discussion, one must have noticed that although $\epsilon_{\pm}^* = \epsilon_{\mp}$, we still define two different basis vectors as ϵ_+ and ϵ_-. The advantage of this notation is that all real quantities appearing as dot products of the basis states can be written as the dot product of one vector with the complex conjugate of the other. This convention makes it clear that the dot products are real. Using the new basis, one can now write the most general form of the electric field in a plane electromagnetic wave as

$$\mathbf{E}(\mathbf{x}, t) = (E_+\,\epsilon_+ + E_-\,\epsilon_-)\,e^{i(\mathbf{k}\cdot\mathbf{x}-\omega t)}\,, \tag{4.64}$$

where E_+ and E_- are complex amplitudes. The above equation contains all the relevant information about plane electromagnetic waves.

Next, we analyze the properties of the above general expression. We assume initially that E_+ and E_- have different magnitudes but have the same phase as

$$E_+ = E_0\,e^{i\delta}\,, \qquad E_- = E_0'\,e^{i\delta}\,, \tag{4.65}$$

where E_0, E_0' and δ are all real numbers. In this case, we can write

$$E_+\,\epsilon_+ + E_-\,\epsilon_- = \frac{E_+}{\sqrt{2}}(\epsilon_1 + i\epsilon_2) + \frac{E_-}{\sqrt{2}}(\epsilon_1 - i\epsilon_2)$$
$$= \left(\frac{E_+ + E_-}{\sqrt{2}}\right)\epsilon_1 + i\left(\frac{E_+ - E_-}{\sqrt{2}}\right)\epsilon_2\,. \tag{4.66}$$

Using the above expression, we now have

$$\mathbf{E}(\mathbf{x}, t) = \left[\left(\frac{E_+ + E_-}{\sqrt{2}}\right)\epsilon_1 + i\left(\frac{E_+ - E_-}{\sqrt{2}}\right)\epsilon_2\right]e^{i(\mathbf{k}\cdot\mathbf{x}-\omega t)}\,, \tag{4.67}$$

as the most general expression of the electric field in a plane electromagnetic wave when E_+ and E_- are as given in Eq. (4.65). This expression can be rewritten as

$$\mathbf{E}(\mathbf{x}, t) = \left(\frac{E_+ + E_-}{\sqrt{2}}\right)\left[\epsilon_1 + i\left(\frac{E_+ - E_-}{E_+ + E_-}\right)\epsilon_2\right]e^{i(\mathbf{k}\cdot\mathbf{x}-\omega t)}\,. \tag{4.68}$$

The above expression of the electric field reveals the nature of the various real components of the electric field. To make things apparent, we again assume that the wave is propagating in the z direction and ϵ_1 and ϵ_2 are along the x and y direction. From the nature of E_+ and E_- in Eq. (4.65), we can now write

$$\mathbf{E}(\mathbf{x}, t) = \left(\frac{E_0 + E_0'}{\sqrt{2}}\right)\left[\epsilon_1 + i\left(\frac{E_0 - E_0'}{E_0 + E_0'}\right)\epsilon_2\right]e^{i(kz-\omega t+\delta)}\,. \tag{4.69}$$

Now, this expression is very similar to the expression we had in the case of circular polarization. We will soon see that although the above expression looks similar to the expression of the circularly polarized electric field, in reality, the present case is very different from the previous one. To see the big difference, we first write down the real components of the electric field as:

$$E_1(\mathbf{x}, t) = \left(\frac{E_0 + E_0'}{\sqrt{2}}\right) \cos(kz - \omega t + \delta), \tag{4.70}$$

$$E_2(\mathbf{x}, t) = -\left(\frac{E_0 - E_0'}{\sqrt{2}}\right) \sin(kz - \omega t + \delta). \tag{4.71}$$

One can now easily see the difference if one compares these results with the previous results. Defining new constants as:

$$a \equiv \left(\frac{E_0 + E_0'}{\sqrt{2}}\right), \qquad b \equiv \left(\frac{E_0 - E_0'}{\sqrt{2}}\right), \tag{4.72}$$

we see that the field components satisfy the equation

$$\frac{E_1^2}{a^2} + \frac{E_2^2}{b^2} = 1. \tag{4.73}$$

From the above equation, it can be seen that the field components satisfy the equation of an ellipse. This shows that, in general, the tip of the electric field vector in a plane electromagnetic wave can traverse an elliptic path, and when this happens, we say that the wave is elliptically polarized. The principal axes of the ellipse, in this case, are along ϵ_1 and ϵ_2. Defining the parameter r as

$$r \equiv \frac{E_-}{E_+}, \tag{4.74}$$

it is observed that the ratio of the length of the semimajor axis to the length of the semiminor axis is given as

$$\frac{a}{b} = \frac{1+r}{1-r}. \tag{4.75}$$

For $r = \pm 1$, we get back linear polarization. All these results are true when E_+ and E_- have the same phase.

One can have a more complicated situation where

$$\frac{E_-}{E_+} = r\, e^{i\theta}, \tag{4.76}$$

where r is real. In this case, both the E_+ and E_- do not have the same phase. In this case, the calculation will be more involved, but straightforward. The electric field will still be moving on an elliptic path, but the axes of the ellipse will not coincide with the directions of ϵ_1 and ϵ_2. The axes will be rotated by some angle with respect to the ϵ_1, ϵ_2 system.

4.2.3 Stokes Parameters

The polarization content of the plane electromagnetic wave can be known if we can write it down in the form

$$\mathbf{E}(\mathbf{x}, t) = (\epsilon_1 \, E_1 + \epsilon_2 \, E_2) \, e^{i(\mathbf{k} \cdot \mathbf{x} - \omega t)} \,, \tag{4.77}$$

showing that the wave is plane polarized, or in the form

$$\mathbf{E}(\mathbf{x}, t) = (E_+ \, \epsilon_+ + E_- \, \epsilon_-) \, e^{i(\mathbf{k} \cdot \mathbf{x} - \omega t)} \,, \tag{4.78}$$

showing that it is elliptically polarized. If we know the pairs (E_1, E_2) or (E_+, E_-), then we know fully the polarization state of the electromagnetic wave. In reality, the converse problem arises. From astrophysical sources or some other sources, we only know that the electric field is of the form

$$\mathbf{E}(\mathbf{x}, t) = \mathcal{E} \, e^{i(\mathbf{k} \cdot \mathbf{x} - \omega t)} \,, \qquad \mathbf{B}(\mathbf{x}, t) = \mathcal{B} \, e^{i(\mathbf{k} \cdot \mathbf{x} - \omega t)} \,, \tag{4.79}$$

which are not written in the polarization basis. The problem now is how do we know about the polarization state of light from this information. We require to measure some quantities related to the incoming light beam to understand the proper polarization basis of the incoming plane waves. We know that light can be linearly polarized or elliptically polarized, but we do not know exactly in which state of polarization the light beam is.

To tackle this problem, the Stokes parameters become useful. These parameters are quadratic in the field strength and can be determined through the intensity measurements and polarizers. Stokes parameters are important as they help us to opine about the vectorial nature of light from intensity measurements. Suppose we are looking at light traveling along z direction, then the Stokes parameters are related to the basic quantities

$$\epsilon_1 \cdot \mathbf{E} \,, \qquad \epsilon_2 \cdot \mathbf{E} \,, \qquad \epsilon_+^* \cdot \mathbf{E} \,, \qquad \epsilon_-^* \cdot \mathbf{E} \,. \tag{4.80}$$

Looking at the above quantities, one can easily see that they specify the amplitude of radiation

1. with linear polarization in the x direction,
2. with linear polarization in the y direction,
3. with elliptical polarization with negative helicity,
4. with elliptical polarization with positive helicity.

We do not know about the actual polarization state of the light beam and so all these quantities have to be known to see the actual polarization state of light. The squares of the above quantities actually give us important information about the distribution of light intensity into various polarization modes. The actual Stokes parameters are built out from the above kind of quantities.

In reality, the linear polarization basis (ϵ_1, ϵ_2) and the elliptical polarization basis (ϵ_+, ϵ_-) are not linearly independent. For this reason, the Stokes parameters are, in general, presented on different bases. Mixing up the basis vectors can produce confusion as one set of basis can always be expressed in terms of the other set. First, we specify the Stokes parameters in the linear polarization basis, where we assume

$$E_1 = a_1\, e^{i\theta_1}\,, \qquad E_2 = a_2\, e^{i\theta_2}\,, \tag{4.81}$$

where a_1, a_2 are real numbers and

$$\mathbf{E}(\mathbf{x}, t) = (\epsilon_1\, E_1 + \epsilon_2\, E_2)\, e^{i(\mathbf{k}\cdot\mathbf{x}-\omega t)}\,.$$

In terms of the linear polarization basis (ϵ_1, ϵ_2), the first Stokes parameter is given as

$$s_0 \equiv |\epsilon_1 \cdot \mathbf{E}|^2 + |\epsilon_1 \cdot \mathbf{E}|^2 = a_1^2 + a_2^2\,. \tag{4.82}$$

This parameter is more like the total intensity of the light beam. The next Stokes parameter is

$$s_1 \equiv |\epsilon_1 \cdot \mathbf{E}|^2 - |\epsilon_2 \cdot \mathbf{E}|^2 = a_1^2 - a_2^2\,. \tag{4.83}$$

This parameter specifies the difference in intensity of the two polarizations. The third Stokes parameter is

$$\begin{aligned}
s_2 &\equiv 2\mathrm{Re}\left[(\epsilon_1 \cdot \mathbf{E})^*(\epsilon_2 \cdot \mathbf{E})\right] = 2\mathrm{Re}\left[\left(a_1\, e^{i(\mathbf{k}\cdot\mathbf{x}-\omega t+\theta_1)}\right)^* \left(a_2\, e^{i(\mathbf{k}\cdot\mathbf{x}-\omega t+\theta_2)}\right)\right]\\
&= 2a_1 a_2 \mathrm{Re}\left[e^{i(\theta_2-\theta_1)}\right]\\
&= 2a_1 a_2\, \cos(\theta_2 - \theta_1)\,.
\end{aligned} \tag{4.84}$$

This parameter specifies the cosine of the phase difference between the two linear polarization modes. The other Stokes parameter in the linear polarization basis is

$$s_3 \equiv 2\mathrm{Im}\left[(\epsilon_1 \cdot \mathbf{E})^*(\epsilon_2 \cdot \mathbf{E})\right] = 2a_1 a_2\, \sin(\theta_2 - \theta_1)\,. \tag{4.85}$$

The last two parameters are related to the relative phase difference of the different linear polarization modes. The polarization states of the electric field expression

$$\mathbf{E}(\mathbf{x}, t) = (\epsilon_1 E_1 + \epsilon_2 E_2)\, e^{i(\mathbf{k}\cdot\mathbf{x}-\omega t)} ,$$

as given by

$$E_1 = a_1\, e^{i\theta_1} , \qquad E_2 = a_2\, e^{i\theta_2}$$

actually has three degrees of freedom, related to a_1, a_2 and $\theta_2 - \theta_1$. It should be noted that only the phase difference between E_1 and E_2 matters, we do not care about common phase. In other words, you can always write the electric field expression as

$$\mathbf{E}(\mathbf{x}, t) = e^{i\theta_1}(\epsilon_1 a_1 + \epsilon_2 a_2\, e^{i(\theta_2-\theta_1)})\, e^{i(\mathbf{k}\cdot\mathbf{x}-\omega t)} = (\epsilon_1 a_1 + \epsilon_2 a_2\, e^{i(\theta_2-\theta_1)})\, e^{i(\mathbf{k}\cdot\mathbf{x}-\omega t+\theta_1)} ,$$

which implies that the polarization part is only sensitive to the phase difference of E_1 and E_2 where the other phase part $e^{i\theta_1}$ modifies the overall phase of the wave and does not affect the relative status of the polarization modes. This observation has an important effect on the definition of the Stokes parameters. Due to this fact, out of the four parameters a_1, a_2, θ_1 and θ_2, only three parameters a_1, a_2 and $\theta_2 - \theta_1$ independently affects the polarization amplitudes. As a result of these, we can anticipate that the four Stokes parameters must not be independent of each other. One can easily check from the above results that

$$s_0^2 = s_1^2 + s_2^2 + s_3^2 , \qquad (4.86)$$

showing that only three of the above parameters are independent of each other. This is a very general result and even holds true for the Stokes parameters represented in terms of the circular polarization basis. The above discussion shows that we can get fair amount of information about the polarization state of the incoming plane waves if we can measure the above four Stokes parameters.

Next, we represent the Stokes parameters in the circular polarization basis. In this basis, the electric field is represented as

$$\mathbf{E}(\mathbf{x}, t) = (E_+ \epsilon_+ + E_- \epsilon_-)\, e^{i(\mathbf{k}\cdot\mathbf{x}-\omega t)} , \qquad (4.87)$$

where we assume the polarization amplitudes as

$$E_+ = a_+\, e^{i\delta_+} , \qquad E_- = a_-\, e^{i\delta_-} . \qquad (4.88)$$

Here, a_+, a_-, δ_+ and δ_- are all real quantities. Using the orthonormality of the circular polarization basis vectors we have the Stokes parameters in this basis as

$$s_0 \equiv |\epsilon_+^* \cdot \mathbf{E}|^2 + |\epsilon_-^* \cdot \mathbf{E}|^2 = a_+^2 + a_-^2 , \tag{4.89}$$

$$s_1 \equiv 2\mathrm{Re}\left[(\epsilon_+^* \cdot \mathbf{E})^*(\epsilon_-^* \cdot \mathbf{E})\right] = 2a_+ a_- \cos(\delta_- - \delta_+) , \tag{4.90}$$

$$s_2 \equiv 2\mathrm{Im}\left[(\epsilon_+^* \cdot \mathbf{E})^*(\epsilon_-^* \cdot \mathbf{E})\right] = 2a_+ a_- \sin(\delta_- - \delta_+) , \tag{4.91}$$

$$s_3 \equiv |\epsilon_+^* \cdot \mathbf{E}|^2 - |\epsilon_-^* \cdot \mathbf{E}|^2 = a_+^2 - a_-^2 . \tag{4.92}$$

In this case also, the parameters can be intuitively interpreted. Except for some minor relabelling of the parameters, the interpretation of them remains similar to the previous case.

In experiments with light beam, the Stokes parameters can be measured by using a waveplate and a linear polarizer. The exact experimental technique to measure the Stokes parameters is beyond the scope of this book, readers who want to know more about it can actually read an interesting pedagogical paper on Stokes parameters. This paper appeared in the American Journal of Physics. The title of the paper is "Measuring the Stokes polarization parameters". The authors of the paper are Beth Schaefer, Edward Collett, Robert Smyth, Daniel Barrett, and Beth Fraher. The citation of the paper is Am. J. Phys. 75 (2), February 2007 page number 163.

4.3   Radiation from Localized Charge and Current Distributions

Till now we were discussing about a particular solution of Maxwell's equations in vacuum, this solution was the plane-wave solution. The source of radiation was not discussed till now. Next, we will discuss a particular case where radiation is produced from moving charges in a localized region of space. The charges are assumed to be moving non-relativistically. We will see that the moving charges produce spherical electromagnetic waves. Far away from the source, these spherical waves behave as plane waves. Before we proceed, we want to present the Green function solution of the inhomogeneous Maxwell's equations as we will require the formal expression of the Green function to derive the results in this section. At first, we present the Green function solution of Maxwell's equation in the Lorenz gauge.

4.3.1   Green Function for Inhomogeneous Maxwell's Equation in Lorenz Gauge

If we look at the inhomogeneous Maxwell equations in the Lorentz gauge, as discussed in the first chapter, we see that their general structure is like

$$\nabla^2 \Psi - \frac{1}{c^2}\frac{\partial^2 \Psi}{\partial t^2} = -4\pi f(\mathbf{x}, t) . \tag{4.93}$$

Here, Ψ stands for the scalar electrodynamic potential in Lorenz gauge. The source function $f(\mathbf{x}, t)$ may be related to charge density or components of the total current. To proceed we Fourier expand $\Psi(\mathbf{x}, t)$ and $f(\mathbf{x}, t)$ as

$$\Psi(\mathbf{x}, t) = \frac{1}{\sqrt{2\pi}} \int_{-\infty}^{\infty} \Psi(\mathbf{x}, \omega) \, e^{-i\omega t} \, d\omega \qquad (4.94)$$

$$f(\mathbf{x}, t) = \frac{1}{\sqrt{2\pi}} \int_{-\infty}^{\infty} f(\mathbf{x}, \omega) \, e^{-i\omega t} \, d\omega, \qquad (4.95)$$

with the inverse Fourier transforms as

$$\Psi(\mathbf{x}, \omega) = \frac{1}{\sqrt{2\pi}} \int_{-\infty}^{\infty} \Psi(\mathbf{x}, t) \, e^{i\omega t} \, d\omega \qquad (4.96)$$

$$f(\mathbf{x}, t) = \frac{1}{\sqrt{2\pi}} \int_{-\infty}^{\infty} f(\mathbf{x}, \omega) \, e^{i\omega t} \, d\omega. \qquad (4.97)$$

From these equations, we see that

$$\frac{\partial^2 \Psi}{\partial t^2} = \frac{1}{\sqrt{2\pi}} \int_{-\infty}^{\infty} \Psi(\mathbf{x}, \omega) \frac{d^2(e^{-i\omega t})}{dt^2} \, d\omega = -\frac{1}{\sqrt{2\pi}} \int_{-\infty}^{\infty} \Psi(\mathbf{x}, \omega)\omega^2 \, e^{-i\omega t} \, d\omega.$$

As a result, the equation for Ψ becomes

$$\frac{1}{\sqrt{2\pi}} \int_{-\infty}^{\infty} (\nabla^2 \Psi(\mathbf{x}, \omega)) \, e^{-i\omega t} \, d\omega + \frac{1}{\sqrt{2\pi c^2}} \int_{-\infty}^{\infty} \Psi(\mathbf{x}, \omega)\omega^2 \, e^{-i\omega t} \, d\omega$$
$$= -4\pi \left[\frac{1}{\sqrt{2\pi}} \int_{-\infty}^{\infty} f(\mathbf{x}, \omega) \, e^{-i\omega t} \, d\omega \right],$$

which can also be written as

$$\frac{1}{\sqrt{2\pi}} \int_{-\infty}^{\infty} \left[\nabla^2 \Psi(\mathbf{x}, \omega) + k^2 \Psi(\mathbf{x}, \omega) + 4\pi f(\mathbf{x}, \omega) \right] e^{-i\omega t} \, d\omega = 0,$$

where

$$k = \frac{\omega}{c}. \qquad (4.98)$$

If the integral vanish, then generally

$$(\nabla^2 + k^2)\Psi(\mathbf{x}, \omega) = -4\pi f(\mathbf{x}, \omega), \qquad (4.99)$$

for each value of ω. Now we can write down the Green function corresponding to this inhomogeneous equation. For a fixed ω, we can write the Green function as $G_\omega(\mathbf{x}, \mathbf{x}')$ and the Green function equation becomes

$$(\nabla^2 + k^2)G_\omega(\mathbf{x}, \mathbf{x}') = -4\pi\delta^3(\mathbf{x} - \mathbf{x}') . \tag{4.100}$$

From isotropy and homogeneity we know that $G_\omega(\mathbf{x}, \mathbf{x}')$ can only be a function of $R = |\mathbf{R}|$ where $\mathbf{R} = \mathbf{x} - \mathbf{x}'$. As a result, we can actually write

$$(\nabla^2 + k^2)G(R, \omega) = -4\pi\delta^3(\mathbf{R}) , \tag{4.101}$$

where we have introduced ω as one of the variables on which the Green function can depend. In the present case, the above equation holds for only a particular ω value. Using spherical polar coordinates, where the origin is at $\mathbf{x}' = 0$, we have the Green equation as

$$\frac{1}{R}\frac{d^2}{dR^2}(RG) + k^2 G = -4\pi\delta^3(\mathbf{R}) . \tag{4.102}$$

To solve the above equation, we first notice that, for $R \neq 0$, the above equation becomes

$$\frac{1}{R}\frac{d^2}{dR^2}(RG) + k^2 G = 0, \quad \Longrightarrow \quad \frac{d^2}{dR^2}(RG) + k^2(RG) = 0, \tag{4.103}$$

whose solution is

$$RG(R, \omega) = Ae^{ikR} + Be^{-ikR} , \tag{4.104}$$

where A, B are constants whose values will be found out shortly.

Next, we see that in the $R \to 0$ limit, the above Green function equation becomes

$$\frac{1}{R}\frac{d^2}{dR^2}(RG) \sim -4\pi\delta^3(\mathbf{R}) , \tag{4.105}$$

which is exactly like the Poisson equation

$$\nabla^2\left(\frac{1}{|\mathbf{x} - \mathbf{x}'|}\right) = -4\pi\delta^3(\mathbf{x} - \mathbf{x}') .$$

where the similarity becomes clear if one represents the above Laplacian in spherical polar coordinates and write $R = |\mathbf{x} - \mathbf{x}'|$. Using the above result, we can now write

$$\lim_{kR \to 0} G(R, \omega) = \frac{1}{R} . \tag{4.106}$$

Previously, we knew for $R \neq 0$,

$$RG(R, \omega) = Ae^{ikR} + Be^{-ikR} \quad \Longrightarrow \quad G(R, \omega) = A\frac{e^{ikR}}{R} + B\frac{e^{-ikR}}{R} .$$

Comparing the above two results in the two limits, we can write

$$G(R, \omega) = AG^+(R, \omega) + BG^-(R, \omega), \tag{4.107}$$

where

$$G^\pm(R, \omega) = \frac{e^{\pm ikR}}{R}. \tag{4.108}$$

As $kR \to 0$ we know $G(R, \omega) = \frac{A+B}{R} = \frac{1}{R}$ and so we have the condition $A + B = 1$. In our convention, $G^+(R, \omega)$ represents the spatial part of a diverging spherical wave, whereas $G^-(R, \omega)$ represents the spatial part of a converging spherical wave. Here, the constants A and B depend upon the conditions we impose upon the Green function. We have

$$G(R, \omega) = AG^+(R, \omega) + BG^-(R, \omega),$$

where $A + B = 1$ and

$$G^\pm(R, \omega) = \frac{e^{\pm ikR}}{R}.$$

If the flow of time is perceived in such a way that we are looking forward in time, the boundary conditions are specified in the past, then of course, $G^+(R, \omega)$ only contributes and $B = 0$ and $A = 1$. On the other hand, if the boundary condition of the problem is specified at some future time, then the relevant Green function is $G^-(R, \omega)$, where $A = 0$ and $B = 1$.

Next, we now try to find out the Green function $G(R, \tau)$, where $\tau = t - t'$. We have assumed the Green function to be a function of τ and not a function of t and t' separately because we have assumed time translation symmetry of the problem. This function appears in the solution of the basic inhomogeneous equation we are trying to solve,

$$\nabla^2 \Psi - \frac{1}{c^2} \frac{\partial^2 \Psi}{\partial t^2} = -4\pi f(\mathbf{x}, t).$$

Previously, we have found the Green function for single ω, now we want the Green function as functions of time. The Green equation in time domain is written as

$$\left(\nabla^2 - \frac{1}{c^2} \frac{\partial^2}{\partial t^2} \right) G(R, \tau) = -4\pi \delta^3(\mathbf{R})\delta(\tau). \tag{4.109}$$

This is the actual problem we want to solve. Our previous discussion will help us to solve the actual problem. To solve it, we first note that

$$G(R, \tau) = \frac{1}{\sqrt{2\pi}} \int_{-\infty}^{\infty} G(R, \omega) e^{-i\omega\tau} d\omega \tag{4.110}$$

and the Dirac delta function can be written as

$$\delta(\tau) = \frac{1}{2\pi} \int_{-\infty}^{\infty} e^{-i\omega\tau} d\omega \,. \tag{4.111}$$

One can easily check that using the Fourier expansion of the Green function and the expression of the Dirac delta function in Eq. (4.109) one gets

$$(\nabla^2 + k^2)G(R, \omega) = -4\pi\delta^3(\mathbf{R}) \,, \tag{4.112}$$

whose solution we already know. Finally, the relevant expression of the Green function $G^{\pm}(R, \tau)$ is then given by the Fourier expansion

$$G^{\pm}(R, \tau) = \frac{1}{2\pi} \int_{-\infty}^{\infty} G^{\pm}(R, \omega) e^{-i\omega\tau} d\omega = \frac{1}{2\pi} \int_{-\infty}^{\infty} \frac{e^{\pm ikR}}{R} e^{-i\omega\tau} d\omega$$
$$= \frac{1}{R} \left[\frac{1}{2\pi} \int_{-\infty}^{\infty} e^{-i\omega(\tau \mp \frac{R}{c})} d\omega \right] \,.$$

From the integral definition of the Dirac delta function

$$\delta(\tau) = \frac{1}{2\pi} \int_{-\infty}^{\infty} e^{-i\omega\tau} d\omega \,,$$

one can now easily see that the Green function in our specific case is

$$G^{\pm}(R, \tau) = \frac{1}{R}\delta\left(\tau \mp \frac{R}{c}\right) \,. \tag{4.113}$$

Here, $G^+(R, \tau)$ is called the retarded Green function and $G^-(R, \tau)$ is called the advanced Green function. The retarded Green function affects the field point when $c\tau = R$ or when $c(t - t') = R$. Here, t' is the time at the source point, and the signal from the source point reaches the field point at time t, where $t = t' + \frac{R}{c}$. This makes sense as the electromagnetic signal traveling at the speed c affects the field point, which is $R = |\mathbf{x} - \mathbf{x}'|$ distance away, after R/c amount of time. If the boundary conditions are given in the past, then this is the Green function with which one should work. On the other hand, the advanced Green function becomes active when $c\tau = -R$ or when $t = t' - \frac{R}{c}$, which means the signal reaches the field point R/c times before. This form of the Green function becomes important when the boundary condition is given in the future.

With the knowledge of the Green function in the present case, we can now write the solution of the inhomogeneous equation as

$$\Psi^{\pm}(\mathbf{x}, t) = \Psi_{\text{hom}}^{\text{in/out}}(\mathbf{x}, t) + \int_{-\infty}^{\infty} d^3x' \int_{-\infty}^{\infty} dt' \, G^{\pm}(|\mathbf{x} - \mathbf{x}'|, t - t') f(\mathbf{x}', t') \,, \tag{4.114}$$

where $\Psi_{\text{hom}}^{\text{in/out}}(\mathbf{x}, t)$ is the solution of the corresponding homogeneous differential equation (in this case, one obtains the homogeneous differential equation from the inhomogeneous Maxwell equations). To be specific, if at $t \to -\infty$ the solution of the homogeneous partial differential equation is called $\Psi_{\text{hom}}^{\text{in}}(\mathbf{x}, t)$ then the solution is

$$\Psi^+(\mathbf{x}, t) = \Psi_{\text{hom}}^{\text{in}}(\mathbf{x}, t) + \int_{-\infty}^{\infty} d^3 x' \int_{-\infty}^{\infty} dt' \; G^+(|\mathbf{x} - \mathbf{x}'|, t - t') \, f(\mathbf{x}', t') .$$
(4.115)

It is generally assumed that the source term $f(\mathbf{x}', t')$ is localized in space and time and vanishes as $t \to -\infty$. In the far past the only solution was $\Psi_{\text{hom}}^{\text{in}}(\mathbf{x}, t)$. At some finite time, the source term starts to contribute and the solution becomes as shown in the above equation. In a similar way, if at $t \to \infty$, the solution of the homogeneous differential equation is $\Psi_{\text{hom}}^{\text{out}}(\mathbf{x}, t)$, then the general solution is

$$\Psi^-(\mathbf{x}, t) = \Psi_{\text{hom}}^{\text{out}}(\mathbf{x}, t) + \int_{-\infty}^{\infty} \int_{-\infty}^{\infty} dt' \; G^-(|\mathbf{x} - \mathbf{x}'|, t - t') \, f(\mathbf{x}', t') .$$
(4.116)

Later, in this book, we will re-derive the Green function in the full relativistic setting. Next, we will see an application of the above solutions in an important problem, the problem related to electromagnetic radiation.

4.3.2 Radiation from Localized Charges

Now we know how to handle time-dependent situations in electrodynamics giving rise to radiation. We will apply our knowledge in an important problem. First of all, we recall that the Maxwell equations in the Lorenz gauge were

$$\nabla^2 \Phi - \frac{1}{c^2} \frac{\partial^2 \Phi}{\partial t^2} = -4\pi\rho, \qquad \nabla^2 \mathbf{A} - \frac{1}{c^2} \frac{\partial^2 \mathbf{A}}{\partial t^2} = -\frac{4\pi}{c} \mathbf{J} . \qquad (4.117)$$

For time-varying charges and currents, we can use their Fourier expansions as

$$\mathbf{J}(\mathbf{x}, t) = \frac{1}{\sqrt{2\pi}} \int_{-\infty}^{\infty} \mathbf{J}(\mathbf{x}, \omega) \, e^{-i\omega t} \, d\omega , \qquad \rho(\mathbf{x}, t) = \frac{1}{\sqrt{2\pi}} \int_{-\infty}^{\infty} \rho(\mathbf{x}, \omega) \, e^{-i\omega t} \, d\omega .$$

Similar Fourier expansion can also be done for Φ and $\mathbf{A}$. As the differential equations are linear in nature, we can actually work with one Fourier component alone. Henceforth, we work with the $\omega > 0$ component and write the source functions as

$$\rho(\mathbf{x}, t) = \rho(\mathbf{x}) e^{-i\omega t} , \qquad \mathbf{J}(\mathbf{x}, t) = \mathbf{J}(\mathbf{x}) e^{-i\omega t} .$$

In our present case, all the relevant time varying fields can be written as

$$\mathbf{A}(\mathbf{x}, t) = \mathbf{A}(\mathbf{x})e^{-i\omega t}, \qquad \mathbf{E}(\mathbf{x}, t) = \mathbf{E}(\mathbf{x})e^{-i\omega t} \qquad \mathbf{B}(\mathbf{x}, t) = \mathbf{B}(\mathbf{x})e^{-i\omega t}.$$

This amounts to saying that the whole system is oscillating with time at the same frequency and the only thing to know is the spatial dependence of the fields. In general, all the fields and source functions which are independent of time can be complex. Time dependence introduces another complex phase. To connect with reality, we must remember that we have to take the real part of the complex solutions as physically relevant.

We know that once the gauge potentials are known, we can evaluate the physical fields as

$$\mathbf{B} = \nabla \times \mathbf{A}, \qquad \mathbf{E} = -\nabla \Phi - \frac{1}{c} \frac{\partial \mathbf{A}}{\partial t}. \tag{4.118}$$

In our study of radiation, we will be interested to find the fields $\mathbf{E}$ and $\mathbf{B}$ in regions where the source charge density or current density is absent. The source charges are localized in some region of space. When there is no source current density, we can bypass the evaluation of the scalar electrodynamical potential Φ by using the homogeneous Maxwell equation

$$\nabla \times \mathbf{B} = \frac{1}{c} \frac{\partial \mathbf{E}}{\partial t}. \qquad \text{(When } \mathbf{J} = 0) \tag{4.119}$$

As we are working in the single frequency mode, we know all time derivatives of fields can be replaced by $-i\omega$ times the field. From the above equation, we see that in our case

$$\mathbf{E} = \frac{i}{k} \nabla \times \mathbf{B}, \tag{4.120}$$

where

$$k = \frac{\omega}{c}. \tag{4.121}$$

So if we have found out the value of $\mathbf{B}$ using $\mathbf{B} = \nabla \times \mathbf{A}$, then we can calculate $\mathbf{E}$ from the above equation. If the current density is non-zero, such a procedure does not work. Fortunately, at the field point, away from localized charge distribution, the current is always zero. In the absence of any electric or magnetic field in the infinite past, we know that the retarded solution for the vector potential solution in the Lorenz gauge is given by

$$\mathbf{A}^+(\mathbf{x}, t) = \int_{-\infty}^{\infty} d^3 x' \int_{-\infty}^{\infty} dt' \; G^+(|\mathbf{x} - \mathbf{x}'|, t - t') \, \mathbf{f}(\mathbf{x}', t'), \tag{4.122}$$

where $\mathbf{f} = (1/c)\mathbf{J}$ and $G^+(|\mathbf{x} - \mathbf{x}'|, t - t')$ is the retarded Green function. Henceforth we will simply write $\mathbf{A}^+$ as $\mathbf{A}$. We know

$$G^+(|\mathbf{x} - \mathbf{x}'|, t - t') = \frac{1}{R}\delta\left(\tau - \frac{R}{c}\right) = \frac{1}{R}\delta\left(t - t' - \frac{|\mathbf{x} - \mathbf{x}'|}{c}\right), \quad (4.123)$$

and as a result

$$\mathbf{A}(\mathbf{x}, t) = \frac{1}{c}\int_{-\infty}^{\infty} d^3x' \int_{-\infty}^{\infty} dt' \frac{\mathbf{J}(\mathbf{x}', t')}{|\mathbf{x} - \mathbf{x}'|}\delta\left(t - t' - \frac{|\mathbf{x} - \mathbf{x}'|}{c}\right), \quad (4.124)$$

where the spatial and temporal integration spans all spacetime, there are no boundaries. We can write the above equation as

$$\begin{aligned}
\mathbf{A}(\mathbf{x}, t) &= \frac{1}{c}\int_{-\infty}^{\infty} d^3x' \int_{-\infty}^{\infty} dt' \frac{\mathbf{J}(\mathbf{x}')e^{-i\omega t'}}{|\mathbf{x} - \mathbf{x}'|}\delta\left(t - t' - \frac{|\mathbf{x} - \mathbf{x}'|}{c}\right) \\
&= \frac{1}{c}\int_{-\infty}^{\infty} d^3x'\, \mathbf{J}(\mathbf{x}')\frac{e^{-i\omega\left(t - \frac{|\mathbf{x} - \mathbf{x}'|}{c}\right)}}{|\mathbf{x} - \mathbf{x}'|} = \left[\frac{1}{c}\int_{-\infty}^{\infty} d^3x'\, \mathbf{J}(\mathbf{x}')\frac{e^{i\frac{\omega}{c}|\mathbf{x} - \mathbf{x}'|}}{|\mathbf{x} - \mathbf{x}'|}\right]e^{-i\omega t}.
\end{aligned}$$

As $\mathbf{A}(\mathbf{x}, t) = \mathbf{A}(\mathbf{x})e^{-i\omega t}$, we have

$$\mathbf{A}(\mathbf{x}) = \frac{1}{c}\int_{-\infty}^{\infty} \mathbf{J}(\mathbf{x}')\frac{e^{ik|\mathbf{x} - \mathbf{x}'|}}{|\mathbf{x} - \mathbf{x}'|}\, d^3x'. \quad (4.125)$$

This is the expression of the vector potential which we will like to evaluate for a specific source configuration.

Formally now we have all the tools to calculate how time-varying sources produce electric and magnetic fields. To proceed further, we require some simplifying assumptions regarding the physical dimensions of the source region and the wavelengths of radiation. First of all, we assume that the source current is confined to a small region in space. The word small implies that the typical length dimension d specifying the source region is much smaller than the wavelength $\lambda = 2\pi/k$ of radiation. Mathematically, our approximation states

$$d \ll \lambda. \quad (4.126)$$

After this assumption, we can proceed to find out the radiation produced by the oscillating source current. Using this assumption, we can broadly categorize three spatial regions of interest:

1. The near or static zone. This region is specified by the range of $r = |\mathbf{x}|$ which satisfies $d \ll r \ll \lambda$.

2. The intermediate or induction zone. This region is specified by the condition $d \ll r \sim \lambda$.
3. The far or radiation zone. This region is specified by $d \ll \lambda \ll r$.

The time dependence of the field $\mathbf{A}(\mathbf{x}, t)$ is the same in all three regions as we have assumed a harmonic time variation for all the fields and the sources. Except for the pure time dependence, the spatial field expressions are different in these three regions. In the near zone, $\mathbf{A}(\mathbf{x})$ is like the vector potential from magnetostatics and so we call this region the static zone. In the far zone, we will see that the $\mathbf{A}(\mathbf{x}, t)$ solution is like spherical waves, and in this region, we will see that energy is transmitted out via these waves. This is the reason why this region is also called the radiation zone. Compared to the solutions of the vector potential in the near or far zones, the solution in the intermediate zone is more complicated, and in this book, we will not work out the physics of the intermediate zone.

4.3.2.1 The Near Zone

We first consider the near zone where $d \ll \lambda$ and $r \ll \lambda$ and $r \gg d$. We have to evaluate

$$\mathbf{A}(\mathbf{x}) = \frac{1}{c} \int_{-\infty}^{\infty} \mathbf{J}(\mathbf{x}') \frac{e^{ik|\mathbf{x}-\mathbf{x}'|}}{|\mathbf{x} - \mathbf{x}'|} \, d^3x' \, ,$$

in this limit. To evaluate the above integral, we first expand the exponential

$$e^{ik|\mathbf{x}-\mathbf{x}'|} = 1 + ik|\mathbf{x} - \mathbf{x}'| + \cdots \tag{4.127}$$

In our case, we have $|\mathbf{x}'| \leq d$ as $\mathbf{x}'$ specifies source points. As $|\mathbf{x}| = r \gg d$, we can safely write the above expansion as

$$e^{ik|\mathbf{x}-\mathbf{x}'|} = 1 + ik|\mathbf{x}| + \cdots = 1 + ikr + \cdots \tag{4.128}$$

Moreover, as

$$kr = 2\pi \frac{r}{\lambda} \sim 0 \tag{4.129}$$

as $r \ll \lambda$ in the near zone, we can replace the exponential function above by unity in the integral. So, in the near zone, we have

$$\mathbf{A}(\mathbf{x}) = \frac{1}{c} \int_{-\infty}^{\infty} \frac{\mathbf{J}(\mathbf{x}')}{|\mathbf{x} - \mathbf{x}'|} \, d^3x' \, . \tag{4.130}$$

This is just like the vector potential produced by a static current distribution $\mathbf{J}(\mathbf{x}')$. In reality, the vector potential is not really static as

$$A(\mathbf{x}, t) = \frac{1}{c} \int_{-\infty}^{\infty} \frac{\mathbf{J}(\mathbf{x}')}{|\mathbf{x} - \mathbf{x}'|} e^{-i\omega t} d^3 x' . \tag{4.131}$$

One can now calculate the electric and magnetic field from this expression in the near zone. Here, the integration is actually over the source volume (where the integrand does not vanish). As this region does not give rise to radiation, we will not calculate the electric and magnetic fields here. The above result does not predict any propagating wave as we do not have any oscillations in space. A factor of $e^{\pm ikr}$ in conjunction with $e^{-i\omega t}$, in general, specifies a wave-like solution, here we do not have any factor as $e^{\pm ikr}$. We will see that such wave-like features appear in the radiation zone. We skip the calculations in the intermediate zone and directly move over to the radiation zone.

4.3.2.2   The Zone Where $r \gg d$ and Radiation Phenomenon

We have the basic expression for the vector potential as

$$A(\mathbf{x}) = \frac{1}{c} \int_{-\infty}^{\infty} \mathbf{J}(\mathbf{x}') \frac{e^{ik|\mathbf{x}-\mathbf{x}'|}}{|\mathbf{x} - \mathbf{x}'|} d^3 x' ,$$

and in the radiation zone or the far zone we have to find how it looks when $r \gg \lambda \gg d$. In this limit $kr \gg 1$. In this case, we cannot neglect the exponential, it affects the calculations. First, we notice

$$|\mathbf{x} - \mathbf{x}'| = \left[(\mathbf{x} - \mathbf{x}') \cdot (\mathbf{x} - \mathbf{x}') \right]^{1/2} = (r^2 - 2\mathbf{x} \cdot \mathbf{x}' + r'^2)^{1/2}$$

$$= r \left[1 - \frac{2\mathbf{x} \cdot \mathbf{x}'}{r^2} + \left(\frac{r'}{r} \right)^2 \right]^{1/2} . \tag{4.132}$$

Defining the unit vector $\hat{\mathbf{n}}$, which is an unit vector along $\mathbf{x}$, as

$$\hat{\mathbf{n}} \equiv \frac{\mathbf{x}}{r} , \tag{4.133}$$

we can write

$$|\mathbf{x} - \mathbf{x}'| = r \left[1 - \frac{2\hat{\mathbf{n}} \cdot \mathbf{x}'}{r} + \left(\frac{r'}{r} \right)^2 \right]^{1/2} . \tag{4.134}$$

We can expand the above term to second order in r'/r and write

$$|\mathbf{x} - \mathbf{x}'| = r \left[1 - \frac{\hat{\mathbf{n}} \cdot \mathbf{x}'}{r} + \frac{1}{2} \left(\frac{r'}{r} \right)^2 - \frac{1}{2} \left(\frac{\hat{\mathbf{n}} \cdot \mathbf{x}'}{r} \right)^2 \cdots \right]^{1/2} . \tag{4.135}$$

Due to our assumption $(r'/r) \ll 1$, the above expansion becomes simply

$$|\mathbf{x} - \mathbf{x}'| = r - \hat{\mathbf{n}} \cdot \mathbf{x}'. \tag{4.136}$$

In our scheme of approximation, it is consistent to take $|\mathbf{x} - \mathbf{x}'|^{-1} = (1/r)$. With all these results, we can now write down the expression of the vector potential in the radiation zone as

$$\mathbf{A}(\mathbf{x}) = \frac{e^{ikr}}{cr} \int_{-\infty}^{\infty} \mathbf{J}(\mathbf{x}') \, e^{-ik\hat{\mathbf{n}}\cdot\mathbf{x}'} \, d^3x', \tag{4.137}$$

and the full time-dependent potential becomes

$$\mathbf{A}(\mathbf{x}, t) = \frac{e^{i(kr-\omega t)}}{cr} \int_{-\infty}^{\infty} \mathbf{J}(\mathbf{x}') \, e^{-ik\hat{\mathbf{n}}\cdot\mathbf{x}'} \, d^3x'. \tag{4.138}$$

The above expression gives the form of the vector potential when $r \gg d$ and kr is non-zero. One must remember that even in the near zone, we have $r \gg d$, but there we neglect kr. This gives us a hint that the above form of the potential can produce the physics in the far zone (where $kr \gg 1$) and in the near zone by setting $kr \sim 0$. We will show how this happens in the later part of this chapter. The above expression shows that in the zone where kr cannot be neglected the vector potential has the form of an outgoing spherical wave. We can also write the above equation as

$$\mathbf{A}(\mathbf{x}, t) = \frac{e^{i(kr-\omega t)}}{r} \, \mathbf{f}(\theta, \phi), \tag{4.139}$$

where

$$\mathbf{f}(\theta, \phi) \equiv \frac{1}{c} \int_{-\infty}^{\infty} \mathbf{J}(\mathbf{x}') \, e^{-ik\hat{\mathbf{n}}\cdot\mathbf{x}'} \, d^3x'. \tag{4.140}$$

Using the above expression, we will now attempt to calculate the magnetic field. In the present case

$$\mathbf{B}(\mathbf{x}, t) = \nabla \times \mathbf{A} = \nabla \times \left(\frac{e^{i(kr-\omega t)}}{r} \, \mathbf{f}(\theta, \phi) \right) = \frac{e^{i(kr-\omega t)}}{r} \nabla \times \mathbf{f}(\theta, \phi)$$
$$+ \left[\nabla \frac{e^{i(kr-\omega t)}}{r} \right] \times \mathbf{f}(\theta, \phi).$$

It must be noted here and henceforth that the nabla operator only acts on the field point. We know

$$\nabla \frac{e^{i(kr-\omega t)}}{r} = \frac{1}{r} \nabla e^{i(kr-\omega t)} + e^{i(kr-\omega t)} \nabla \left(\frac{1}{r}\right) = ik \frac{e^{i(kr-\omega t)}}{r} \hat{\mathbf{n}} - e^{i(kr-\omega t)} \frac{\hat{\mathbf{n}}}{r^2}$$

$$= ik \frac{e^{i(kr-\omega t)}}{r} \hat{\mathbf{n}} \left(1 - \frac{1}{ikr}\right).$$

The calculation shows the first signature of a spherical wave. We will see later that in the radiation zone both the magnetic and the electric fields propagate as spherical waves. To find the magnetic field, we require to evaluate the curl of $\mathbf{f}(\theta, \phi)$. We evaluate it as

$$\nabla \times \mathbf{f}(\theta, \phi) = \frac{1}{c} \int_{-\infty}^{\infty} \nabla \times \left[\mathbf{J}(\mathbf{x}') e^{-ik\hat{\mathbf{n}}\cdot\mathbf{x}'}\right] d^3 x' = \frac{1}{c} \int_{-\infty}^{\infty} \left[(\nabla e^{-ik\hat{\mathbf{n}}\cdot\mathbf{x}'}) \times \mathbf{J}(\mathbf{x}')\right] d^3 x'.$$

To evaluate the above integral, we have to calculate $\nabla e^{-ik\hat{\mathbf{n}}\cdot\mathbf{x}'}$. It is

$$\nabla e^{-ik\hat{\mathbf{n}}\cdot\mathbf{x}'} = -ik\nabla(\hat{\mathbf{n}}\cdot\mathbf{x}') - \frac{k^2}{2}\nabla(\hat{\mathbf{n}}\cdot\mathbf{x}')^2 + \cdots \qquad (4.141)$$

Next we calculate

$$\nabla(\hat{\mathbf{n}}\cdot\mathbf{x}') = \nabla\left(\frac{\mathbf{x}}{r}\cdot\mathbf{x}'\right) = (\mathbf{x}'\cdot\nabla)\frac{\mathbf{x}}{r} + \mathbf{x}' \times \left(\nabla \times \frac{\mathbf{x}}{r}\right). \qquad (4.142)$$

Without doing further calculations, one can see that the right-hand side of the above equation will have terms which are dependent on $\mathbf{x}'/r$. As we have assumed $d \ll r$, these terms do not contribute. In a similar way, $\nabla(\hat{\mathbf{n}}\cdot\mathbf{x}')^2$ and the other terms in $\nabla e^{-ik\hat{\mathbf{n}}\cdot\mathbf{x}'}$ also do not contribute. As a result of this, when $kr \gg 1$, we can write

$$\mathbf{B}(\mathbf{x}, t) = \left[\nabla \frac{e^{i(kr-\omega t)}}{r}\right] \times \mathbf{f}(\theta, \phi) = ik \frac{e^{i(kr-\omega t)}}{r} \left[\hat{\mathbf{n}} \times \mathbf{f}(\theta, \phi)\right]. \qquad (4.143)$$

We know that the electric field can be obtained from the magnetic field as

$$\mathbf{E} = \frac{i}{k} \nabla \times \mathbf{B},$$

which gives

$$\mathbf{E}(\mathbf{x}, t) = \frac{i}{k} \nabla \times \left\{ik \frac{e^{i(kr-\omega t)}}{r} \left[\hat{\mathbf{n}} \times \mathbf{f}(\theta, \phi)\right]\right\} = -\nabla \left(\frac{e^{i(kr-\omega t)}}{r}\right) \times \left[\hat{\mathbf{n}} \times \mathbf{f}(\theta, \phi)\right]$$

$$- \frac{e^{i(kr-\omega t)}}{r} \nabla \times \left[\hat{\mathbf{n}} \times \mathbf{f}(\theta, \phi)\right], \qquad (4.144)$$

when $kr \gg 1$. Actual calculation shows that the last term on the right-hand side of the above equation does not contribute in the region where $r \gg d$. As a consequence we have

$$\mathbf{E}(\mathbf{x}, t) = -ik\frac{e^{i(kr-\omega t)}}{r}\left[\hat{\mathbf{n}} \times \left(\hat{\mathbf{n}} \times \mathbf{f}(\theta, \phi)\right)\right] = \mathbf{B}(\mathbf{x}, t) \times \hat{\mathbf{n}}. \quad (4.145)$$

These expressions show clearly that both the electric and magnetic fields in the radiation zone fall off as $1/r$ and both the electric and magnetic fields are transverse to the direction of wave propagation specified by $\hat{\mathbf{n}}$. The electric and magnetic fields are themselves orthogonal.

4.3.3   The Vector Potential in the Electric Dipole Approximation

We have seen that in the region where $r \gg d$ the vector potential is given by

$$\mathbf{A}(\mathbf{x}, t) = \frac{e^{i(kr-\omega t)}}{r} \mathbf{f}(\theta, \phi),$$

where

$$\mathbf{f}(\theta, \phi) = \frac{1}{c}\int_{-\infty}^{\infty} \mathbf{J}(\mathbf{x}')\, e^{-ik\hat{\mathbf{n}}\cdot\mathbf{x}'}\, d^3x'.$$

As the source dimension is much small compared to the wavelength of radiation we can actually expand $e^{-ik\hat{\mathbf{n}}\cdot\mathbf{x}'}$ in a power series and write

$$\mathbf{f}(\theta, \phi) = \frac{1}{c}\int_{-\infty}^{\infty} \mathbf{J}(\mathbf{x}')\left[1 - ik(\mathbf{n}\cdot\mathbf{x}') - \frac{1}{2}k^2(\mathbf{n}\cdot\mathbf{x}')^2 + \cdots\right] d^3x'. \quad (4.146)$$

The successive terms in the bracket inside the integral contribute decreasingly. If we take the first term of the above series, we get

$$\mathbf{A}(\mathbf{x}, t) = \frac{e^{i(kr-\omega t)}}{cr}\int_{-\infty}^{\infty} \mathbf{J}(\mathbf{x}')\, d^3x'. \quad (4.147)$$

Working with the first term of the series gives rise to the electric dipole approximation. Why we call it the dipole approximation will become clear later. This approximation plays an important part in physics related to radiation. The reader must note that one does not get other relevant radiation patterns, as the magnetic dipole radiation or electric quadrupole radiation, by simply taking higher order terms inside the bracket on the right-hand side of the above integral. To obtain the other radiation patterns, one has to be a bit more careful. We will show how one obtains the other radiation patterns later in this chapter.

All the above discussion is done in a region where $r \gg d$. We can get both the radiation zone and the near zone results from our work. To appreciate this idea, one must first of all notice that the expression of the vector potential in the near zone was given by

$$\mathbf{A}(\mathbf{x}, t) = \frac{e^{-i\omega t}}{c} \int_{-\infty}^{\infty} \frac{\mathbf{J}(\mathbf{x}')}{|\mathbf{x} - \mathbf{x}'|} d^3 x' \,.$$

In the near zone $r \gg d$ so we can also write the above result as

$$\mathbf{A}(\mathbf{x}, t) = \frac{e^{-i\omega t}}{cr} \int_{-\infty}^{\infty} \mathbf{J}(\mathbf{x}') \, d^3 x' \,. \tag{4.148}$$

From Eq. (4.147), we see that if we first evaluate $\mathbf{A}(\mathbf{x}, t)$ for $r \gg d$ and then if we take the limit $kr \to 0$, we obtain the above expression of the potential in the near zone. This says that our calculation of $\mathbf{A}(\mathbf{x}, t)$ using the first term of the series in $\mathbf{f}(\theta, \phi)$ can produce both the radiation zone (r is greater than the wavelength, kr is also large) and the near zone (r is smaller than the wavelength, kr is approximately zero)[1] results. This is certainly an interesting observation and we will use this important bit of information.

Taking the first term in the expansion in $\mathbf{f}(\theta, \phi)$, we have

$$\mathbf{f}(\theta, \phi) = \frac{1}{c} \int_{-\infty}^{\infty} \mathbf{J}(\mathbf{x}') \, d^3 x' \,. \tag{4.149}$$

To proceed further, we use a general mathematical technique. Let $h(\mathbf{x}')$ and $g(\mathbf{x}')$ be two well-behaved functions of $\mathbf{x}'$. We can now define an integral as

$$K \equiv \int (h\mathbf{J} \cdot \nabla' g + g\mathbf{J} \cdot \nabla' h + hg\nabla' \cdot \mathbf{J}) \, d^3 x' \,, \tag{4.150}$$

and try to see its properties. In the present case, the nabla operator acts on the source points. We note that

$$\nabla' \cdot [g(h\mathbf{J})] = g\nabla' \cdot (h\mathbf{J}) + h\mathbf{J} \cdot \nabla' g = gh\nabla' \cdot \mathbf{J} + g\mathbf{J} \cdot \nabla' h + h\mathbf{J} \cdot \nabla' g \,.$$

From this relation, we see that

$$K = \int \nabla' \cdot [g(h\mathbf{J})] \, d^3 x' = \oint_{\partial V} [g(h\mathbf{J})] \cdot \hat{\mathbf{n}} da \,, \tag{4.151}$$

where we have used Gauss theorem. Here ∂V is the boundary surface which contains the current distribution and $\hat{\mathbf{n}} da$ is an infinitesimal outward oriented area vector on ∂V. Assuming $\mathbf{J}$ to be localized inside the boundary, we have $\mathbf{J} = 0$ on ∂V giving $K = 0$ and consequently:

---

[1] Although in both cases $r \gg d$.

$$\int (h\mathbf{J} \cdot \nabla' g + g\mathbf{J} \cdot \nabla' h + hg\nabla' \cdot \mathbf{J}) \, d^3x' = 0. \tag{4.152}$$

Now, we take $h = 1$ and $g = x'^i$ and get

$$\int (J^i + x'^i \nabla' \cdot \mathbf{J}) \, d^3x' = 0 \quad \Longrightarrow \quad \int J^i \, d^3x' = -\int x'^i \nabla' \cdot \mathbf{J} \, d^3x'.$$

Written as a vector equation, it says

$$\int \mathbf{J} \, d^3x' = -\int \mathbf{x}' \nabla' \cdot \mathbf{J} \, d^3x'. \tag{4.153}$$

The right-hand side now contains explicit derivatives operating on functions of the source coordinates. From the definition of $\mathbf{f}(\theta, \phi)$, we can now write

$$\mathbf{f}(\theta, \phi) = -\frac{1}{c} \int_{-\infty}^{\infty} [\nabla' \cdot \mathbf{J}(\mathbf{x}')] \, \mathbf{x}' \, d^3x'. \tag{4.154}$$

The source charge density and current density satisfies the continuity equation. The continuity condition on the sources is given by

$$\frac{\partial \rho(\mathbf{x}', t)}{\partial t} + \nabla' \cdot \mathbf{J}(\mathbf{x}', t) = 0. \tag{4.155}$$

As both the charge density and current density have the same time dependence as $e^{-i\omega t}$, we can recast the above equation as

$$\rho(\mathbf{x}') = -i\frac{\nabla' \cdot \mathbf{J}(\mathbf{x}')}{\omega}. \tag{4.156}$$

Using this equation, we can now write

$$\mathbf{f}(\theta, \phi) = -ik \int_{-\infty}^{\infty} \mathbf{x}' \rho(\mathbf{x}') \, d^3x'. \tag{4.157}$$

The volume integral on the right-hand side of the above equation is now recognizable, it is the electric dipole moment of the source charge distribution. Defining the electric dipole moment of the source charge distribution in the usual way as

$$\mathbf{p} \equiv \int_{-\infty}^{\infty} \mathbf{x}' \rho(\mathbf{x}') \, d^3x', \tag{4.158}$$

we have

$$\mathbf{f}(\theta, \phi) = -ik\mathbf{p}, \tag{4.159}$$

and as a result we wee that $\mathbf{f}(\theta, \phi)$ is independent of θ and ϕ in the dipole approximation. As we had

$$\mathbf{A}(\mathbf{x}, t) = \frac{e^{i(kr-\omega t)}}{r} \, \mathbf{f}(\theta, \phi) \, ,$$

we see that in our scheme of approximation, where we have only taken the first term in the expansion of $e^{-ik\hat{\mathbf{n}} \cdot \mathbf{x}'}$ in $\mathbf{f}(\theta, \phi)$, we have

$$\mathbf{A}(\mathbf{x}, t) = -ik\mathbf{p} \, \frac{e^{i(kr-\omega t)}}{r} \, . \tag{4.160}$$

Due to the explicit presence of the electric dipole term in the vector potential, we call our approximation scheme as the electric dipole approximation. One must note that the dipole moment term, $\mathbf{p}$, can be complex in general. Later, we will explicitly show an example where $\mathbf{p}$ is complex. The radiation generated from this kind of field is called electric dipole radiation.

4.3.4   The Magnetic Field and the Electric Field in the Electric Dipole Approximation

Now we will calculate the magnetic field and the electric field in the dipole approximation. We work with the time independent fields. We can introduce time dependence at any moment as time dependence of the quantities is purely harmonic in nature. The magnetic field is given by

$$\mathbf{B}(\mathbf{x}) = \nabla \times \mathbf{A}(\mathbf{x}) = -ik\nabla \left(\frac{e^{ikr}}{r} \right) \times \mathbf{p}$$
$$= -ik \left(ik - \frac{1}{r} \right) \frac{e^{ikr}}{r} (\hat{\mathbf{n}} \times \mathbf{p}) \, .$$

From this expression, we can write

$$\mathbf{B}(\mathbf{x}) = k^2 \left(1 - \frac{1}{ikr} \right) \frac{e^{ikr}}{r} (\hat{\mathbf{n}} \times \mathbf{p}) \, . \tag{4.161}$$

In the above calculation, we have still not used the fact that $kr \gg 1$ in the radiation zone. This limit will be taken later. In the above expression, if $kr \sim 0$, we will get the near zone expression of the magnetic field. Next, we will calculate the electric field in the dipole approximation. We know that

$$\mathbf{E}(\mathbf{x}) = \frac{i}{k}\nabla \times \mathbf{B}(\mathbf{x}) = ik\nabla\left[\frac{e^{ikr}}{r}\left(1 - \frac{1}{ikr}\right)\right] \times (\hat{\mathbf{n}} \times \mathbf{p})$$

$$+ ik\frac{e^{ikr}}{r}\left(1 - \frac{1}{ikr}\right)\nabla \times (\hat{\mathbf{n}} \times \mathbf{p}). \tag{4.162}$$

As the above calculation is a bit long, we do it in parts. First, we see that

$$ik\nabla\left[\frac{e^{ikr}}{r}\left(1 - \frac{1}{ikr}\right)\right] \times (\hat{\mathbf{n}} \times \mathbf{p}) = ik\left[\left(\frac{ik}{r} - \frac{1}{r^2}\right)\left(1 - \frac{1}{ikr}\right)\right.$$

$$\left. + \frac{1}{ikr^3}\right]e^{ikr}\,\hat{\mathbf{n}} \times (\hat{\mathbf{n}} \times \mathbf{p}).$$

Using the above step, we can now write the first term on the right-hand side of Eq. (4.162) as

$$ik\nabla\left[\frac{e^{ikr}}{r}\left(1 - \frac{1}{ikr}\right)\right] \times (\hat{\mathbf{n}} \times \mathbf{p}) = \frac{k^2}{r}\left(-1 + \frac{2}{ikr} + \frac{2}{k^2r^2}\right)e^{ikr}\,\hat{\mathbf{n}} \times (\hat{\mathbf{n}} \times \mathbf{p}).$$

To tackle the next term on the right-hand side of Eq. (4.162), we must use the vector analysis result

$$\nabla \times (\hat{\mathbf{n}} \times \mathbf{p}) = \hat{\mathbf{n}}(\nabla \cdot \mathbf{p}) - \mathbf{p}(\nabla \cdot \hat{\mathbf{n}}) + (\mathbf{p} \cdot \nabla)\hat{\mathbf{n}} - (\hat{\mathbf{n}} \cdot \nabla)\mathbf{p}.$$

As $\mathbf{p}$ is a constant vector the above result simplifies to

$$\nabla \times (\hat{\mathbf{n}} \times \mathbf{p}) = (\mathbf{p} \cdot \nabla)\hat{\mathbf{n}} - \mathbf{p}(\nabla \cdot \hat{\mathbf{n}}) = (\mathbf{p} \cdot \nabla)\frac{\mathbf{x}}{r} - \mathbf{p}\nabla \cdot \left(\frac{\mathbf{x}}{r}\right).$$

One can easily show, using spherical polar coordinates, that

$$(\mathbf{p} \cdot \nabla)\frac{\mathbf{x}}{r} = \frac{\mathbf{p}}{r} - \frac{(\mathbf{p} \cdot \hat{\mathbf{n}})\hat{\mathbf{n}}}{r}, \qquad \text{and} \qquad \mathbf{p}\nabla \cdot \left(\frac{\mathbf{x}}{r}\right) = \frac{2\mathbf{p}}{r}. \tag{4.163}$$

Using these results in the second term on the right-hand side of Eq. (4.162), we get

$$ik\frac{e^{ikr}}{r}\left(1 - \frac{1}{ikr}\right)\nabla \times (\hat{\mathbf{n}} \times \mathbf{p}) = -ik\frac{e^{ikr}}{r}\left(1 - \frac{1}{ikr}\right)\left[\frac{(\mathbf{p} \cdot \hat{\mathbf{n}})\hat{\mathbf{n}} + \mathbf{p}}{r}\right].$$

$$\tag{4.164}$$

We can use a simple manipulation of terms to write down the above expression in a different way. Noting that

$$\hat{\mathbf{n}} \times (\hat{\mathbf{n}} \times \mathbf{p}) = (\mathbf{p} \cdot \hat{\mathbf{n}})\hat{\mathbf{n}} - \mathbf{p}, \tag{4.165}$$

we can write

$$-ik\frac{e^{ikr}}{r}\left(1-\frac{1}{ikr}\right)\left[\frac{(\mathbf{p}\cdot\hat{\mathbf{n}})\hat{\mathbf{n}}+\mathbf{p}}{r}\right] = -ik\frac{e^{ikr}}{r^2}\left(1-\frac{1}{ikr}\right)\left[3(\mathbf{p}\cdot\hat{\mathbf{n}})\hat{\mathbf{n}}-\mathbf{p}\right]$$
$$+ 2ik\frac{e^{ikr}}{r^2}\left(1-\frac{1}{ikr}\right)e^{ikr}\hat{\mathbf{n}}\times(\hat{\mathbf{n}}\times\mathbf{p}).$$

The expression looks a bit complicated at this stage. We have written the above term in this particular way for some purpose, the purpose will become clear shortly. These complicated expressions give the form of the electric field, in dipole approximation, in both the far zone and the near zone.

Now we can combine our bits of calculation and write down the electric field as

$$\mathbf{E}(\mathbf{x}) = \left\{-\frac{k^2}{r}\hat{\mathbf{n}}\times(\hat{\mathbf{n}}\times\mathbf{p})+\frac{1}{r^3}(1-ikr)\left[3(\mathbf{p}\cdot\hat{\mathbf{n}})\hat{\mathbf{n}}-\mathbf{p}\right]\right\}e^{ikr}. \quad (4.166)$$

From the expressions of the magnetic field and the electric field, we immediately see that these fields are mutually perpendicular to each other. To appreciate the form of the above expression, we refer to the expression of the electric field in the radiation zone as given in Eq. (4.145). In the dipole approximation, we have

$$\mathbf{f}(\theta, \phi) = -ik\mathbf{p},$$

and as a result we see that the electric field in the dipole approximation must be

$$\mathbf{E}(\mathbf{x}) = -k^2\frac{e^{ikr}}{r}\hat{\mathbf{n}}\times(\hat{\mathbf{n}}\times\mathbf{p}), \quad (4.167)$$

exactly matching with the first term on the right-hand side of Eq. (4.166). This implies the first term on the right-hand side of Eq. (4.166) stands for the electric field in the radiation zone. We expected this as we can always write $k^2/r = (kr)^2/r^4$ and as r is large in the radiation zone so is kr and consequently the first term on the right-hand side of Eq. (4.166) remains. On the other hand, as r is large in the radiation zone the second term on the right-hand side of Eq. (4.166) becomes relatively less important as there is r^3 in the denominator. Actually the second term on the right-hand side of the above expression stands for the electric field in the near zone where $kr \ll 1$. In the near zone, the first term on the right-hand side of Eq. (4.166) becomes negligible as we are looking in the small r limit. The splitting of the electric field expression has been done in such a way that the far zone and the near zone results become explicit.

Now, we separately write down the electric and magnetic fields in the far zone and the near zone. The results can now be easily derived from the expressions of the fields. In the radiation zone, we have $kr \gg 1$ and r is large. The fields in the dipole approximation turns out to be:

$$\mathbf{B}(\mathbf{x}) = k^2\frac{e^{ikr}}{r}(\hat{\mathbf{n}}\times\mathbf{p}), \qquad \mathbf{E}(\mathbf{x}) = -k^2\frac{e^{ikr}}{r}\hat{\mathbf{n}}\times(\hat{\mathbf{n}}\times\mathbf{p}). \quad (4.168)$$

From the above expression, one can easily see that in the radiation zone

$$\mathbf{E} = \mathbf{B} \times \hat{\mathbf{n}}, \qquad (4.169)$$

showing that dipole radiation is transverse in nature. We did see that the plane electromagnetic wave was also transverse, here the wave is not a plane wave but they are spherical waves. The electric field is polarized in the plane defined by $\hat{\mathbf{n}}$ and $\mathbf{p}$. The actual time-dependent fields can easily be obtained from the above expressions by introducing the $e^{-i\omega t}$ factor.

On the other hand, in the near zone, where $kr \ll 1$ and r is relatively small, we have

$$\mathbf{B}(\mathbf{x}) = \frac{ik}{r^2}(\hat{\mathbf{n}} \times \mathbf{p}), \qquad \mathbf{E}(\mathbf{x}) = \left[3(\mathbf{p} \cdot \hat{\mathbf{n}})\hat{\mathbf{n}} - \mathbf{p}\right]\frac{1}{r^3}. \qquad (4.170)$$

Here, we have assumed $e^{ikr} \sim 1$ in the near zone. When we introduce the time-dependent factors, we see that these fields only oscillate in time but there is no oscillation in spatial coordinates. The above expression of the electric field shows that apart from its temporal oscillations it is just the static dipole field as we calculated in electrostatics. The fields are not transverse in the near zone. In the near zone, we see roughly that the magnitude of the magnetic field is kr times that of the electric field and as $kr \ll 1$ in the near zone we have mainly electric field in the near zone.

One can now go on to calculate the contributions from the magnetic dipole terms. We will briefly describe the magnetic dipole term later in this chapter.

4.3.5   The Electric Monopole Term

Till now we were only discussing the dipole term without mentioning the monopole term. We will now discuss the monopole term. We had seen that the simplest assumption in the expansion of the vector potential in the Lorenz gauge gave us the dipole fields. The electric monopole term does not arise from the vector potential, it arises from the scalar potential which we have not used till now. We know that, in Lorenz gauge, the scalar electrodynamic potential satisfies

$$\nabla^2 \Phi - \frac{1}{c^2}\frac{\partial^2 \Phi}{\partial t^2} = -4\pi\rho,$$

where ρ is the charge density in the source region. This equation can be directly solved using the Green function. We know if we use the retarded Green function, then we obtain

$$\Phi(\mathbf{x}, t) = \int_{-\infty}^{\infty} d^3 x' \int_{-\infty}^{\infty} dt' \frac{\rho(\mathbf{x}', t')}{|\mathbf{x} - \mathbf{x}'|}\delta\left(t - t' - \frac{|\mathbf{x} - \mathbf{x}'|}{c}\right). \qquad (4.171)$$

The time integration can easily be done using the Dirac delta function. Assuming $r \gg d$, we can safely approximate $|\mathbf{x} - \mathbf{x}'|^{-1} \sim 1/r$. Using this fact in the time integral, we obtain

$$\Phi(\mathbf{x}, t) = \frac{1}{r} \int_{-\infty}^{\infty} d^3 x' \, \rho \left(\mathbf{x}', t' = t - \frac{r}{c} \right). \tag{4.172}$$

Now,

$$\int_{-\infty}^{\infty} d^3 x' \, \rho \left(\mathbf{x}', t' = t - \frac{r}{c} \right) = Q \left(t' = t - \frac{r}{c} \right), \tag{4.173}$$

where Q is the total charge of the source configuration. As a result of this, we can now write the scalar potential as

$$\Phi(\mathbf{x}, t) = \frac{Q \left(t' = t - \frac{r}{c} \right)}{r}. \tag{4.174}$$

This shows that the electromagnetic scalar potential depends on total charge of the source configuration. As the total charge is conserved, so the scalar potential does not oscillate in time in the radiation zone and consequently does not radiate energy. For obvious reasons we call this form of the scalar potential the monopole term and as it does not radiate, we in general ignore its effect.

4.3.6 Power Radiated in the Electric Dipole Approximation

We know the fields in the radiation zone in the electric dipole approximation, they are given as

$$\mathbf{B}(\mathbf{x}) = k^2 \frac{e^{ikr}}{r} (\hat{\mathbf{n}} \times \mathbf{p}), \qquad \mathbf{E}(\mathbf{x}) = -k^2 \frac{e^{ikr}}{r} \hat{\mathbf{n}} \times (\hat{\mathbf{n}} \times \mathbf{p}).$$

where $\mathbf{p}$ is the dipole moment of the source charge distribution and $\hat{\mathbf{n}}$ is an unit vector in the direction of $\mathbf{x}$. These fields are complex quantities just like the fields we were dealing with when we were discussing plane electromagnetic waves. We want to find out the power radiated in such dipole radiation. We imagine a closed spherical surface enclosing the localized charge and current distribution. The radius of the spherical surface is $r = |\mathbf{x}|$ where $r \gg d$ where d is the typical source dimension. The power emitted across an infinitesimal surface area $d\mathbf{a} = \hat{\mathbf{n}} r^2 d\Omega$ is written as

$$dP = \text{Re} \left[\mathbf{S}(\mathbf{x}) \cdot d\mathbf{a} \right] = \text{Re} \left[r^2 d\Omega \, \mathbf{S}(\mathbf{x}) \cdot \hat{\mathbf{n}} \right], \tag{4.175}$$

where $\mathbf{S}$ is the appropriate Poynting vector for complex fields as we defined when we were discussing plane waves. Although we are calling P the power radiated in the electric dipole approximation, we actually mean P to be the time average of radiated power over one cycle (of field oscillations) as the fields are oscillating quantities. The form of the Poynting vector we are using is in reality a time average, as discussed previously in this chapter. Here, $d\Omega$ is the solid angle subtended by the infinitesimal area element through which energy is flowing normally. From the above expression, we can write the differential power emitted or the power emitted per unit solid angle as

$$\frac{dP}{d\Omega} = \text{Re}\left[r^2\,\mathbf{S(x)} \cdot \hat{\mathbf{n}}\right] = \frac{c}{8\pi}\text{Re}\left[r^2\hat{\mathbf{n}} \cdot (\mathbf{E} \times \mathbf{B}^*)\right]. \qquad (4.176)$$

In the radiation zone, we have

$$\mathbf{E} \times \mathbf{B}^* = \frac{k^4}{r^2}\left\{(\hat{\mathbf{n}} \times \mathbf{p}^*) \times \left[\hat{\mathbf{n}} \times (\hat{\mathbf{n}} \times \mathbf{p})\right]\right\}$$

$$= \frac{k^4}{r^2}\left\{(\hat{\mathbf{n}} \times \mathbf{p}^*) \times \left[\hat{\mathbf{n}}(\hat{\mathbf{n}} \cdot \mathbf{p}) - \mathbf{p}\right]\right\}. \qquad (4.177)$$

One can easily simplify the right-hand side of the above equation and write

$$\mathbf{E} \times \mathbf{B}^* = \frac{k^4}{r^2}\left[|\mathbf{p}|^2 - (\hat{\mathbf{n}} \cdot \mathbf{p})(\hat{\mathbf{n}} \cdot \mathbf{p}^*)\right]\hat{\mathbf{n}}. \qquad (4.178)$$

In our notation $|\mathbf{p}|^2 \equiv |\mathbf{p}||\mathbf{p}^*|$, where the modulus sign specifies the length of the vector. This shows that $\mathbf{E} \times \mathbf{B}^*$ is actually a real quantity. If the angle between $\hat{\mathbf{n}}$ and $\mathbf{p}$ is θ then $|\mathbf{p}|^2 - (\hat{\mathbf{n}} \cdot \mathbf{p})(\hat{\mathbf{n}} \cdot \mathbf{p}^*) = |\mathbf{p}|^2 - |\mathbf{p}|^2 \cos\theta^2 = |\mathbf{p}|^2 \sin^2\theta$. We can choose θ to be the polar angle by rotating the coordinate axes in such a way that the z-axis coincides with $\mathbf{p}$. Now, we have all the ingredients to write down the differential power emitted

$$\frac{dP}{d\Omega} = \frac{c}{8\pi}\text{Re}\left[r^2\hat{\mathbf{n}} \cdot (\mathbf{E} \times \mathbf{B}^*)\right],$$

by a localized source in the dipole approximation. In our case, the differential power is

$$\frac{dP}{d\Omega} = \frac{c}{8\pi}k^4|\mathbf{p}|^2 \sin^2\theta. \qquad (4.179)$$

The total power emitted in the radiation zone is simply

$$P = \int \frac{dP}{d\Omega}\,d\Omega = \frac{ck^4|\mathbf{p}|^2}{8\pi}\int_{\phi=0}^{2\pi}\int_{\theta=0}^{\pi} d\phi\,d\theta\,\sin\theta\,\sin^2\theta = \frac{ck^4|\mathbf{p}|^2}{3}. \quad (4.180)$$

This expression gives the total time averaged power emitted by the localized source in the electric dipole approximation.

4.3.7 Short, Center-Fed, Linear Antenna

Till now we have been talking theoretically, here we will present a practical discussion about radiation. Let us now see a simple example where our theory can be practiced and radiation can be created from oscillating currents. We have seen oscillating charges can produce electric dipole radiation, here we will see that indeed in a linear center-fed antenna such kind of radiation is produced. This kind of antenna is produced by a short and thin metallic rod whose length is d placed along the z-axis as shown in Fig. 4.3. The length of the antenna is much smaller than the wavelength. There is a small gap in the middle of the antenna from where external power cables pass current to the system. This process of charging the antenna is called feeding, and consequently the system has the name center-fed antenna. As shown in Fig. 4.3, the antenna extends from $z = d/2$ to $z = -d/2$. The oscillating current is set in such a manner that its magnitude is symmetric about $z = 0$. As this is a linear system, there is no associated magnetic moment. The current is set in such a way that it is maximal at $z = 0$ and fades to zero at $z = \pm d/2$. With this input, we can now write the current in the system as

$$I(z')e^{-i\omega t} = I_0 \left(1 - \frac{2|z'|}{d}\right)e^{-i\omega t} . \tag{4.181}$$

Fig. 4.3 A short, center-fed linear antenna of length d. The observation point direction where the fields are calculated is shown by the orientation of the unit vector

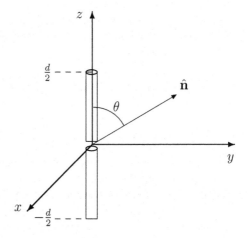

The peak current in the center is I_0. To find out the dipole moment of the system, we first require to know the linear charge density, $\rho(z')$, in the antenna. To find out the charge density, we use the continuity equation in one dimension

$$\rho(z')\frac{\partial(e^{-i\omega t})}{\partial t} + \frac{\partial I(z')}{\partial z'}e^{-i\omega t} = 0\,,$$

which yields

$$-i\omega\rho(z') \mp \frac{2I_0}{d} = 0\,.$$

The $\mp$ signs correspond to cases where $z' > 0$ and $z' < 0$, respectively. We can now write down the linear charge density on the antenna as

$$\rho(z') = \pm\frac{2i\,I_0}{\omega d}\,, \tag{4.182}$$

where now the plus sign corresponds to the case when $z' > 0$ and the minus sign corresponds to the case when $z' < 0$. We see that the spatial part of the linear charge density (amplitude) is constant on the antenna although the charge density is oscillating in time. The time-dependent charge density looks like

$$\rho(z',t) = \pm\frac{2i\,I_0}{\omega d}e^{-i\omega t} = \pm\frac{2i\,I_0}{\omega d}(\cos\omega t - i\sin\omega t)\,, \tag{4.183}$$

whose real part is

$$\mathrm{Re}[\rho(z',t)] = \pm\frac{2I_0}{\omega d}\sin\omega t\,.$$

The charge density expression directly shows that this configuration of charges have a definite electric dipole moment which oscillates with time.

To find out the dipole moment of the system, along the z-axis, we again suppress time variable and write

$$p = \int_{-d/2}^{d/2} z'\rho(z')dz' = -\frac{2i\,I_0}{\omega d}\int_{-d/2}^{0} z'dz' + \frac{2i\,I_0}{\omega d}\int_{0}^{d/2} z'dz'$$

$$= \frac{i\,I_0 d}{2\omega}\,. \tag{4.184}$$

In reality, the time-dependent dipole moment vector along the positive z-axis is oscillating in time. This becomes apparent as soon as we include the factor $e^{-i\omega t}$. From the above equation, one can see that $\mathbf{p}$ in the present case is purely imaginary.

Now, we can write down the radiation power generated from this kind of a dipole antenna. Using Eq. (4.180), we ultimately get

$$P = \frac{I_0^2(kd)^2}{12c}\,. \tag{4.185}$$

From Joule heating theory, where power emitted in the form of heat is given by $P \propto I^2 R$, R being the resistance in the circuit we immediately see that kd has the dimension of resistance. This resistance is in general called radiation resistance. We see from the above formula that for the short, center-fed antenna the power is proportional to the square of the frequency of the radiated waves.

4.3.8 Magnetic Dipole Radiation

In our initial phase of discussion about the radiation zone, we showed that up to second order in $1/r^2$ we have

$$|\mathbf{x} - \mathbf{x}'| = r \left[1 - \frac{\hat{\mathbf{n}} \cdot \mathbf{x}'}{r} + \frac{1}{2} \left(\frac{r'}{r} \right)^2 - \frac{1}{2} \left(\frac{\hat{\mathbf{n}} \cdot \mathbf{x}'}{r} \right)^2 \cdots \right]^{1/2}.$$

There we obtained $|\mathbf{x} - \mathbf{x}'| = r - \hat{\mathbf{n}} \cdot \mathbf{x}'$. If we expand $|\mathbf{x} - \mathbf{x}'|^{-1}$ up to $1/r^3$ we get

$$\frac{1}{|\mathbf{x} - \mathbf{x}'|} = \frac{1}{r} \left[1 - \frac{2\hat{\mathbf{n}} \cdot \mathbf{x}'}{r} + \left(\frac{r'}{r} \right)^2 \right]^{-1/2}$$

$$= \frac{1}{r} + \frac{\hat{\mathbf{n}} \cdot \mathbf{x}'}{r^2} + \frac{1}{2r^3} \left[3(\hat{\mathbf{n}} \cdot \mathbf{x}')^2 - r'^2 \right]. \qquad (4.186)$$

Up to order $1/r^2$ we can now write

$$\frac{e^{ik|\mathbf{x} - \mathbf{x}'|}}{|\mathbf{x} - \mathbf{x}'|} = e^{ik(r - \hat{\mathbf{n}} \cdot \mathbf{x}')} \left[\frac{1}{r} + \frac{\hat{\mathbf{n}} \cdot \mathbf{x}'}{r^2} \right]$$

$$= e^{ikr} (1 - ik\hat{\mathbf{n}} \cdot \mathbf{x}') \left[\frac{1}{r} + \frac{\hat{\mathbf{n}} \cdot \mathbf{x}'}{r^2} \right]$$

$$= \frac{e^{ikr}}{r} \left[1 + \frac{\hat{\mathbf{n}} \cdot \mathbf{x}'}{r} - ik\hat{\mathbf{n}} \cdot \mathbf{x}' - ik\frac{(\hat{\mathbf{n}} \cdot \mathbf{x}')^2}{r} \right].$$

As primarily we will be looking at regions where $(r'/r)^2$ is negligible, we omit the last term and write

$$\frac{e^{ik|\mathbf{x} - \mathbf{x}'|}}{|\mathbf{x} - \mathbf{x}'|} = \frac{e^{ikr}}{r} \left[1 + \left(\frac{1}{r} - ik \right) \hat{\mathbf{n}} \cdot \mathbf{x}' \right]. \qquad (4.187)$$

Using this result, we can now write the basic expression of the vector potential up to order $1/r^2$, in the region, where there are no source charges or currents, as given in Eq. (4.125)

$$\mathbf{A}(\mathbf{x}) = \frac{e^{ikr}}{cr}\left(\frac{1}{r} - ik\right)\int_{-\infty}^{\infty}\mathbf{J}(\mathbf{x}')\,(\hat{\mathbf{n}}\cdot\mathbf{x}')\,d^3x'\,. \tag{4.188}$$

This form of the vector potential gives rise to magnetic dipole radiation and electric quadrupole radiation. To get the magnetic dipole contribution, we, at first, write the above integrand as a sum of two terms. One of these terms is symmetric in $\hat{\mathbf{n}}$ and $\mathbf{x}'$ and the other term is antisymmetric under the exchange of those two vectors. We note

$$\begin{aligned}\nabla\cdot(x_i'x_j'\mathbf{J}') &= x_j'(\nabla x_i'\cdot\mathbf{J}') + x_i'(\nabla x_j'\cdot\mathbf{J}') + x_i'x_j'\nabla\cdot\mathbf{J}'\\ &= x_j'J_i' + x_i'J_j' + x_i'x_j'(i\omega\rho')\,.\end{aligned}$$

Here, we have used the equation of continuity to replace the third term on the right-hand side. Taking the volume integral on both sides over a region enclosing the localized source distribution, it is easy to show that the left-hand side vanishes when converted into surface term using divergence theorem. Then

$$\int_V \left[x_j'J_i' + x_i'J_j' + i\omega\rho'x_i'x_j'\right]d^3x' = 0\,. \tag{4.189}$$

Now, let us rewrite the integral in Eq. (4.188) as follows:

$$\int\mathbf{J}(\mathbf{x}')\,(\hat{\mathbf{n}}\cdot\mathbf{x}')\,d^3x' = \frac{1}{r}\int\mathbf{J}(\mathbf{x}')\,(\mathbf{x}\cdot\mathbf{x}')\,d^3x'\,. \tag{4.190}$$

We pick up the ith component of the integral on the right-hand side and represent it as

$$\sum_{j=1}^{3}x_j\int x_j'J_i\,d^3x' = -\sum_{j=1}^{3}x_j\int(x_i'J_j + i\omega\rho'x_i'x_j')\,d^3x'\,. \tag{4.191}$$

In the last step, we have used Eq. (4.189). Now we rewrite the first term in the integrand on the right-hand side as a sum of symmetric and antisymmetric parts.

$$\sum_{j=1}^{3}x_j\int x_j'J_i\,d^3x' = -\sum_{j=1}^{3}x_j\int\left[\frac{1}{2}\left(x_i'J_j + x_j'J_i\right) + \frac{1}{2}\left(x_i'J_j - x_j'J_i\right) + i\omega\rho'x_i'x_j'\right]d^3x' \tag{4.192}$$

Using Eq. (4.189) for the symmetric part, we have,

$$\sum_{j=1}^{3}x_j\int x_j'J_i\,d^3x' = -\sum_{j=1}^{3}x_j\int\left[\frac{1}{2}\left(-i\omega\rho'x_i'x_j'\right) + \frac{1}{2}\left(x_i'J_j - x_j'J_i\right) + i\omega\rho'x_i'x_j'\right]d^3x' \tag{4.193}$$

Note that the antisymmetric part in the integrand on the right-hand side is actually kth component of the cross-product $\mathbf{x}' \times \mathbf{J}'$. Taking the unprimed variable inside the integrand, we can now write the right hand side in terms of a triple cross product as:

$$\sum_{j=1}^{3} x_j \int x_j' J_i \, d^3x' = -\frac{1}{2} \int \left[\{ \mathbf{x} \times (\mathbf{x}' \times \mathbf{J}) \}_i + i\omega\rho' x_i' (\mathbf{x} \cdot \mathbf{x}') \right] d^3x' . \quad (4.194)$$

It is now easy to write down Eq. (4.190) in the following form

$$\int \mathbf{J}(\mathbf{x}') (\hat{\mathbf{n}} \cdot \mathbf{x}') \, d^3x' = \frac{1}{2} \int \left[(\mathbf{x}' \times \mathbf{J}) \times \hat{\mathbf{n}} - i\omega\rho' \mathbf{x}'(\hat{\mathbf{n}} \cdot \mathbf{x}') \right] d^3x' . \quad (4.195)$$

The first integrand on the right-hand side is antisymmetric in $\hat{\mathbf{n}}$ and $\mathbf{x}'$ and gives rise to magnetic dipole radiation, the second part of the above expression is symmetric in $\hat{\mathbf{n}}$ and $\mathbf{x}'$ and gives rise to electric quadrupole radiation. We shall discuss about quadrupole radiation later. For the moment, we focus our attention on the first term, the antisymmetric term, which gives rise to magnetic dipole radiation. We first note that the antisymmetric part contains the magnetization of the source region as

$$\mathbf{M} = \frac{1}{2c} (\mathbf{x} \times \mathbf{J}) .$$

This magnetization is caused due to the source current $\mathbf{J}$. The magnetic dipole moment contribution in the vector potential can now be written as

$$\mathbf{A}(\mathbf{x}) = \frac{e^{ikr}}{r} \left(\frac{1}{r} - ik \right) \int_{-\infty}^{\infty} \mathbf{M} \times \hat{\mathbf{n}} \, d^3x' . \quad (4.196)$$

In the above integral, $\hat{\mathbf{n}}$ is independent of x' and the total dipole moment, $\mathbf{m}$, of the source region is

$$\mathbf{m} = \int_{-\infty}^{\infty} \mathbf{M} \, d^3x' .$$

Using this information, we can now write the vector potential, giving rise to magnetic dipole radiation, as

$$\mathbf{A}(\mathbf{x}) = ik(\hat{\mathbf{n}} \times \mathbf{m}) \frac{e^{ikr}}{r} \left(1 - \frac{1}{ikr} \right) . \quad (4.197)$$

Once we have the vector potential, we can calculate the magnetic field $\mathbf{B}(\mathbf{x}) = \nabla \times \mathbf{A}(\mathbf{x})$. In the present case, we do not require to do a full calculation to obtain the magnetic field as we can get the result by noting some similarities of the present quantities with those we calculated previously. If one looks at the expression of the magnetic field in the case of electric dipole radiation, as given in Eq. (4.161), one can immediately notice the similarity of that expression and the vector potential written above. Instead of k^2 in Eq. (4.161), we have ik in the present case and the magnetic dipole moment replaces the electric dipole moment. Except these minor differences, both the expressions are similar. Multiplying the previous magnetic field expression by i/k and replacing $\mathbf{p}$ with $\mathbf{m}$ there, yields the vector potential. In the previous case, the electric field was given by

$$\mathbf{E} = \frac{i}{k}\nabla \times \mathbf{B} = \nabla \times \left(\frac{i}{k}\mathbf{B}\right).$$

From this equation, we see that the previous electric field expression gives the present magnetic field by the substitution of the dipole moment by the magnetic moment in the previous expression of the electric field. We can now directly write down the result as

$$\mathbf{B}(\mathbf{x}) = \left\{-\frac{k^2}{r}\hat{\mathbf{n}} \times (\hat{\mathbf{n}} \times \mathbf{m}) + \frac{1}{r^3}(1 - ikr)\left[3(\mathbf{m} \cdot \hat{\mathbf{n}})\hat{\mathbf{n}} - \mathbf{m}\right]\right\}e^{ikr}. \quad (4.198)$$

To find out the electric field, in the present case, we use another technique which will help us to bypass a long and cumbersome calculation. We know that the Lorenz gauge condition is given by

$$\frac{1}{c}\frac{\partial \Phi}{\partial t} + \nabla \cdot \mathbf{A} = 0.$$

Suppressing the temporal part, we can write the above equation as

$$\nabla \cdot \mathbf{A}(\mathbf{x}) - ik\Phi(\mathbf{x}) = 0. \quad (4.199)$$

We know that the vector potential $\mathbf{A}(\mathbf{x})$ with which we are working resembles the magnetic field in the case of electric dipole radiation (with some minor changes). As a magnetic field always has zero divergence, we expect $\nabla \cdot \mathbf{A}(\mathbf{x}) = 0$ in our present case. This gives $\Phi(\mathbf{x}) = 0$ in the our case. The electric field is given by

$$\mathbf{E}(\mathbf{x}) = -\frac{1}{c}\frac{\partial \mathbf{A}(\mathbf{x})}{\partial t} - \nabla \Phi(\mathbf{x}),$$

and as the scalar potential is zero we simply have

$$\mathbf{E}(\mathbf{x}) = ik\mathbf{A}(\mathbf{x}). \quad (4.200)$$

As a result, we can now write the electric field expression, obtained from the vector potential in Eq. (4.197), as

$$\mathbf{E}(\mathbf{x}) = -k^2(\hat{\mathbf{n}} \times \mathbf{m})\frac{e^{ikr}}{r}\left(1 - \frac{1}{ikr}\right). \quad (4.201)$$

The electric and magnetic fields thus obtained contain information about the far zone as well as the near zone. We can now easily find out the far zone results as we have done in the previous case. We observe that we obtain the above field expressions from the expressions in the previous case by replacing $\mathbf{E} \rightarrow \mathbf{B}$, $\mathbf{B} \rightarrow -\mathbf{E}$ and $\mathbf{p} \rightarrow \mathbf{m}$. In the present case, one can easily check that the magnetic dipole radiation is polarized in the plane defined by the vectors $\hat{\mathbf{n}}$ and $\mathbf{m}$. The time averaged power emitted from the magnetic dipole can easily be obtained from Eq. (4.180) if we replace the electric dipole moment vector $\mathbf{p}$ by the magnetic dipole moment vector $\mathbf{m}$.

Magnetic dipole radiation requires antennas that are not purely linear, as in that case, the magnetic dipole moment vanishes. For producing magnetic dipole radiation, one requires antennas that have a loop where current can flow producing magnetic dipole moment.

4.3.9   Electric Quadrupole Radiation

We had already mentioned that Eq. (4.188) contains both magnetic dipole and electric quadrupole terms. Let us write down explicitly the quadrupole contribution to the vector potential using Eq. (4.195) as:

$$
\begin{aligned}
\mathbf{A}(\mathbf{x}) &= \frac{e^{ikr}}{cr}\left(\frac{1}{r} - ik\right)\int_{-\infty}^{\infty}\frac{1}{2}\left[-i\omega\rho'\mathbf{x}'(\hat{\mathbf{n}}\cdot\mathbf{x}')\right]d^3x' \\
&= \frac{-k^2}{2}\frac{e^{ikr}}{r}\left(1 - \frac{1}{ikr}\right)\int_{-\infty}^{\infty}\left[\mathbf{x}'(\hat{\mathbf{n}}\cdot\mathbf{x}')\rho'\right]d^3x' .
\end{aligned} \tag{4.202}
$$

We consider the radiation zone ($kr \gg 1$). Keeping only $\frac{e^{ikr}}{r}$ term, we write down the radiation fields as

$$
\begin{aligned}
\mathbf{B} &= \nabla\times\mathbf{A} = ik\hat{\mathbf{n}}\times\mathbf{A} \\
&= ik\left(\frac{-k^2}{2}\right)\frac{e^{ikr}}{r}\int_{-\infty}^{\infty}\left[(\hat{\mathbf{n}}\times\mathbf{x}')(\hat{\mathbf{n}}\cdot\mathbf{x}')\rho(\mathbf{x}')\right]d^3x' ,
\end{aligned} \tag{4.203}
$$

and,

$$
\begin{aligned}
\mathbf{E} &= \frac{i}{k}\nabla\times\mathbf{B} \\
&= \frac{i}{k}\left[ik\hat{\mathbf{n}}\times\left(ik\hat{\mathbf{n}}\times\mathbf{A}\right)\right] = ik\left[(\hat{\mathbf{n}}\times\mathbf{A})\times\hat{\mathbf{n}}\right] .
\end{aligned} \tag{4.204}
$$

We concentrate on the integral in Eq. (4.203). We take the unprimed $\hat{\mathbf{n}}$ involved in the cross product outside the integral over primed variable and consider only the ith component of the remaining integrand and expand it as follows:

$$
\sum_{j=1}^{3}x_i'(n_j x_j')\rho(\mathbf{x}') = \frac{1}{3}\sum_{j=1}^{3}\left(3x_i'x_j' - r'^2\delta_{ij}\right)\rho(\mathbf{x}')n_j + \frac{1}{3}\sum_{j=1}^{3}r'^2\delta_{ij}\rho(\mathbf{x}')n_j .
$$
$$ \tag{4.205} $$

We have already defined the components Q_{ij} of the quadrupole moment tensor in the previous chapter.

$$Q_{ij} = \int \left(3x_i' x_j' - r'^2 \delta_{ij} \right) \rho(\mathbf{x}') \, d^3 x' . \tag{4.206}$$

We now construct the components of a vector $\mathbf{Q}$ as follows:

$$Q_i = \sum_{j=1}^{3} Q_{ij} n_j . \tag{4.207}$$

It follows from Eq. (4.203),

$$
\begin{aligned}
\mathbf{B} &= ik \left(\frac{-k^2}{2} \right) \frac{e^{ikr}}{r} \hat{\mathbf{n}} \times \int_{-\infty}^{\infty} \mathbf{x}' (\hat{\mathbf{n}} \cdot \mathbf{x}') \rho(\mathbf{x}') \, d^3 x' \\
&= ik \left(\frac{-k^2}{2} \right) \frac{e^{ikr}}{r} \hat{\mathbf{n}} \times \left(\frac{1}{3} \mathbf{Q} \right) .
\end{aligned} \tag{4.208}
$$

The above step follows from the fact that the second term on the right-hand side of Eq. (4.205) is actually proportional to $\hat{\mathbf{n}}$. Similarly, from Eq. (4.204)

$$
\begin{aligned}
\mathbf{E} &= ik \left(\frac{-k^2}{2} \right) \frac{e^{ikr}}{r} \left\{ \hat{\mathbf{n}} \times \left[\int_{-\infty}^{\infty} \left[\mathbf{x}' (\hat{\mathbf{n}} \cdot \mathbf{x}') \rho' \right] d^3 x' \right] \right\} \times \hat{\mathbf{n}} \\
&= \frac{ik}{3} \left(\frac{-k^2}{2} \right) \frac{e^{ikr}}{r} \left[(\hat{\mathbf{n}} \times \mathbf{Q}) \times \hat{\mathbf{n}} \right] .
\end{aligned} \tag{4.209}
$$

The above expressions give the electric field and the magnetic field for electric quadrupole radiation.

The power radiated per unit solid angle is given by

$$
\begin{aligned}
\frac{dP}{d\Omega} &= \text{Re} \left[r^2 \mathbf{S}(\mathbf{x}) \cdot \hat{\mathbf{n}} \right] = \frac{c}{8\pi} \text{Re} \left[r^2 \hat{\mathbf{n}} \cdot (\mathbf{E} \times \mathbf{B}^*) \right] \\
&= \frac{c}{8\pi} \text{Re} \left[r^2 \hat{\mathbf{n}} \cdot ((\mathbf{B} \times \hat{\mathbf{n}}) \times \mathbf{B}^*) \right] \\
&= \frac{c}{8\pi} r^2 \text{Re} \left[(\mathbf{B} \times \hat{\mathbf{n}}) \cdot (\mathbf{B}^* \times \hat{\mathbf{n}}) \right] ,
\end{aligned} \tag{4.210}
$$

where, in the second step, we have used the result from Eq. (4.204) which states $\mathbf{E} = \mathbf{B} \times \hat{\mathbf{n}}$. In the last step, we have used the standard vector identity $\mathbf{a} \cdot (\mathbf{b} \times \mathbf{c}) = \mathbf{b} \cdot (\mathbf{c} \times \mathbf{a})$. We see that $(\mathbf{B} \times \hat{\mathbf{n}}) \cdot (\mathbf{B}^* \times \hat{\mathbf{n}})$ is actually a real quantity and we will write it as $|\mathbf{B} \times \hat{\mathbf{n}}|^2$. Finally, using Eq. (4.208)

$$\frac{dP}{d\Omega} = \frac{c}{8\pi} r^2 |\mathbf{B} \times \hat{\mathbf{n}}|^2 = \left(\frac{c}{8\pi} \right) \frac{k^6}{36} |(\hat{\mathbf{n}} \times \mathbf{Q}) \times \hat{\mathbf{n}}|^2 , \tag{4.211}$$

where,

$$
\begin{aligned}
|(\hat{\mathbf{n}} \times \mathbf{Q}) \times \hat{\mathbf{n}}|^2 &= \left[\mathbf{Q} - \hat{\mathbf{n}}(\mathbf{Q} \cdot \hat{\mathbf{n}})\right] \cdot \left[\mathbf{Q}^* - \hat{\mathbf{n}}(\mathbf{Q}^* \cdot \hat{\mathbf{n}})\right] \\
&= \mathbf{Q} \cdot \mathbf{Q}^* - (\mathbf{Q} \cdot \hat{\mathbf{n}})(\mathbf{Q}^* \cdot \hat{\mathbf{n}}) \\
&= \sum_{ijk} Q_{ij}^* n_j Q_{ik} n_k - \sum_{ijkl} Q_{ij}^* n_i n_j Q_{kl} n_k n_l .
\end{aligned}
\tag{4.212}
$$

Thus, for a given Q_{ij} and $\hat{\mathbf{n}}$, we can calculate the angular distribution of radiated power for sources with non-zero electric quadrupole moments. Here, n_i's are the components of $\hat{\mathbf{n}} = (\sin\theta \cos\varphi, \sin\theta \sin\varphi, \cos\theta)$, obeying the following identities:

$$
\int n_i n_j \, d\Omega = \frac{4\pi}{3} \delta_{ij} ,
\tag{4.213}
$$

$$
\int n_i n_j n_k n_l \, d\Omega = \frac{4\pi}{15} \left(\delta_{ij}\delta_{kl} + \delta_{ik}\delta_{jl} + \delta_{il}\delta_{jk}\right) .
\tag{4.214}
$$

Therefore, using Eqs. (4.212), (4.213) and (4.214), we can calculate the total radiated power by integrating over all directions as

$$
\begin{aligned}
\int |(\hat{\mathbf{n}} \times \mathbf{Q}) \times \hat{\mathbf{n}}|^2 \, d\Omega &= \sum_{ijk} Q_{ij}^* Q_{ik} \frac{4\pi}{3} \delta_{jk} - \sum_{ijkl} Q_{ij}^* Q_{kl} \frac{4\pi}{15} \left(\delta_{ij}\delta_{kl} + \delta_{ik}\delta_{jl} + \delta_{il}\delta_{jk}\right) \\
&= \frac{4\pi}{3} \sum_{ij} Q_{ij}^* Q_{ij} - \frac{4\pi}{15} \left(\sum_{ik} Q_{ii}^* Q_{kk} + \sum_{ij} |Q_{ij}|^2 + \sum_{ij} Q_{ij}^* Q_{ji}\right) .
\end{aligned}
$$

Since Q_{ij} is traceless and symmetric,[2] we obtain

$$
\int |(\hat{\mathbf{n}} \times \mathbf{Q}) \times \hat{\mathbf{n}}|^2 \, d\Omega = \frac{4\pi}{5} \sum_{ij} |Q_{ij}|^2 .
\tag{4.215}
$$

Using Eqs. (4.211) and (4.215), the total time averaged radiated power turns out to be

$$
P = \int \frac{dP}{d\Omega} \, d\Omega = \frac{ck^6}{360} \sum_{ij} |Q_{ij}|^2 .
\tag{4.216}
$$

Thus, the total radiated power varies as sixth power of frequency for quadrupole radiation.

A simple example of radiating quadrupole source is an oscillating spheroidal charge distribution. A spheroid is also known as an ellipsoid of revolution and a spheroid is, in general, not a sphere. A spheroid is obtained by rotating an ellipse around one of its principal axes. If the ellipse is rotated about its major axis we obtain a prolate spheroid. On the other hand, if the ellipse is rotated about its minor axis,

---

[2] See Chap. 3.

Fig. 4.4 A typical angular
distribution of radiated
power from electric
quadrupole source

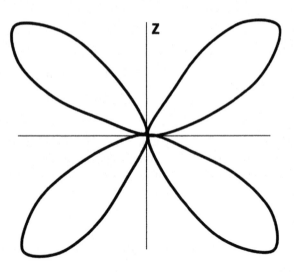

the result is an oblate spheroid. An oblate spheroid is flattened along the equatorial
plane and resembles the shape of the Earth, whereas a prolate spheroid is an elongated
structure. The angular distribution of radiated power for such a quadrupole source
exhibits a 4-lobed pattern, which is typical of quadrupole radiation as shown in
Fig. 4.4. In the coordinate system, where the charge distribution is spheroidal, the
quadrupole moment tensor is in general given by $Q_{ij} = 0$ if $i \neq j$; $Q_{zz} = Q_0$; $Q_{xx} =
Q_{yy} = -\frac{1}{2}Q_0$. Clearly, Q_{ij} is symmetric and traceless as required. We can readily
calculate, $\sum |Q_{ij}|^2 = \frac{3}{2}|Q_0|^2$. The total power radiated is then

$$P = \frac{ck^6}{360}\frac{3}{2}|Q_0|^2 = \frac{ck^6}{240}|Q_0|^2. \tag{4.217}$$

We now evaluate the following expression for the spheroidal charge distribution.

$$|(\hat{\mathbf{n}} \times \mathbf{Q}) \times \hat{\mathbf{n}}|^2 = \mathbf{Q} \cdot \mathbf{Q}^* - |\mathbf{Q} \cdot \hat{\mathbf{n}}|^2. \tag{4.218}$$

We use the Cartesian components of $\hat{\mathbf{n}}$ and the given form of Q_{ij} and obtain:

$$\mathbf{Q} = Q_{xx}n_x\hat{x} + Q_{yy}n_y\hat{y} + Q_{zz}n_z\hat{z},$$
$$\mathbf{Q} \cdot \mathbf{Q}^* = |Q_{xx}|^2 n_x^2 + |Q_{yy}|^2 n_y^2 + |Q_{zz}|^2 n_z^2,$$
$$\mathbf{Q} \cdot \hat{\mathbf{n}} = Q_{xx}n_x^2 + Q_{yy}n_y^2 + Q_{zz}n_z^2.$$

The expression of $\mathbf{Q}$ is simplified since Q_{ij} does not have any off-diagonal components in the present case. Therefore,

$$\mathbf{Q} \cdot \mathbf{Q}^* - |\mathbf{Q} \cdot \hat{\mathbf{n}}|^2 = |Q_{xx}|^2 n_x{}^2 + |Q_{yy}|^2 n_y{}^2 + |Q_{zz}|^2 n_z{}^2 - \left|Q_{xx}n_x{}^2 + Q_{yy}n_y{}^2 + Q_{zz}n_z{}^2\right|^2.$$

Plugging in the values of Q_{xx}, Q_{yy} and Q_{zz} on the right-hand side, we have

$$\mathbf{Q} \cdot \mathbf{Q}^* - |\mathbf{Q} \cdot \hat{\mathbf{n}}|^2 = |Q_0|^2 \left[\frac{1}{4}n_x^2 + \frac{1}{4}n_y^2 + n_z^2 - \right.$$
$$\left. \left(\frac{1}{4}n_x^4 + \frac{1}{4}n_y^4 + n_z^4 + \frac{1}{2}n_x^2 n_y^2 - n_x^2 n_z^2 - n_y^2 n_z^2 \right) \right].$$

As we know $\hat{\mathbf{n}} = (\sin\theta\cos\varphi, \sin\theta\sin\varphi, \cos\theta)$, we can write

$$\mathbf{Q} \cdot \mathbf{Q}^* - |\mathbf{Q} \cdot \hat{\mathbf{n}}|^2 = |Q_0|^2 \left[\frac{1}{4}\sin^2\theta + \cos^2\theta - \frac{1}{4}\sin^4\theta(\cos^4\varphi + \sin^4\varphi) - \cos^4\theta \right.$$
$$\left. -\frac{1}{2}\sin^4\theta\cos^2\varphi\sin^2\varphi + \sin^2\theta\cos^2\theta \right]$$
$$= |Q_0|^2 \left[\frac{1}{4}\sin^2\theta + \cos^2\theta - \frac{1}{4}\sin^4\theta - \cos^4\theta + \sin^2\theta\cos^2\theta \right]$$
$$= |Q_0|^2 \left[\frac{9}{4}\sin^2\theta\cos^2\theta \right]. \tag{4.219}$$

Therefore

$$|(\hat{\mathbf{n}} \times \mathbf{Q}) \times \hat{\mathbf{n}}|^2 = \frac{9}{4}|Q_0|^2 \sin^2\theta\cos^2\theta. \tag{4.220}$$

Then using Eq. (4.211), the angular distribution of radiated power is given by

$$\frac{dP}{d\Omega} = \frac{c}{8\pi}\frac{k^6}{36}|(\hat{\mathbf{n}} \times \mathbf{Q}) \times \hat{\mathbf{n}}|^2 = \frac{ck^6}{128\pi}|Q_0|^2 \sin^2\theta\cos^2\theta. \tag{4.221}$$

The angular distribution displays a 4-lobed pattern having azimuthal symmetry with a maxima in each quadrant.

4.4   Problems

1. Show that the function $\psi(x \pm vt)$ satisfies the one dimensional wave equation. Here, v is a constant.
2. Suppose a plane electromagnetic wave is traveling along the z-direction and the independent orthogonal oscillating electric field components are given by

$$E_x(z,t) = E_{0x}\cos(\omega t - kz + \delta_x), \quad E_y(z,t) = E_{0y}\cos(\omega t - kz + \delta_y),$$

where E_{0x} and E_{0y} specify the amplitude of oscillations and δ_x and δ_y are phase constants. Eliminate $\cos(\omega t - kz)$ from the above expressions and obtain an equation containing only E_x, E_y, E_{0x}, E_{0y} and $\delta \equiv \delta_y - \delta_x$. The equation you will obtain is called the polarization ellipse. From this equation, figure out what happens when either of E_{0x}, E_{0y} is zero. What kind of equation do you get when $\delta = 0$ or $\delta = \pi$ and $E_{0x} = E_{0y} = E_0$? Finally, what kind of equation do you get when $\delta = \pi/2$ or $\delta = 3\pi/2$ and $E_{0x} = E_{0y} = E_0$?

3. Suppose there are two nonmagnetic ($\mu = 1$) dielectric media separated by an interface at the $z = 0$ plane. The medium on the left of $z = 0$ plane is called medium one and the medium on the right of the interface is called medium two. The dielectric constants in the two media are ϵ_1 and ϵ_2. A plane wave with wave vector parallel to the z-axis is incident on the interface. The wave is polarized along the x-axis. Write down the incident and reflected electric and magnetic field expressions in medium one. Write down the form of the transmitted electric and magnetic field expressions in medium two. Apply the matching conditions of the fields on the interface where you know that there is no free charge or current on the interface. Express the amplitude of the reflected and transmitted electric field in terms of the amplitude of the incident electric field. From this result show that if $\epsilon_1 = \epsilon_2$ there is no reflection and the whole wave is transmitted.

4. Suppose again we have two media as proposed in our previous problem, but instead of normal incidence, now we have oblique incidence. The incident wave makes an angle θ_i with the normal on the interface ($z = 0$ surface) and the transmitted wave makes an angle θ_t with the normal. You can assume that the incident angle is equal to the angle of reflection. Apply all the matching conditions of the fields on the interface (devoid of charge or current). Particularly focus on the matching of the oscillating magnetic field at $z = 0$. From the last matching condition, can you show that the x and y components of the incident, reflected and transmitted wave vector are all equal. This result shows that the incident wave, reflected wave and the transmitted wave can all lie in the same plane.

5. Evaluate the electric field and the magnetic field in the near zone from the expression of the vector potential in Eq. (4.131).

6. Show that the last term on the extreme right-hand side of Eq. (4.144) does not contribute in the radiation zone.

7. Two point charges, q each, are attached to the ends of a rigid rod of length ℓ. The rod is rotated with angular speed ω about an axis through the center of the rod and this axis is perpendicular to the rod. Calculate the electric and magnetic dipole moment of the system taking the center of the rod as the origin. What is the total power radiated via electric dipole radiation and magnetic dipole radiation?

8. A wire of length ℓ is bent in a circular loop to form a magnetic dipole antenna. The current in the dipole varies as $I = I_0 \sin \omega_0 t$ where ω_0 and I_0 are constants. Determine the electric and magnetic field as seen by an observer at a large distance r from the dipole at an angle θ to the axis of the loop (passing through its center and perpendicular to the plane of the loop). Calculate the total power radiated by such an antenna.

9. Assuming that the instantaneous (not the time average) power emitted in the electric dipole approximation is given by

$$P = \frac{2}{3c^3}\ddot{\mathbf{p}}^2 \,,$$

where $\mathbf{p} = \mathbf{p}_0 \cos \omega t$ represents an oscillating dipole moment, show that the time averaged power obtained from the above equation is the same as that given in Eq. (4.180).

10. A pulsar is a rapidly rotating neutron star. Generally, pulsars have magnetic fields attached to them. Assume the star to be spherical in shape and it is rotating with angular speed Ω about the z-axis. The magnetic moment of the neutron star $\mathbf{m}$ makes an angle α with the rotation axis and rotates with the star. From this information, write down the components of $\mathbf{m}$ in terms of Cartesian unit vectors. Calculate the power emitted by such a time varying magnetic dipole moment. To calculate the power emitted, you can use the expression of power in the previous question where you replace the electric dipole moment with the magnetic dipole moment. If I is the moment of inertia of the star about one of its diameter then write down the rotational energy E of the pulsar. Can you now relate P and $\dot{E}$?

11. Derive the results given in Eqs. (4.213) and (4.214).

12. A quadrupole consists of a square of sides ℓ with charges $\pm q$ at alternate corners. The square rotates with angular speed ω about an axis normal to the plane of the square and through its center. Calculate the quadrupole moments, the radiation fields, the differential power and the total power radiated by such a rotating quadrupole.

Chapter 5
Long Wavelength Scattering

In this chapter, we discuss the scenario where the plane electromagnetic wave incident on a source (a scatterer), typically a single charge-neutral object, triggers oscillation of the dipole moments in the source with the same frequency as that of the incident wave, thus producing new secondary radiation in all directions. This mechanism is called dipole scattering. We shall limit ourselves to dealing with dipole scattering in the so-called long wavelength limit,[1] where the dimension d of the scatterer is small compared to the wavelength λ of the incident wave.

5.1 General Formulation of the Dipole Scattering Problem

Let us consider a plane electromagnetic wave traveling along $\hat{\mathbf{n}}_0$ direction and having wavelength λ is incident on a single scatterer of dimension d such that $d \ll \lambda$. Since d is small compared to λ, the phase of the incident wave is effectively constant on the scatterer. The incident wave, with fields

$$\mathbf{E}_{\text{inc}}(\mathbf{x}) = \epsilon_0 E_0 \, e^{ik\hat{\mathbf{n}}_0 \cdot \mathbf{x}} \,, \tag{5.1}$$

$$\mathbf{H}_{\text{inc}}(\mathbf{x}) = \hat{\mathbf{n}}_0 \times \mathbf{E}_{\text{inc}} \,, \tag{5.2}$$

induces oscillating electric dipoles (with moment $\mathbf{p} \, e^{-i\omega t}$) and (or) magnetic dipoles (with moment $\mathbf{m} \, e^{-i\omega t}$) having the same phase relationship with the incident wave. These induced dipoles reradiate along a direction $\hat{\mathbf{n}}$. It is assumed for simplicity that the medium surrounding the scatterer has $\epsilon \simeq \mu \simeq 1$. Far away from the scatterer, in

---

[1] It is also called Rayleigh scattering.

© Springer Nature Singapore Pte Ltd. 2021
K. Bhattacharya and S. Mukhopadhyay, *Introduction to Advanced Electrodynamics*,
https://doi.org/10.1007/978-981-16-7802-8_5

the lowest order approximation, the induced electric and magnetic dipoles produce scattered (radiated) fields as follows[2]

$$\mathbf{E}_{sc}(\mathbf{x}) = k^2 \frac{e^{ikr}}{r} \left[(\hat{\mathbf{n}} \times \mathbf{p}) \times \hat{\mathbf{n}} - \hat{\mathbf{n}} \times \mathbf{m} \right] . \tag{5.3}$$

$$\mathbf{B}_{sc}(\mathbf{x}) = \hat{\mathbf{n}} \times \mathbf{E}_{sc} . \tag{5.4}$$

The question we ask is the following: how much radiation is scattered along a given direction, with what polarization? The power radiated in the direction $\hat{\mathbf{n}}$ with polarization ϵ, per unit solid angle (Ω), per unit incident flux of radiation along $\hat{\mathbf{n}}_0$ with the incident field polarized along ϵ_0 is called the differential scattering cross section[3]

$$\frac{d\sigma}{d\Omega} = \frac{\operatorname{Re} (r^2 \hat{\mathbf{n}} \cdot \mathbf{E}_{sc} \times \mathbf{H}_{sc}^*)}{\operatorname{Re} (\hat{\mathbf{n}}_0 \cdot \mathbf{E}_{inc} \times \mathbf{H}_{inc}^*)} . \tag{5.5}$$

For generality, we assume that polarization vectors are complex quantities.[4] Then

$$\hat{\mathbf{n}}_0 \cdot \left(\mathbf{E}_{inc} \times \mathbf{H}_{inc}^* \right) = \hat{\mathbf{n}}_0 \cdot \left[\left(\epsilon_0^* \cdot \mathbf{E}_{inc} \right) \epsilon_0^* \times \{ \hat{\mathbf{n}}_0 \times \left(\epsilon_0^* \cdot \mathbf{E}_{inc} \right)^* \epsilon_0^* \} \right] . \tag{5.6}$$

The triple product on the right-hand side can be simplified using standard vector identity and keeping in mind that $\epsilon_0 \cdot \hat{\mathbf{n}}_0 = 0$.

$$\hat{\mathbf{n}}_0 \cdot \left(\mathbf{E}_{inc} \times \mathbf{H}_{inc}^* \right) = \hat{\mathbf{n}}_0 \cdot \left[\left(\epsilon_0^* \cdot \mathbf{E}_{inc} \right) \left(\epsilon_0^* \cdot \mathbf{E}_{inc} \right)^* \left(\epsilon_0^* \cdot \epsilon_0^* \right) \hat{\mathbf{n}}_0 \right]$$
$$= | \left(\epsilon_0^* \cdot \mathbf{E}_{inc} \right) |^2 . \tag{5.7}$$

Similarly, the numerator in Eq. (5.5) can be evaluated, and we can write

$$\frac{d\sigma}{d\Omega} = r^2 \frac{| \left(\epsilon^* \cdot \mathbf{E}_{sc} \right) |^2}{| \left(\epsilon_0^* \cdot \mathbf{E}_{inc} \right) |^2} . \tag{5.8}$$

If E_0 is the incident electric field amplitude, then $| \left(\epsilon_0^* \cdot \mathbf{E}_{inc} \right) |^2 = E_0^2$. Using Eq. (5.3), we finally obtain

$$\frac{d\sigma}{d\Omega} = \frac{k^4}{E_0^2} \left| \epsilon^* \cdot \{ (\hat{\mathbf{n}} \times \mathbf{p}) \times \hat{\mathbf{n}} \} - \epsilon^* \cdot \hat{\mathbf{n}} \times \mathbf{m} \right|^2$$
$$= \frac{k^4}{E_0^2} \left| \epsilon^* \cdot \{ \mathbf{p} - (\hat{\mathbf{n}} \cdot \mathbf{p}) \hat{\mathbf{n}} \} - \epsilon^* \cdot \hat{\mathbf{n}} \times \mathbf{m} \right|^2$$
$$= \frac{k^4}{E_0^2} \left| \epsilon^* \cdot \mathbf{p} + (\hat{\mathbf{n}} \times \epsilon^*) \cdot \mathbf{m} \right|^2 . \tag{5.9}$$

---

[2] Recall the far fields due to electric and magnetic dipole radiation discussed in Chap. 4.

[3] It has dimension of area.

[4] Such as in case of circular polarization.

The differential scattering cross section varies as k^4. This is a universal characteristic of long wavelength scattering by any finite system.[5] We shall now utilize Eq. (5.9) to calculate the differential scattering cross section for a few specific cases.

5.1.1 Scattering by a Dielectric Sphere

Consider a small dielectric sphere of radius R, with $\mu = 1$ and uniform, isotropic dielectric constant $\epsilon = \epsilon(\omega)$. We already know what should be the expression for the electric dipole moment produced by the incident electric field.[6] If the incident electric field is given by, $\mathbf{E}_{\text{inc}} = \epsilon_0 E_0 \, e^{i(k\hat{\mathbf{n}}_0 \cdot \mathbf{x} - \omega t)}$, then

$$\mathbf{p} = \frac{\epsilon - 1}{\epsilon + 2} R^3 \mathbf{E}_{\text{inc}} . \tag{5.10}$$

We assume that there is no induced magnetic moment ($\mathbf{m} = 0$). Therefore

$$\frac{d\sigma}{d\Omega} = \frac{k^4}{E_0^2} \left| \epsilon^* \cdot \left(\frac{\epsilon - 1}{\epsilon + 2} \right) R^3 \mathbf{E}_{\text{inc}} \right|^2$$

$$= k^4 R^6 \left(\frac{\epsilon - 1}{\epsilon + 2} \right)^2 \left| \epsilon^* \cdot \epsilon_0 \right|^2 . \tag{5.11}$$

From Eq. (5.3), it is clear that the scattered radiation is linearly polarized in the plane defined by the direction of dipole moment $\mathbf{p}$ (or ϵ_0) and the unit vector $\hat{\mathbf{n}}$. Now, suppose the incident wave is unpolarized. We want to calculate the fraction of scattered radiation with polarization in the scattering plane defined by $\hat{\mathbf{n}}$ and $\hat{\mathbf{n}}_0$ and the fraction perpendicular to it (Fig. 5.1). We can average the effects of two orthogonal incident incoherent waves to produce an unpolarized incident beam of radiation. The random polarization of the incident beam can then be expressed as a mixture of two polarizations: ϵ_0^1 in the scattering plane and ϵ_0^2 perpendicular to the scattering plane as

$$\epsilon_0 = \epsilon_0^1 \cos \varphi + \epsilon_0^2 \sin \varphi . \tag{5.12}$$

where φ is the instantaneous angle between ϵ_0 and ϵ_0^1 in the xy plane. The components of polarization vectors corresponding to incident and scattered radiation which are in the scattering plane are ϵ_0^1 and ϵ^1, respectively, while the components ϵ_0^2 and ϵ^2 are perpendicular to the scattering plane (Fig. 5.1). Note that θ is the angle between ϵ_0^1 and ϵ^1. Then, $\epsilon^1 \cdot \epsilon_0 = \cos \theta \cos \varphi$.

---

[5] Such dependence on wave number is also known as Rayleigh's law.

[6] See Eq. (3.79) in Chap. 3.

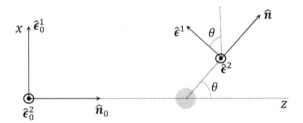

Fig. 5.1 Propagation vectors of the incident and scattered waves ($\hat{n}_0$ and $\hat{n}$) are in the scattering (xz) plane. The components of polarization vectors of the incident and scattered radiation in the xz plane and perpendicular to it (along positive y-direction) are also shown

The scattering cross section parallel to the scattering plane is obtained by replacing ϵ^* by ϵ^1 in Eq. (5.11) and averaging over φ

$$\frac{d\sigma_\parallel}{d\Omega} = k^4 R^6 \left(\frac{\epsilon-1}{\epsilon+2}\right)^2 \frac{1}{2\pi} \int_0^{2\pi} \left|\epsilon^1 \cdot \epsilon_0\right|^2 d\varphi = \frac{k^4 R^6}{2} \left(\frac{\epsilon-1}{\epsilon+2}\right)^2 \cos^2\theta \tag{5.13}$$

Similarly, $\epsilon^2 \cdot \epsilon_0 = \sin\varphi$. Therefore, the scattering cross section perpendicular to the scattering plane is

$$\frac{d\sigma_\perp}{d\Omega} = k^4 R^6 \left(\frac{\epsilon-1}{\epsilon+2}\right)^2 \frac{1}{2\pi} \int_0^{2\pi} \left|\epsilon^2 \cdot \epsilon_0\right|^2 d\varphi = \frac{k^4 R^6}{2} \left(\frac{\epsilon-1}{\epsilon+2}\right)^2 . \tag{5.14}$$

The total differential scattering cross section is

$$\frac{d\sigma}{d\Omega} = \frac{d\sigma_\parallel}{d\Omega} + \frac{d\sigma_\perp}{d\Omega} = \frac{k^4 R^6}{2} \left(\frac{\epsilon-1}{\epsilon+2}\right)^2 \left(1 + \cos^2\theta\right) . \tag{5.15}$$

The differential scattering cross section peaks in the forward ($\theta = 0$) and backward ($\theta = \pi$) directions. The total scattering cross section is obtained by integrating over all directions

$$\sigma = \int \frac{d\sigma}{d\Omega} d\Omega = \frac{k^4 R^6}{2} \left(\frac{\epsilon-1}{\epsilon+2}\right)^2 \int_0^{2\pi} d\varphi \int_{-1}^{1} d\cos\theta \left(1 + \cos^2\theta\right)$$

$$= \frac{8\pi k^4 R^6}{3} \left(\frac{\epsilon-1}{\epsilon+2}\right)^2 . \tag{5.16}$$

We finally define the polarization factor of the scattered radiation as

$$\Pi(\theta) = \frac{\frac{d\sigma_\perp}{d\Omega} - \frac{d\sigma_\parallel}{d\Omega}}{\frac{d\sigma_\perp}{d\Omega} + \frac{d\sigma_\parallel}{d\Omega}} . \tag{5.17}$$

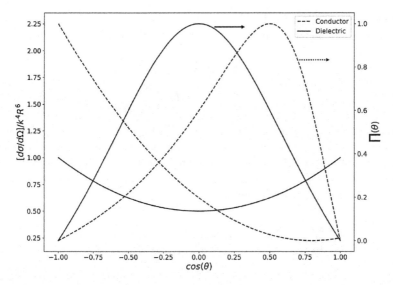

Fig. 5.2 Angle dependence of the differential scattering cross section and the polarization factor for the dielectric and perfectly conducting sphere

Using Eqs. (5.13) and (5.14) in Eq. (5.17), we obtain

$$\Pi(\theta) = \frac{\sin^2 \theta}{1 + \cos^2 \theta}. \tag{5.18}$$

The polarization factor $\Pi(\theta)$ reaches the maximum value of unity at $\theta = \frac{\pi}{2}$. It means that the radiation scattered in the transverse direction is 100% linearly polarized perpendicular to the scattering plane. Whereas the differential scattering cross section is maximum in the forward and backward direction, the forward and back scattered radiation remains unpolarized. The scattered radiation is linearly polarized significantly on either side of $\theta = \frac{\pi}{2}$ direction (Fig. 5.2).

5.1.2 Scattering by a Perfectly Conducting Sphere

We now discuss the scattering of a plane electromagnetic wave incident on a perfectly conducting sphere of radius R. In the long wavelength limit ($\lambda \gg R$), the incident fields are essentially uniform over the sphere. A perfectly conducting sphere is equivalent to a sphere of infinite dielectric constant and zero permeability. The last statement can be intuitively understood in the following way. The electric field inside a perfect conductor is zero. From Eq. (3.43), this is possible only when ϵ approaches infinity. On the other hand, if a time-varying external magnetic field is applied on a perfect conductor, it cannot penetrate the conductor since otherwise, it will lead to induced electric field inside. Keeping these in mind, we write down the induced electric dipole moment in presence of incident electric field $\mathbf{E}_{\text{inc}}$ as follows:

$$\mathbf{p} = R^3 \mathbf{E}_{\text{inc}}, \tag{5.19}$$

$$\mathbf{m} = -\frac{R^3}{2} \mathbf{H}_{\text{inc}}. \tag{5.20}$$

To get the first result, one must take $\epsilon \to \infty$ in Eq. (3.79) of Chap. 3. To get the second result, one must take the limit $\mu \to 0$ in Eq. (3.171) of Chap. 3 and calculate the moment $\mathbf{m} = \frac{4\pi}{3} R^3 \mathbf{M}$. We replace $\mathbf{p}$ and $\mathbf{m}$ in Eq. (5.9) by Eqs. (5.19) and (5.20), respectively, and noting that $\mathbf{H}_{\text{inc}} = \mathbf{n}_0 \times \mathbf{E}_{\text{inc}}$, we write down the corresponding differential scattering cross section as

$$\frac{d\sigma}{d\Omega} = k^4 R^6 \left| \boldsymbol{\epsilon}^* \cdot \boldsymbol{\epsilon}_0 - \frac{1}{2}(\hat{\mathbf{n}} \times \boldsymbol{\epsilon}^*) \cdot (\mathbf{n}_0 \times \boldsymbol{\epsilon}_0) \right|^2. \tag{5.21}$$

As earlier, we label the components of scattered radiation polarization, parallel and perpendicular to the scattering plane as $\boldsymbol{\epsilon}^1$ and $\boldsymbol{\epsilon}^2$, respectively. Assuming that the incident wave is unpolarized, we follow the same procedure used in case of dielectric sphere and write

$$\boldsymbol{\epsilon}_0 = \boldsymbol{\epsilon}_0^1 \cos \varphi + \boldsymbol{\epsilon}_0^2 \sin \varphi. \tag{5.22}$$

Then,

$$\boldsymbol{\epsilon}^1 \cdot \boldsymbol{\epsilon}_0 - \frac{1}{2}(\hat{\mathbf{n}} \times \boldsymbol{\epsilon}^1) \cdot (\mathbf{n}_0 \times \boldsymbol{\epsilon}_0) = \left(\cos\theta - \frac{1}{2}\right) \cos\varphi, \tag{5.23}$$

$$\boldsymbol{\epsilon}^2 \cdot \boldsymbol{\epsilon}_0 - \frac{1}{2}(\hat{\mathbf{n}} \times \boldsymbol{\epsilon}^2) \cdot (\mathbf{n}_0 \times \boldsymbol{\epsilon}_0) = \left(1 - \frac{1}{2}\cos\theta\right) \sin\varphi. \tag{5.24}$$

After averaging over φ, it follows that

$$\frac{d\sigma_{\parallel}}{d\Omega} = \frac{1}{2} k^4 R^6 \left(\cos\theta - \frac{1}{2}\right)^2, \tag{5.25}$$

$$\frac{d\sigma_{\perp}}{d\Omega} = \frac{1}{2} k^4 R^6 \left(1 - \frac{1}{2}\cos\theta\right)^2. \tag{5.26}$$

Total differential scattering cross section is obtained after summing over both polarization states[7] as

---

[7] Note that the result will be the same for incident circularly polarized radiation.

$$\frac{d\sigma}{d\Omega} = \frac{d\sigma_\parallel}{d\Omega} + \frac{d\sigma_\perp}{d\Omega} = k^4 R^6 \left[\frac{5}{8} \left(1 + \cos^2 \theta\right) - \cos\theta \right]. \qquad (5.27)$$

A plot of the differential scattering cross section shows that the backward scattering is much stronger compared to forward scattering, in case of perfectly conducting sphere. As we have already found out, such a behavior is in contrast to the case for dielectric sphere. The total scattering cross section is

$$\sigma = \int \frac{d\sigma}{d\Omega} d\Omega = k^4 R^6 \frac{5}{8} 2\pi \int_{-1}^{1} (1 + \cos^2\theta)\, d(\cos\theta) = \frac{10\pi}{3} k^4 R^6. \qquad (5.28)$$

The polarization is

$$\Pi(\theta) = \frac{3 \sin^2 \theta}{5(1 + \cos^2 \theta) - 8\cos\theta}. \qquad (5.29)$$

Note that the polarization factor Π is 100% at $\theta = \frac{\pi}{3}$. This is understandable as $\frac{d\sigma_\parallel}{d\Omega}$ is always non-zero, but $\frac{d\sigma_\perp}{d\Omega} = 0$ for $\theta = \frac{\pi}{3}$. Therefore, the scattered radiation is linearly polarized parallel to the scattering plane for arbitrary incident polarization at $\theta = \frac{\pi}{3}$.

5.1.3   Scattering by a Collection of Scatterers

Now suppose, instead of a single scatterer, the plane wave radiation is incident on a collection of scatterers having fixed spatial separation between each other with the detector being far away from the scatterers. This is a common problem encountered in solid state physics. The incident fields as well as the scattered fields depend on the position of the scatterers.

The incident and the scattered electric fields for the jth scatterer positioned at $\mathbf{x_j}'$ with the detector (or the observation point) at $\mathbf{x}$ is given by

$$\mathbf{E}_{\text{inc}}(\mathbf{x_j}') = \boldsymbol{\epsilon}_0 E_0 e^{ik\hat{\mathbf{n}}_0 \cdot \mathbf{x_j}'} \qquad (5.30)$$

$$\mathbf{E}_{\text{sc}}(\mathbf{x}) = k^2 \frac{e^{ikr}}{r} e^{-ik\hat{\mathbf{n}} \cdot \mathbf{x_j}'} \left[(\hat{\mathbf{n}} \times \mathbf{p}_j) \times \hat{\mathbf{n}} - \hat{\mathbf{n}} \times \mathbf{m}_j \right]. \qquad (5.31)$$

Here, $\mathbf{E}_{\text{sc}}(\mathbf{x})$ is the scattered field at $\mathbf{x}$ due to the jth scatterer at position $\mathbf{x_j}'$. The induced moments acquire a position dependent phase in the present case. The phases appear since in the present case we should replace r by $|\mathbf{x} - \mathbf{x_j}'| = r - \mathbf{n} \cdot \mathbf{x_j}'$ as the position of the scatterers matter now. Moreover, as r is greater than the distance scale specifying the position of the localized scatterers, we assume $|\mathbf{x} - \mathbf{x_j}'|^{-1} = 1/r$.

We sum over fields at $\mathbf{x}$ due to all scatterers at $\mathbf{x}_j$ with $j = 1, 2, 3, \ldots$ and obtain the total differential scattering cross section as follows:

$$\frac{d\sigma}{d\Omega} = r^2 \frac{|(\boldsymbol{\epsilon}^* \cdot \mathbf{E}_{sc})|^2}{|(\boldsymbol{\epsilon}_0^* \cdot \mathbf{E}_{inc})|^2} = \frac{k^4}{E_0^2} \left| \sum_j [\boldsymbol{\epsilon}^* \cdot \mathbf{p}_j + (\hat{\mathbf{n}} \times \boldsymbol{\epsilon}^*) \cdot \mathbf{m}_j] e^{ik(\hat{\mathbf{n}}_0 - \hat{\mathbf{n}}) \cdot \mathbf{x}_j'} \right|^2 . \qquad (5.32)$$

Note that this is a case of elastic scattering as the magnitude of $\mathbf{k}$ is unaltered. We define $\mathbf{q} = k(\hat{\mathbf{n}}_0 - \hat{\mathbf{n}})$ as the change in wave vector due to scattering. The scattering cross section clearly depends on the spatial arrangement of scatterers. If all the scatterers are identical (i.e., with same $\mathbf{p}_j$ and $\mathbf{m}_j$), then the cross section can be written as the product of single particle scattering cross section and an additional factor which encapsulates the phase information at the observation point due to the particular spatial arrangement of scatterers as follows:

$$\frac{d\sigma}{d\Omega} = \frac{k^4}{E_0^2} \left| \boldsymbol{\epsilon}^* \cdot \mathbf{p}_j + (\hat{\mathbf{n}} \times \boldsymbol{\epsilon}^*) \cdot \mathbf{m}_j \right|^2 \left| \sum_j e^{i\mathbf{q} \cdot \mathbf{x}_j'} \right|^2 . \qquad (5.33)$$

We define the so-called 'structure factor'

$$\mathcal{F}(\mathbf{q}) = \left| \sum_j e^{i\mathbf{q} \cdot \mathbf{x}_j'} \right|^2 = \sum_j \sum_i e^{i\mathbf{q} \cdot (\mathbf{x}_j' - \mathbf{x}_i')} . \qquad (5.34)$$

We can think of two different scenarios. First, if there is a random distribution of scatterers, contribution by the terms $j \neq i$ is negligible. Consequently, $\mathcal{F}(\mathbf{q}) = N$, if there are total N scatterers and we have the somewhat uninteresting case of incoherent superposition of scattered fields. Second, we discuss the case of a regular arrangement of scatterers as observed in crystal lattices. Let us consider, as a simple illustration, the case of scattering in a large cubic array of scatterers.

Let us assume that N_1, N_2 and N_3 are the number of lattice sites (scattering centers) along the three Cartesian axes of the cubic array such that the total number of scatterers in the collection is $N = N_1 N_2 N_3$. The lattice spacing along each direction is a. Let us also assume that the Cartesian components of $\mathbf{q}$ along the three axes are q_1, q_2 and q_3. The structure factor for such a simple cubic lattice is given by

$$\mathcal{F}(\mathbf{q}) = \left| \sum_{l=0}^{N_1-1} e^{iq_1 la} \sum_{m=0}^{N_2-1} e^{iq_2 ma} \sum_{n=0}^{N_3-1} e^{iq_3 na} \right|^2 . \qquad (5.35)$$

Here, we have used the fact that the spacing between any two scattering centers along a particular direction is integer multiple of a. Now, we can write down $\mathcal{F}$ in a simplified form

$$\mathcal{F}(\mathbf{q}) = \left| \left(\frac{e^{iq_1 N_1 a} - 1}{e^{iq_1 a} - 1} \right) \left(\frac{e^{iq_2 N_2 a} - 1}{e^{iq_2 a} - 1} \right) \left(\frac{e^{iq_3 N_3 a} - 1}{e^{iq_3 a} - 1} \right) \right|^2,$$

$$= \left(\frac{\sin^2 \frac{q_1 N_1 a}{2}}{\sin^2 \frac{q_1 a}{2}} \right) \left(\frac{\sin^2 \frac{q_2 N_2 a}{2}}{\sin^2 \frac{q_2 a}{2}} \right) \left(\frac{\sin^2 \frac{q_3 N_3 a}{2}}{\sin^2 \frac{q_3 a}{2}} \right). \tag{5.36}$$

In the long wavelength limit, $ka \ll 1$ or alternatively, $q_i a \ll 1$, and therefore, the structure factor can be written as

$$\mathcal{F}(\mathbf{q}) = N^2 \left(\frac{\sin \frac{q_1 N_1 a}{2}}{\frac{q_1 N_1 a}{2}} \right)^2 \left(\frac{\sin \frac{q_2 N_2 a}{2}}{\frac{q_2 N_2 a}{2}} \right)^2 \left(\frac{\sin \frac{q_3 N_3 a}{2}}{\frac{q_3 N_3 a}{2}} \right)^2. \tag{5.37}$$

Clearly, the structure factor peaks at $q_i a = 0$. Since $|\mathbf{q}| = k|\hat{\mathbf{n}}_0 - \hat{\mathbf{n}}|$, such a situation corresponds to the scattering in the forward direction, i.e., for $\hat{\mathbf{n}} = \hat{\mathbf{n}}_0$. This is, in general, true for scattering in crystal lattices in the long wavelength limit[8]: the structure factor is negligible everywhere except along the forward direction. This is the reason a large number of single crystals are transparent to visible light.

5.2   Perturbation Theory of Scattering by Liquids and Gases

In this section, we show that spatial and temporal variations in a medium lead to scattering. We assume that the variation is small enough so that the problem can be treated perturbatively. We consider a medium with dielectric constant $\epsilon(\mathbf{x})$ and permeability $\mu(\mathbf{x})$ expressed as

$$\epsilon(\mathbf{x}) = <\epsilon> \pm \delta\epsilon(\mathbf{x}), \tag{5.38}$$

$$\mu(\mathbf{x}) = <\mu> \pm \delta\mu(\mathbf{x}). \tag{5.39}$$

Here, $<\epsilon>$ and $<\mu>$ are the spatial averages while $\delta\epsilon(\mathbf{x})$ and $\delta\mu(\mathbf{x})$ are the corresponding fluctuations in space. The Maxwell equations in absence of sources are given by

$$\nabla \cdot \mathbf{D} = 0, \tag{5.40}$$

$$\nabla \cdot \mathbf{B} = 0, \tag{5.41}$$

$$\nabla \times \mathbf{E} = -\frac{1}{c} \frac{\partial \mathbf{B}}{\partial t}, \tag{5.42}$$

---

[8] For cubic lattice, in the short wavelength limit, i.e., for $ka > 1$, $\mathcal{F}(\mathbf{q})$ peaks at $\frac{1}{2} q_i a = 0, \pi, 2\pi, \ldots$ etc. We may write $\mathbf{q} = k(\hat{\mathbf{n}}_0 - \hat{\mathbf{n}}) = \Delta \mathbf{k}$. Moreover q_i is the component of $\mathbf{q}$ along ith Cartesian component of the lattice vector. This is the familiar Bragg scattering condition.

$$\nabla \times \mathbf{H} = \frac{1}{c} \frac{\partial \mathbf{D}}{\partial t} \,. \tag{5.43}$$

The equations connecting $\mathbf{D}$ and $\mathbf{E}$ on one hand and $\mathbf{H}$ and $\mathbf{B}$ on the other are given by

$$\mathbf{D} = \mathbf{E} + 4\pi \mathbf{P} = \epsilon \mathbf{E}\,, \tag{5.44}$$

$$\mathbf{B} = \mathbf{H} + 4\pi \mathbf{M} = \mu \mathbf{H}\,. \tag{5.45}$$

Using the Maxwell equations, we construct a wave equation in $\mathbf{D}$ as follows. First, we rewrite Eq. (5.42) as

$$\frac{1}{\epsilon} \nabla \times [\mathbf{D} - (\mathbf{D} - \epsilon \mathbf{E})] = -\frac{1}{c} \frac{\partial}{\partial t} [\mu \mathbf{H} + (\mathbf{B} - \mu \mathbf{H})] \,. \tag{5.46}$$

Taking curl of both sides, we get

$$\nabla \times \nabla \times [\mathbf{D} - (\mathbf{D} - \epsilon \mathbf{E})] = -\frac{\epsilon}{c} \frac{\partial}{\partial t} [\mu \nabla \times \mathbf{H} + \nabla \times (\mathbf{B} - \mu \mathbf{H})] \,. \tag{5.47}$$

Since $\nabla \cdot \mathbf{D} = 0$, it follows that

$$-\nabla^2 \mathbf{D} - \nabla \times \nabla \times (\mathbf{D} - \epsilon \mathbf{E}) = -\frac{\epsilon \mu}{c^2} \frac{\partial^2 \mathbf{D}}{\partial t^2} - \frac{\epsilon}{c} \frac{\partial}{\partial t} [\nabla \times (\mathbf{B} - \mu \mathbf{H})] \,. \tag{5.48}$$

We rearrange the terms to obtain the following:

$$\nabla^2 \mathbf{D} - \frac{\epsilon \mu}{c^2} \frac{\partial^2 \mathbf{D}}{\partial t^2} = -\nabla \times \nabla \times (\mathbf{D} - \epsilon \mathbf{E}) + \frac{\epsilon}{c} \frac{\partial}{\partial t} [\nabla \times (\mathbf{B} - \mu \mathbf{H})] \,. \tag{5.49}$$

This is an inhomogeneous wave equation in $\mathbf{D}$ with the right hand side describing the source term. The reader will immediately notice that the source term is trivially zero as $\mathbf{D} = \epsilon \mathbf{E}$ and $\mathbf{B} = \mu \mathbf{H}$ and consequently the above equation is not really interesting. Here, one must notice that we can actually transform the above equation into a perturbation equation where the source term can become meaningful. To see how one can reinterpret the last equation, we remind the reader that the initial equations in this section give the fluctuations of the dielectric constant and the permeability of the medium. In presence of such fluctuations, one can show that

$$\mathbf{D} - <\epsilon> \mathbf{E} = \frac{\delta \epsilon}{<\epsilon>} \mathbf{D}_0\,, \quad \text{and} \quad \mathbf{B} - <\mu> \mathbf{H} = \frac{\delta \mu}{<\mu>} \mathbf{B}_0\,,$$

where $\mathbf{D}_0 = <\epsilon> \mathbf{E}$ and $\mathbf{B}_0 = <\mu> \mathbf{H}$ are the fields in the absence of any fluctuations in dielectric constant and the magnetic permeability. We will discuss about this approximation scheme more vividly in the next subsection. Using this knowledge, we can now actually replace $\mathbf{D} - \epsilon \mathbf{E}$ and $\mathbf{B} - \mu \mathbf{H}$ on the right-hand side of the last equation by $\mathbf{D} - <\epsilon> \mathbf{E}$ and $\mathbf{B} - <\mu> \mathbf{H}$, respectively. These terms are proportional to $\delta\epsilon$ and $\delta\mu$. With these replacements, we can now solve $\mathbf{D}$ in presence of fluctuating medium properties. We assume harmonic time dependence $(e^{-i\omega t})$ of the fields. To[9] first order in perturbations we can write

$$\left(\nabla^2 + k^2\right)\mathbf{D} = -\nabla \times \nabla \times (\mathbf{D} - <\epsilon> \mathbf{E}) - i\frac{<\epsilon>\omega}{c}\left[\nabla \times (\mathbf{B} - <\mu> \mathbf{H})\right],$$

(5.50)

where $k^2 = \frac{\epsilon\mu\omega^2}{c^2}$. The 'source term' on the right-hand side is treated as a perturbation over the 'unperturbed' homogeneous wave solution. We can readily write down the solution provided the perturbative source term is known[10]:

$$\mathbf{D} = \mathbf{D}_0 + \frac{1}{4\pi}\int d^3x' \frac{e^{ik|\mathbf{x}-\mathbf{x}'|}}{|\mathbf{x}-\mathbf{x}'|}\left[\nabla' \times \nabla' \times (\mathbf{D} - <\epsilon> \mathbf{E})\right.$$
$$\left. +i\frac{<\epsilon>\omega}{c}\nabla' \times (\mathbf{B} - <\mu> \mathbf{H})\right]. \quad (5.51)$$

Here, $\mathbf{D}_0$ is the solution to the homogeneous wave equation (when the source term or the perturbation term is zero), which is interpreted as an incident wave propagating along certain direction. As stated already, the second term on the right-hand side is non-zero only over certain finite regions in space, interpreted as the scattering regions. The radiation field measured far away from the scattering regions is expressed as the combination of the incident wave and the 'scattered' waves, as follows:

$$\mathbf{D} = \mathbf{D}_0 + \frac{e^{ikr}}{4\pi r}\int d^3x' e^{-ik\hat{\mathbf{n}}\cdot\mathbf{x}'}\left[\nabla' \times \nabla' \times (\mathbf{D} - <\epsilon> \mathbf{E})\right.$$
$$\left. +i\frac{<\epsilon>\omega}{c}\nabla' \times (\mathbf{B} - <\mu> \mathbf{H})\right]. \quad (5.52)$$

We define the 'scattering amplitude'

$$\mathbf{A}_{sc} = \frac{1}{4\pi}\int d^3x' e^{-ik\hat{\mathbf{n}}\cdot\mathbf{x}'}\left[\nabla' \times \nabla' \times (\mathbf{D} - <\epsilon> \mathbf{E}) + i\frac{<\epsilon>\omega}{c}\nabla' \times (\mathbf{B} - <\mu> \mathbf{H})\right].$$

(5.53)

Since the observation point is far away from the source, we can simplify the above expression further by replacing the ∇ operators with $ik\hat{\mathbf{n}}$, appropriate for vector field operations on a plane wave. The final expression for the scattering amplitude is

$$\mathbf{A}_{sc} = \frac{k^2}{4\pi}\int d^3x' e^{-ik\hat{\mathbf{n}}\cdot\mathbf{x}'}\left[\{\hat{\mathbf{n}} \times (\mathbf{D} - <\epsilon> \mathbf{E})\} \times \hat{\mathbf{n}} - \frac{<\epsilon>\omega}{kc}\hat{\mathbf{n}} \times (\mathbf{B} - <\mu> \mathbf{H})\right].$$

(5.54)

---

[9] $\frac{\partial}{\partial t} \equiv -i\omega$; $\frac{\partial^2}{\partial t^2} \equiv -\omega^2$.

[10] For the formal solution of an inhomogeneous wave equation, see Chap. 4.

The integrand in Eq. (5.54) makes interesting comparison with the expression of scattered dipole field in Eq. (5.3). When $\epsilon \simeq \mu \simeq 1$, $\frac{\epsilon\omega}{kc} = \sqrt{\frac{\epsilon}{\mu}} \simeq 1$ and except for the term $e^{-ik\hat{n}\cdot x'}$ in Eq. (5.54), these two are equivalent expressions. The additional phase factor in Eq. (5.54) takes care of interference between scattered waves from multiple scattering centers. We rewrite Eq. (5.52) in compact form as

$$\mathbf{D} = \mathbf{D}_0 + \frac{e^{ikr}}{r}\mathbf{A}_{sc}, \qquad (5.55)$$

where $\mathbf{A}_{sc}$ is given by Eq. (5.54). Let the polarization vector of the scattered radiation be ϵ and the same for incident radiation be ϵ_0. Following Eq. (5.8), it is now straightforward to write down the expression for differential scattering cross section as

$$\frac{d\sigma}{d\Omega} = r^2 \frac{\left|\left(\epsilon^* \cdot \left(\frac{e^{ikr}}{r}\right)\mathbf{A}_{sc}\right)\right|^2}{|\left(\epsilon_0^* \cdot \mathbf{D}_0\right)|^2} = \frac{|\left(\epsilon^* \cdot \mathbf{A}_{sc}\right)|^2}{|D_0|^2}. \qquad (5.56)$$

The scattering amplitude cannot be calculated exactly if the fields are only approximately known. However, it is possible to converge to an adequately accurate solution through a series of successive approximations.

5.2.1 Born Approximation

We restrict our discussion to the lowest order approximation for the scattering amplitude and to the linear media. Using Eqs. (5.38) and (5.39), we write

$$\mathbf{D}(\mathbf{x}) = [< \epsilon > \pm \delta\epsilon(\mathbf{x})]\,\mathbf{E}(\mathbf{x})\,, \qquad (5.57)$$

$$\mathbf{B}(\mathbf{x}) = [< \mu > \pm \delta\mu(\mathbf{x})]\,\mathbf{H}(\mathbf{x})\,, \qquad (5.58)$$

where the spatial fluctuations $\delta\epsilon$, $\delta\mu$ are small compared to $< \epsilon >$, $< \mu >$, respectively. For brevity and simplicity, we replace the spatial averages $< \epsilon >$, $< \mu >$ by simply $\epsilon^{(0)}$, $\mu^{(0)}$[11] and only focus on the positive signs of the above fluctuations, respectively, so that

$$\mathbf{D} - \epsilon^{(0)}\mathbf{E} = \frac{\delta\epsilon(\mathbf{x})}{\epsilon^{(0)}}\mathbf{D}_0\,, \qquad (5.59)$$

$$\mathbf{B} - \mu^{(0)}\mathbf{H} = \frac{\delta\mu(\mathbf{x})}{\mu^{(0)}}\mathbf{B}_0\,, \qquad (5.60)$$

---

[11] Not to be confused with free space values of ϵ and μ which are one in our units.

where $\mathbf{D}_0 = \epsilon^{(0)}\mathbf{E}$ and $\mathbf{B}_0 = \mu^{(0)}\mathbf{H}$ are the unperturbed (incident) plane waves propagating along $\hat{\mathbf{n}}_0$. We have

$$\mathbf{D}_0 = \epsilon_0 D_0 e^{ik\hat{\mathbf{n}}_0 \cdot \mathbf{x}}, \tag{5.61}$$

$$\mathbf{B}_0 = \sqrt{\frac{\mu^{(0)}}{\epsilon^{(0)}}} \hat{\mathbf{n}}_0 \times \mathbf{D}_0. \tag{5.62}$$

Here, ϵ_0 is the polarization direction of the unperturbed wave. The scattering amplitude is now given by

$$\mathbf{A}_{sc} = \frac{k^2}{4\pi} \int d^3x' e^{-ik\hat{\mathbf{n}}\cdot\mathbf{x}'} \left[\{\hat{\mathbf{n}} \times (\mathbf{D} - \epsilon^{(0)}\mathbf{E})\} \times \hat{\mathbf{n}} - \frac{\epsilon^{(0)}\omega}{kc} \hat{\mathbf{n}} \times (\mathbf{B} - \mu^{(0)}\mathbf{H}) \right]. \tag{5.63}$$

Combining Eqs. (5.59)–(5.62), we can write

$$\mathbf{D} - \epsilon^{(0)}\mathbf{E} = \frac{\delta\epsilon(\mathbf{x})}{\epsilon^{(0)}} \epsilon_0 D_0 e^{ik\hat{\mathbf{n}}_0 \cdot \mathbf{x}}, \tag{5.64}$$

$$\mathbf{B} - \mu^{(0)}\mathbf{H} = \sqrt{\frac{\mu^{(0)}}{\epsilon^{(0)}}} \frac{\delta\mu(\mathbf{x})}{\mu^{(0)}} \hat{\mathbf{n}}_0 \times \epsilon_0 D_0 e^{ik\hat{\mathbf{n}}_0 \cdot \mathbf{x}}. \tag{5.65}$$

Using Eqs. (5.64) and (5.65) in Eq. (5.63), we obtain the first Born approximation

$$\frac{\epsilon^* \cdot \mathbf{A}_{sc}^{(1)}}{D_0} = \frac{k^2}{4\pi} \int d^3x' e^{-ik(\hat{\mathbf{n}}-\hat{\mathbf{n}}_0)\cdot\mathbf{x}'} \left[\frac{\delta\epsilon(\mathbf{x}')}{\epsilon^{(0)}} \epsilon^* \cdot (\hat{\mathbf{n}} \times \epsilon_0) \times \hat{\mathbf{n}} - \frac{\delta\mu(\mathbf{x}')}{\mu^{(0)}} \epsilon^* \cdot \hat{\mathbf{n}} \times (\hat{\mathbf{n}}_0 \times \epsilon_0) \right].$$

This is an approximate scheme to calculate the scattering amplitude as we have changed the constitutive relations in the medium up to the first order. This is called the Born approximation scheme. We define $\mathbf{q} = k(\hat{\mathbf{n}}_0 - \hat{\mathbf{n}})$ as the change of wave vector due to scattering. Moreover, $\epsilon^* \cdot (\hat{\mathbf{n}} \times \epsilon_0) \times \hat{\mathbf{n}} = \epsilon^* \cdot [(\hat{\mathbf{n}} \cdot \hat{\mathbf{n}})\epsilon_0 - (\hat{\mathbf{n}} \cdot \epsilon_0)\hat{\mathbf{n}}] = \epsilon^* \cdot \epsilon_0$ and $\epsilon^* \cdot \hat{\mathbf{n}} \times (\hat{\mathbf{n}}_0 \times \epsilon_0) = (\epsilon^* \times \hat{\mathbf{n}}) \cdot (\hat{\mathbf{n}}_0 \times \epsilon_0)$. We finally have

$$\frac{\epsilon^* \cdot \mathbf{A}_{sc}^{(1)}}{D_0} = \frac{k^2}{4\pi} \int d^3x' e^{i\mathbf{q}\cdot\mathbf{x}'} \left[\epsilon^* \cdot \epsilon_0 \frac{\delta\epsilon(\mathbf{x}')}{\epsilon^{(0)}} + (\hat{\mathbf{n}} \times \epsilon^*) \cdot (\hat{\mathbf{n}}_0 \times \epsilon_0) \frac{\delta\mu(\mathbf{x}')}{\mu^{(0)}} \right]. \tag{5.66}$$

The differential scattering cross section can be calculated by using Eq. (5.66) in Eq. (5.56).

Let us now reproduce the result for scattering by dielectric sphere using this method. Let us assume that a dielectric sphere of radius R having uniform dielectric constant is placed in vacuum. If we consider the composite medium involving the sphere and its surrounding, then $\delta\epsilon \neq 0$ and constant inside the dielectric sphere and vanishes outside, while $\epsilon^{(0)} = 1$. Note that $\delta\mu = 0$ both inside and outside the sphere. Then from Eq. (5.66)

$$\frac{\epsilon^* \cdot \mathbf{A}_{sc}^{(1)}}{D_0} = \frac{k^2}{4\pi} \epsilon^* \cdot \epsilon_0 \frac{\delta\epsilon}{\epsilon^{(0)}} \int d^3x' e^{i\mathbf{q}\cdot\mathbf{x}'}. \tag{5.67}$$

Without loss of generality, we choose z-axis along $\mathbf{q}$. Using spherical coordinates, we can evaluate the integral

$$\frac{\epsilon^* \cdot \mathbf{A}_{sc}^{(1)}}{D_0} = \frac{k^2}{2}(\epsilon^* \cdot \epsilon_0)\delta\epsilon \int_0^R dr\, r^2 \int_{-1}^1 d\cos\theta\, e^{iqr\cos\theta}, \qquad (5.68)$$

$$= k^2(\epsilon^* \cdot \epsilon_0)\delta\epsilon \left(\frac{\sin qR - qR\cos qR}{q^3} \right). \qquad (5.69)$$

In the long wavelength limit or for scattering in the forward direction, as $q \to 0$, the bracketed term approaches the value $\frac{R^3}{3}$. Thus, the method of Born approximation applied to the problem of scattering by dielectric sphere leads to the following expression for differential scattering cross section:

$$\lim_{q\to 0} \frac{d\sigma}{d\Omega} = k^4 R^6 |\epsilon^* \cdot \epsilon_0)|^2 \left| \frac{\delta\epsilon}{3} \right|^2. \qquad (5.70)$$

For small values of $\delta\epsilon$, the above expression is indeed equivalent to that in Eq. (5.11).

5.2.2 Sunset and Blue Sky: Rayleigh Scattering Law

Lord Rayleigh was the first to present a quantitative explanation of the blueness of the sky away from the Sun and the redness of the sunset. We have already developed the necessary theoretical framework to address this natural phenomenon. We assume that the earth's atmosphere consists of dilute ideal gas and the gas molecules are randomly distributed in space.[12] It is quite reasonable to assume that the magnetic moments of most gas molecules are negligible so that the scattering is purely electric dipolar in character.

The variations in ϵ are due to the granular nature of the distribution of individual molecules in space. The dipole moment of the jth molecule located at $\mathbf{x}_j$ is given by $\mathbf{p}_j = \gamma_{mol}\mathbf{E}(\mathbf{x}_j)$; $\mathbf{E}(\mathbf{x}_j)$ being the incident electric field at $\mathbf{x}_j$ and γ_{mol}, the molecular polarizability.[13] The variation in ϵ in space can be expressed as

$$\delta\epsilon(\mathbf{x}) = 4\pi \sum_j \gamma_{mol}\delta(\mathbf{x} - \mathbf{x_j}). \qquad (5.71)$$

In the first Born approximation, we write down the differential scattering cross section using Eqs. (5.66) and (5.56) as

---

[12] This is only a rough approximation. A more rigorous treatment should take into account the density fluctuations in the gas. For details, see Jackson, Classical Electrodynamics, Third Edition.

[13] The molecular polarizability is defined as the ratio of the average dipole moment of a molecule to the electric field applied at the molecule.

$$\frac{d\sigma}{d\Omega} = \left| k^2 \int d^3x' e^{i\mathbf{q}\cdot\mathbf{x}'} \left[\boldsymbol{\epsilon}^* \cdot \boldsymbol{\epsilon}_0 \sum_j \gamma_{\text{mol}} \delta(\mathbf{x}' - \mathbf{x_j}) \right] \right|^2 ,$$

$$= k^4 |\gamma_{\text{mol}}|^2 |\boldsymbol{\epsilon}^* \cdot \boldsymbol{\epsilon}_0|^2 \sum_i \sum_j e^{i\mathbf{q}\cdot(\mathbf{x_i} - \mathbf{x_j})} . \tag{5.72}$$

We easily identify the double summation term with the structure factor $\mathcal{F}(\mathbf{q})$. Since the molecules are assumed to be randomly distributed in space, $\mathcal{F}(\mathbf{q}) = N$, the total number of molecules, say per unit volume, contributing to the scattering. For dilute gas, the dielectric constant ϵ is related to the molecular polarizability as follows: $\epsilon \simeq 1 + 4\pi N \gamma_{\text{mol}}$. Therefore, the total scattering cross section per molecule is

$$\sigma_1 = k^4 |\gamma_{\text{mol}}|^2 \int d\Omega |\boldsymbol{\epsilon}^* \cdot \boldsymbol{\epsilon}_0|^2 . \tag{5.73}$$

We have already encountered similar integral for electric dipole scattering. Evaluating the integral and replacing γ_{mol}, we have

$$\sigma_1 = \frac{k^4}{2} |\gamma_{\text{mol}}|^2 \int d\Omega \, (1 + \cos^2\theta) \simeq \frac{8\pi}{3} k^4 \frac{|\epsilon - 1|^2}{16\pi^2 N^2} \simeq \frac{k^4}{6\pi N^2} |\epsilon - 1|^2 . \tag{5.74}$$

Radiation incident on the medium is effectively scattered away from the forward direction. In other words, scattering to the sides leads to attenuation of radiation in the forward direction. Since σ_1 represents the power scattered per molecule (or the energy 'loss' per unit time per molecule) per unit incident energy flux, the fractional loss of intensity (I) for traveling a distance dx in the forward direction is $\frac{dI}{I} = -N\sigma_1 \, dx$. The incident beam intensity can be written as $I(x) = I_0 e^{-\alpha x}$, where $\alpha = N\sigma_1$ is called the attenuation coefficient. Therefore

$$\alpha = \frac{k^4}{6\pi N} |\epsilon - 1|^2 . \tag{5.75}$$

Equations (5.74) and (5.75) are known as Rayleigh scattering law, describing the incoherent scattering of radiation by randomly distributed gas molecules, each acting as dipole scatterers. The side scattering or the attenuation in the forward direction is proportional to the fourth power of frequency or inversely proportional to the fourth power of wavelength of incident radiation. This means, for the visible range of radiation, red is attenuated less than blue or in other words, blue is scattered to the side more than red. During the sunset, sun rays travel a longer distance through the atmosphere. The transmitted beam reaching the observer's eyes becomes increasingly red with distance traveled while the observer looking in the transverse direction to the incident beam will detect predominantly (high frequency) blue radiation. That is why the sky is blue when observed sufficiently away from the sun, while the sun looks red during sunset.

5.3 Problems

1. Consider two halves of a perfectly conducting shell of radius R separated by a very thin insulating ring. An alternating voltage is applied between the two halves so that the potentials on the two halves at any instant are $\pm V_0 \cos \omega t$. Find the angular distribution of radiated power and the total radiated power from the sphere, in the long wavelength limit.

2. Find out the long wavelength scattering cross section of a perfectly conducting sphere of radius R, summed over outgoing polarizations for an incident linearly polarized wave as a function of the angle θ between incident and scattering wave propagation directions $\hat{n}_0$ and $\hat{n}$, respectively.

3. One can calculate the differential scattering cross section of a point charge using the electric dipole scattering formula. Consider an electron with charge e and mass m without any 'spin' magnetic moment. Calculate the differential scattering cross section for an incident wave by the electron.

4. Consider an incident electromagnetic wave with elliptic polarization, described by the polarization vector

$$\epsilon = \frac{1}{\sqrt{1+r^2}} \left(\epsilon_+ + r e^{i\alpha} \epsilon_- \right) .$$

The wave is scattered by a perfectly conducting sphere of radius R. Calculate the angular distribution of scattering cross section, in spherical coordinates (r, θ, φ), in the long wavelength limit.

5. Calculate the differential scattering cross section for a plane electromagnetic wave of angular frequency ω incident normally on a perfectly conducting cylinder of radius R, in the following cases: (1) incident polarization parallel to the symmetry axis of the cylinder; (2) incident polarization perpendicular to the symmetry axis of the cylinder.

Chapter 6
Special Relativity and Fourier Transform Theory for Electrodynamics

Any advanced course in electromagnetism requires a fair understanding of the special theory of relativity and various concepts of Fourier analysis. Except for these, one has to work with various kinds of special functions in such a course. This chapter will present a brief introduction to the necessary mathematical concepts and specify some important results whose proof will be left to the readers to verify.

6.1 Lorentz Boosts

Special relativity assumes particular kinds of reference frames called inertial reference frames. The particular property which defines inertial frames is their uniform, linear motion (moving with uniform relative velocity) with respect to other inertial frames. All frames of references that are moving with uniform relative velocity with respect to each other are called a set of inertial frames. Einstein's postulates about special relativity are as follows:

1. identical experiments carried out in different inertial frames produce identical results.
2. The speed of light, propagating in vacua, is independent of inertial frames and its universal value is written as c.

Most of the special relativity can be deduced from the application of the above two postulates in an intelligent way. In this chapter, we will not utilize this method, rather we will propose the important results which we require for advanced electrodynamics.

The Lorentz transformations connect the spacetime coordinates of an event, called P, from two inertial frames traditionally called K and K'. The picture given in Fig. 6.1 illustrates the two reference frames and the event P. The spacetime coordinates of P in the K frame are specified by (x, y, z, t) and the coordinates of the same event

© Springer Nature Singapore Pte Ltd. 2021
K. Bhattacharya and S. Mukhopadhyay, *Introduction to Advanced Electrodynamics*,
https://doi.org/10.1007/978-981-16-7802-8_6

Fig. 6.1 The two inertial frames K and K' and their relative motion. Here, O and O' are the origins of the two reference frames. Here, P specifies the dot near it. The dot stands for an event that can be observed using both the reference frames K and K'

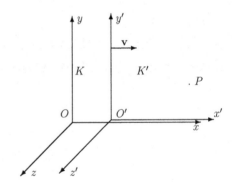

in the K' frame are given by (x', y', z', t'). Suppose the frame K' moves along the direction of common $x - x'$ axes with a speed v. If the frames coincided with each other at $t = t' = 0$, then the transformations for Lorentz boost[1] are:

$$x' = \gamma(x - vt),$$
$$t' = \gamma\left(t - \frac{v}{c^2}x\right),$$
$$y' = y,$$
$$z' = z,$$

$$(6.1)$$

where

$$\gamma \equiv \frac{1}{\sqrt{1 - \beta^2}}, \quad \text{and} \quad \boldsymbol{\beta} \equiv \frac{\mathbf{v}}{c}. \tag{6.2}$$

In the above expression, $\beta^2 \equiv \boldsymbol{\beta} \cdot \boldsymbol{\beta}$. From the principle of relativity, we can easily infer that

$$x = \gamma(x' + vt'),$$
$$t = \gamma\left(t' + \frac{v}{c^2}x'\right),$$
$$y = y',$$
$$z = z',$$

$$(6.3)$$

where we interchange the primed coordinates and the unprimed coordinates from Eq. (6.1) with a simultaneous change in the sign of the relative velocity of the frames. In the present case, the frame K moves with a velocity $-v$ along the common $x - x'$ axes and the frame K' is assumed to be at rest.

---

[1] The specific transformations we are dealing with are called Lorentz boost. Except a Lorentz boost there can be rotations which also is a part of Lorentz transformations. We will briefly discuss about rotations later in this chapter.

Now, we first specify an incomplete definition of 4-vectors in special relativity. The complete definition will appear soon in this chapter. Any four components which transform like the position and time coordinates under a Lorentz boost are supposed to compose a 4-vector. The simplest 4-vector is then written as x^μ where the Greek index μ runs from zero to three. In our notation, we will write

$$x^\mu \equiv (x^0, x^1, x^2, x^3),$$

where $x^0 = ct$ and $x^1 = x$, $x^2 = y$ and $x^3 = z$. Here, x^0 is ct and not purely time due to dimensional reason, as all the 4-vector components must have the same physical dimension. The Lorentz boost formulas can now be written in terms of the x^μ components as:

$$\begin{aligned}
x'^0 &= \gamma \left(x^0 - \beta x^1 \right), \\
x'^1 &= \gamma (x^1 - \beta x^0), \\
x'^2 &= x^2, \\
x'^3 &= x^3.
\end{aligned} \tag{6.4}$$

The inverse Lorentz boost formulas can easily be obtained by the application of the principle of relativity. The above transformations are written generally as

$$\begin{aligned}
x'^\mu &= x'^\mu(x^0, x^1, x^2, x^3), \\
&= x'^\mu(x^\nu),
\end{aligned} \tag{6.5}$$

form an important part of the total transformations which constitute the Lorentz transformations. The other set of Lorentz transformations, rotations, can also be written down as above. The important point to note is that Lorentz transformations are linear coordinate transformations connecting the spacetime coordinates of an event as observed by two inertial observers.

In this book, $P(\mathbf{x}, t)$ will specify an event which happens at spatial position specified by the Cartesian coordinates x, y, z and at time t in frame K. Similarly, $P(\mathbf{x}', t')$ specifies the same event in frame K'. In a manifestly relativistic description, where space and time produces a 4-dimensional world, we can also describe the event as $P(x)$ where x now stands for the four numbers $(\mathbf{x}, t)$. The notation used looks a bit sloppy because x represents different objects simultaneously. Sometimes it means purely the x-coordinate and sometimes it means the four numbers $(\mathbf{x}, t)$. We will still use this convention as it is economical and misunderstanding does not arise because the meaning of the quantity becomes explicit from the context.

6.2   More Formal Way to Define 4-Vectors and Other Tensors

In the previous section, we gave an incomplete definition of 4-vectors. Our definition was purely based on the Lorentz boost formulas. In this section, we formally define 4-vectors. A formal analysis of the general structure yields two kinds of vectors and many other indexed objects called tensors. So we start with the simplest construction which is the scalar function of coordinates. A function of coordinates as $s(x^\mu)$ is said to be a scalar function when it remains the same under a coordinate transformation as given in Eq. (6.5). Although we have only introduced the Lorentz boosts, whatever we discuss henceforth are true for any Lorentz transformations. Ideally, the Lorentz transformations do include rotations in three space, Lorentz boosts and products of boosts and rotations. We will formally define the Lorentz transformations later. Under a Lorentz transformation if $x^\mu \rightarrow x'^\mu$ the scalar function behaves as

$$s(x'^\mu) = s(x^\mu) \,. \tag{6.6}$$

Next, we present the standard definition of contravariant 4-vectors.

6.2.1   Contravariant Vectors

If under a coordinate transformation, as given in Eq. (6.5), the four components $v^\alpha(x)$ (corresponding to $\alpha = 0$, 1, 2, 3) transforms to $v'^\alpha(x)$ as

$$v'^\alpha(x) = \sum_{\beta=0}^{3} \frac{\partial x'^\alpha}{\partial x^\beta} v^\beta(x) \,, \tag{6.7}$$

then $v^\alpha(x)$ are supposed to be components of a contravariant vector. The reader can easily see that during a coordinate transformation the components like dx^μ transforms like a contravariant vector. The first thing to note about the above definition is that both left and right-hand sides are written as functions of x. Usually one expects $v'^\alpha(x')$ on the left-hand side as coordinates have transformed. However in this case we have re-expressed x' as a function of x using Eq. (6.5) on the left-hand side of the above equation and consequently both sides of the above equation are functions of x. In the same spirit, one could have expressed both sides of the last equation as functions of x'. The other point which must be noted down is about the position of the indices β on the right-hand side of the last equation. The index β is repeated once and it is a dummy index, there is no trace of β on the left-hand side. In such cases, we will apply Einstein's summation convention which allows us to sum over repeated indices. In this convention, one does not require to write the summation explicitly, the repetition of an index is enough to specify that a summation over that index must take place. We will specify the summation convention more precisely as we proceed, for the time being, we give the compact form of the last equation

$$v'^{\alpha}(x) = \frac{\partial x'^{\alpha}}{\partial x^{\beta}} v^{\beta}(x) \,, \tag{6.8}$$

using the summation convention. Henceforth, we will apply this convention and we will specify when we do otherwise.

For the Lorentz boost transformations, one can easily calculate the quantities $\frac{\partial x'^{\alpha}}{\partial x^{\beta}}$ as because we know the relation between x'^{α} and x^{β} (as given in Eq. (6.4)). We name this factors as

$$\Lambda^{\mu}{}_{\nu} \equiv \frac{\partial x'^{\mu}}{\partial x^{\nu}} \,, \tag{6.9}$$

where all $\Lambda^{\mu}{}_{\nu}$ factors are independent of spacetime coordinates and are purely functions of the constant relative velocity v along the common $x - x'$ axes. If one chooses to represent the upper index as the row index and the lower index as the column index, one can write down a matrix form of $\Lambda^{\mu}{}_{\nu}$ as

$$\Lambda = \begin{pmatrix} \gamma & -\gamma\beta & 0 & 0 \\ -\gamma\beta & \gamma & 0 & 0 \\ 0 & 0 & 1 & 0 \\ 0 & 0 & 0 & 1 \end{pmatrix} \,, \tag{6.10}$$

which is a 4×4 matrix with constant matrix elements. In the matrix language, the contravariant vector transformation rule becomes

$$v' = \Lambda v \,, \tag{6.11}$$

where v and v' represents 4×1 column matrix where the row number is given by the contravariant index. Most of the 3-vectors which we have used in physics till now, as the radius vector, momentum, electric current etc., can be successfully generalized to four dimensional contravariant 4-vectors.

6.2.2 Covariant Vectors

In physics, we have also come across vectors as $\frac{\partial s}{\partial x^i}$ where s is scalar function. In this case, the Latin index i can take values between 1 and 3. Here, we make a clear distinction between the Greek indices like $\alpha, \beta, \ldots$ which can assume values ranging from 0 to 3. The Latin indices $i, j, \ldots$ will only range from 1 to 3. This convention will be followed throughout the book. Vectors of the form $\frac{\partial s}{\partial x^i}$ generally appears as a component of the gradient of the scalar function s. Gradient operations on scalars produce 3-vectors which transforms under a coordinate transformation as

$$\frac{\partial s}{\partial x'^i} = \frac{\partial x^j}{\partial x'^i} \frac{\partial s}{\partial x^j},$$

where a summation over index j is done according to the summation convention. The above transformation of $\left(\frac{\partial s}{\partial x^j}\right)$ is not how a contravariant vector transforms. These gradient vectors are actually different kind of vectors which can be generalized to covariant vectors. If any four quantities $v_\alpha(x)$ (corresponding to $\alpha = 0, 1, 2, 3$) transforms under a spacetime transformation as given in Eq. (6.5) as

$$v'_\alpha(x) = \frac{\partial x^\beta}{\partial x'^\alpha} v_\beta(x), \tag{6.12}$$

then those four quantities are the components of a covariant vector. In our language, the Lorentz transformation properties of the components of vectors define the nature of the vector. In the present case, one can define a matrix $\Lambda_\nu{}^\mu$ as

$$\Lambda_\nu{}^\mu \equiv \frac{\partial x^\mu}{\partial x'^\nu}, \tag{6.13}$$

where again the upper index is identified with row number and the lower index is identified with column number. One can show that

$$\Lambda_\mu{}^\tau \Lambda^\mu{}_\nu = \delta^\tau_\nu, \tag{6.14}$$

which implies that in the matrix notation one can write the last equation as $\Lambda^{-1}\Lambda = 1$, where

$$(\Lambda^{-1})^\mu{}_\nu = \Lambda_\nu{}^\mu.$$

In the present case

$$\delta^\mu_\nu = \begin{cases} 1 \text{ if } \mu = \nu \\ 0 \text{ if } \mu \neq \nu \end{cases} \tag{6.15}$$

is called the Kronecker delta symbol. In matrix notation, the covariant transformation rule as given in Eq. (6.12) is written as

$$v' = v\Lambda^{-1}, \tag{6.16}$$

where one must notice that the covariant 4-vector is treated as a 1×4 row vector.

6.2.3 A Note on the Einstein Summation Convention

Before we define the other useful quantities in this chapter let us specify the properties of Einstein's summation convention. We had previously said that terms (appearing in products) containing repeated indices, in a formula, must be added up. We have used the convention in Eqs. (6.8), (6.12) and even in Eq. (6.14). We want to specify that the summation convention requires some more specification. The indices which are repeated must be repeated in such a way that one index must stand as a superscript and the other as a subscript, as for the index μ in Eq. (6.14) or the index β in Eq. (6.12). If one of the terms having a repeated index appears in a partial (or ordinary) derivative, then one has to be careful and remember that a covariant vector is defined as $v_\alpha = \frac{\partial s}{\partial x^\alpha}$ where s is a scalar. This implies that $\frac{\partial}{\partial x^\alpha}$ can be interpreted as a vector operator (we can for the time being forget about the scalar as it does not transform under Lorentz transformations) with an index appearing as the subscript and one writes such an operator as $v_\alpha = \partial_\alpha$. If this latter point is taken into account, then we see that the index β in Eq. (6.8) appear as a subscript in the first term (as $\partial_\beta x'^\alpha$) on the right-hand side and the other β appears in the superscript and the summation convention as stated in this subsection holds. Henceforth, we will always follow the above stipulated rules when we employ the Einstein summation convention.

6.3 Tensors

Once we have defined the 4-vectors, it becomes much easier to define the higher rank tensors. The definition of tensors is based on the generalization of the definitions of the vectors. The 4-vectors are rank one tensors.

6.3.1 Contravariant Tensors

If a two indexed quantity $T^{\mu\nu}$ which has sixteen components transform under a coordinate transformation, as given in Eq. (6.5), in the following way

$$T'^{\mu\nu} = \frac{\partial x'^\mu}{\partial x^\alpha} \frac{\partial x'^\nu}{\partial x^\beta} T^{\alpha\beta}, \qquad (6.17)$$

then the sixteen components of $T^{\mu\nu}$ define a second rank contravariant tensor. Here, we are using the summation convention on the right-hand side where α and β are repeated. In a similar fashion, one can define a third rank contravariant tensor as

$$T'^{\mu\nu\tau} = \frac{\partial x'^\mu}{\partial x^\alpha} \frac{\partial x'^\nu}{\partial x^\beta} \frac{\partial x'^\tau}{\partial x^\gamma} T^{\alpha\beta\gamma}. \qquad (6.18)$$

The reader now will be able to define for himself/herself how to define an nth rank contravariant tensor. Now the reader also must be able to see that contravariant vectors are contravariant tensors of rank one and scalars are contravariant tensors of rank zero.

Before we end our discussion on contravariant tensors, let us present an important quantity which we will often use in our book. This is the antisymmetric quantity written $\varepsilon^{\alpha\beta\mu\nu}$. We assume that this four indexed quantity is a contravariant tensor of rank four. The symbol representing the tensor is called the Levi-Civita symbol in four dimensions. Formally, the tensor is defined as

$$
\varepsilon^{\alpha\beta\mu\nu} =
\begin{cases}
1 \text{ if } \alpha, \beta, \mu, \nu \text{ are even permutations of } 0, 1, 2, 3. \\
-1 \text{ if } \alpha, \beta, \mu, \nu \text{ are odd permutations of } 0, 1, 2, 3. \\
0 \text{ if there is any repeatation of indices} .
\end{cases}
\tag{6.19}
$$

One can easily show that under a Lorentz transformation, as given in Eq. (6.5), the transformed tensor in the K' frame have the same components as $\varepsilon^{\alpha\beta\mu\nu}$ had in the K frame. In such cases, we will not call the transformed Levi-Civita in four dimensions as $\varepsilon'^{\alpha\beta\mu\nu}$, we will simply write the transformed Levi-Civita as the old one like $\varepsilon^{\alpha\beta\mu\nu}$.

6.3.2 Covariant Tensors

If under a Lorentz transformation, as given in Eq. (6.5), a two indexed quantity $T_{\mu\nu}$ gets transformed as

$$
T'_{\mu\nu} = \frac{\partial x^\alpha}{\partial x'^\mu} \frac{\partial x^\beta}{\partial x'^\nu} T_{\alpha\beta} ,
\tag{6.20}
$$

then the sixteen components $T_{\mu\nu}$ define a second rank covariant tensor. One can now generalize the above result for nth rank covariant tensors in a mechanical way (following the example of the third rank contravariant tensor in the last subsection). The above definition shows that a covariant vector is a covariant tensor of rank one and one may like to think a scalar as a covariant tensor of rank zero.

A particularly important symmetric, second rank covariant tensor is called the metric tensor, represented by $\eta_{\mu\nu}$. We can motivate the metric tensor in the following way. If any event P has spacetime coordinates as (x, y, z, t) in the K frame, then in special relativity, one can show that the quantity

$$
s^2 = c^2 t^2 - x^2 - y^2 - z^2 ,
$$

remains a scalar invariant under a Lorentz boost, by which we mean that in K' frame the corresponding coordinates of the event P (x', y', z', t') satisfies

$$
s'^2 = c^2 t'^2 - x'^2 - y'^2 - z'^2 = s^2 .
$$

Here c is the speed of light in vacuum. From the above result, we can go one step further. If dt, dx, dy and dz are the coordinate separations of two infinitesimally close events P and $\tilde{P}$ in the K frame then we expect that the quantity defined as ds^2, where

$$ds^2 = c^2 dt^2 - dx^2 - dy^2 - dz^2,$$

will also remain invariant under a Lorentz boost.[2] This fact is trivial to prove if one has shown the invariance of s^2 under Lorentz transformations. The proof relies on the fact that dt and the components of $d\mathbf{x}$ transforms exactly as t and the components of $\mathbf{x}$ under Lorentz transformations. Here, $d\mathbf{x} \equiv (dx, dy, dz)$ and $\mathbf{x} \equiv (x, y, z)$. Using the Einstein summation convention, one can write the square of the infinitesimal separation of two events in special theory of relativity formally as

$$ds^2 = \eta_{\mu\nu} dx^\mu dx^\nu, \tag{6.21}$$

where the components of the tensor $\eta_{\mu\nu}$ are given as the elements of the corresponding matrix as

$$\eta = \begin{pmatrix} 1 & 0 & 0 & 0 \\ 0 & -1 & 0 & 0 \\ 0 & 0 & -1 & 0 \\ 0 & 0 & 0 & -1 \end{pmatrix}. \tag{6.22}$$

In writing the above matrix, we have used the first covariant index as the row number and the second covariant index as the column number. Although for the case of the metric tensor, such fixed identification of row or column is not mandatory, as the tensor is a symmetric one. The reader must note the metric tensor is ideally defined as a covariant tensor and its components remain the same in all inertial frames. In this subsection, we have just claimed that $\eta_{\mu\nu}$ is a covariant tensor of rank two, but we have still not described its transformation properties. The fact that $\eta_{\mu\nu}$ is actually a covariant tensor of rank two can be shown by the application of the quotient theorem of tensor analysis. We present the quotient theorem after the next subsection after discussing mixed tensors.

6.3.3 The Mixed Tensor

One can have tensors that are not purely contravariant or covariant, and these kinds of tensors are called mixed tensors. If under a Lorentz transformation, as given in Eq. (6.5), a two indexed quantity $T^\mu{}_\nu$ transform as

---

[2] Here $ds^2 = (ds)^2$ and similarly for the other coordinates, as $dx^2 = (dx)^2$.

$$T'^{\mu}_{\nu} = \frac{\partial x'^{\mu}}{\partial x^{\alpha}} \frac{\partial x^{\beta}}{\partial x'^{\nu}} T^{\alpha}_{\beta} , \qquad (6.23)$$

then the sixteen components T^{μ}_{ν} define a mixed tensor with one contravariant index and one covariant index. The reader must see how the summation convention is applied for index α and index β in the transformation law. This is perhaps the simplest mixed tensor one can construct. In the next level, we can have a more complicated mixed tensor. If a quantity as $T^{\mu\nu}_{\kappa}$ transforms under the Lorentz transformation as

$$T'^{\mu\nu}_{\kappa} = \frac{\partial x'^{\mu}}{\partial x^{\alpha}} \frac{\partial x'^{\nu}}{\partial x^{\beta}} \frac{\partial x^{\gamma}}{\partial x'^{\kappa}} T^{\alpha\beta}_{\gamma} , \qquad (6.24)$$

then these 64 quantities $T^{\mu\nu}_{\kappa}$ constitute another mixed tensor with two contravariant indices and one covariant index. One important thing to note about the mixed tensors is that the lower index is not written just below the upper indices. The lower index appears a bit removed to the right. The reason for such representation will be made clear later in this chapter. Using the example above, one can now write down the transformation law of a mixed tensor with one contravariant index and two covariant indices, and it should be like

$$T'^{\mu}_{\nu\kappa} = \frac{\partial x'^{\mu}}{\partial x^{\alpha}} \frac{\partial x^{\beta}}{\partial x'^{\nu}} \frac{\partial x^{\gamma}}{\partial x'^{\kappa}} T^{\alpha}_{\beta\gamma} . \qquad (6.25)$$

The more complex mixed tensors with n contravariant indices and m covariant indices can now be made by simply generalizing the rules to construct the above examples. One of the most common mixed tensors is the Kronecker delta symbol as given in Eq. (6.15). In this case, one must note that the covariant index is not pushed to the right it stands below the contravariant index, unlike the way we have defined the mixed tensors in this subsection. The reason for such a rule change will be discussed after we introduce the inner product and raising and lowering operation of indices involving the metric tensor. The matrix representing the Kronecker delta symbol is simply the 4×4 unit matrix, where the contravariant index specifies the rows and the covariant index specifies the columns. The Kronecker delta symbol is a mixed tensor whose components remain the same in all reference frames connected via Lorentz transformations and so we never use δ'^{μ}_{ν} to represent the Kronecker delta in the K' frame. The Kronecker delta tensor is δ^{μ}_{ν} in all inertial frames. The proof is simple and left as an exercise for the readers.

6.3.4   Quotient Theorem and the Metric Tensor

Before we proceed, let us describe the quotient theorem in tensor analysis. The quotient theorem states that if we know $T^{\mu\nu}_{\alpha\beta}$ to be a mixed tensor and we have a mathematical quantity called A, whose transformation properties we do not know a priory, and we also know that $B^\mu{}_\alpha$ is a mixed tensor such that $B^\mu{}_\alpha = AT^{\mu\nu}{}_{\alpha\beta}$ holds true, then A must be a tensor and the indices which specifies it are as $A^\nu{}_\beta$. The reader can easily generalize the theorem by making $T^{\mu\nu}{}_{\alpha\beta}$ more complex by increasing indices in the top and bottom and similarly giving some other index structure to $B^\mu{}_\alpha$. In this case, A will turn out to be a tensor such that the indices on both sides of the tensor equation $B^{\mu\cdots}{}_{\alpha\cdots} = AT^{\mu\nu\cdots}{}_{\alpha\beta\cdots}$ matches. The dots specify possible unspecified indices. The proof of this theorem is simple, one has to keep in mind how the left-hand side and right-hand side transforms under a Lorentz transformation. The transformation laws from both sides will give the index structure of A.

From the definition of the metric in Eq. (6.21), we see that the left-hand side of the equation is a scalar, whereas there is a second rank tensor $dx^\mu dx^\nu$ (it transforms as a second rank contravariant tensor under a Lorentz transformation) on the right-hand side multiplied by $\eta_{\mu\nu}$. These facts compel us to assume (via the quotient theorem) that $\eta_{\mu\nu}$ must be a covariant tensor of rank two.

The metric is by definition a covariant tensor whose matrix representation is given in Eq. (6.22). One can define an inverse of metric tensor $(\eta^{-1})^{\mu\nu}$ via the relation $\eta_{\mu\nu}(\eta^{-1})^{\nu\kappa} = \delta^\kappa_\mu$. The last relation shows that $(\eta^{-1})^{\nu\kappa}$ is a contravariant tensor of rank two. In general, one omits the inverse sign explicitly and expresses the inverse of $\eta_{\mu\nu}$ simply by $\eta^{\nu\kappa}$. The contravariant metric has the same matrix representation as the metric tensor. The contravariant metric is the inverse of the metric tensor.

6.4 Inner Product and Raising and Lowering of Tensor Indices

From previous discussions, it is seen that the 4-vectors are identified by their components, a^μ or a_μ, depending on the way they transform under a Lorentz transformation. In general, the 4-vector can also be specified without any index, as simply a, where one is free to interpret its components as contravariant or covariant depending on a particular situation. This is more like expressing a Euclidean 3-vector as simply $\mathbf{a}$ instead of specifying the 3-vector in its component form as a^i or a_i. When one represents the 3-vector as $\mathbf{a}$, one knows that actually the numbers which define it are a triplet as a^i or a_i where i runs from 1 to 3. In four dimensions, a 4-vector is simply written as a by which one means (a^0, a^i) or (a_0, a_i). In this book, we will often represent a 4-vector without its component form.

The inner product, sometimes also called the scalar product of two 4-vectors, a and b is defined as

$$a \cdot b \equiv a^\mu b_\mu = a_\mu b^\mu . \qquad (6.26)$$

The quantity $(a \cdot b)$ is a Lorentz scalar. To show this, we assume that a' and b' are the transformed 4-vectors in the K' frame and consequently

$$a' \cdot b' = a'^\mu b'_\mu = \frac{\partial x'^\mu}{\partial x^\alpha} \frac{\partial x^\beta}{\partial x'^\mu} a^\alpha b_\beta \,.$$

By chain rule of partial differentiation, we have

$$a' \cdot b' = \frac{\partial x^\beta}{\partial x^\alpha} a^\alpha b_\beta = \delta^\beta_\alpha a^\alpha b_\beta = a^\alpha b_\alpha = a \cdot b \,.$$

The expression in Eq. (6.26) shows that summation of the dummy index μ produces a quantity without any index, hence a scalar. Such omissions of summed up indices are technically called contractions and formally an inner product is the contraction of one tensor index. For higher rank tensors, one can contract multiple indices producing a lower rank tensor. If $A^{\mu\nu}$ is a second rank contravariant tensor and b_μ is covariant 4-vector then a contraction on μ in the product is $A^{\mu\nu} b_\mu = a^\nu$ where a^ν is a contravariant 4-vector. The reader can easily prove the above statement from the transformation property of the tensors.

The invariant ds^2 is a Lorentz scalar and it is made up of two 4-vectors each of which are dx as shown in Eq. (6.21). We have seen that we can obtain a Lorentz scalar from 4-vectors via the inner product, so we can interpret

$$ds^2 = dx \cdot dx = dx^\mu dx_\mu \,.$$

When the above expression is compared with Eq. (6.21), it is seen that one can write $dx_\mu = \eta_{\mu\nu} dx^\nu$, showing that one can actually lower a Lorentz index by contraction with the metric tensor. The above property is a general one and for any arbitrary 4-vector a one can write

$$a_\mu = \eta_{\mu\nu} a^\nu \,. \tag{6.27}$$

From the above relation, one can obtain

$$\eta^{\alpha\mu} a_\mu = \eta^{\alpha\mu} \eta_{\mu\nu} a^\nu = \delta^\alpha_\nu a^\nu = a^\alpha \,,$$

showing that proper contraction of the inverse metric with a covariant 4-vector can raise the Lorentz index of the 4-vector. As a result, $a \cdot b \equiv a^\mu b_\mu$ can also be written as $a \cdot b = \eta_{\mu\nu} a^\mu b^\nu = a_\nu b^\nu$ justifying the form of $a \cdot b$ given in Eq. (6.26). As $a \cdot b = \eta_{\mu\nu} a^\mu b^\nu$, we can write

$$a \cdot b = a^0 b^0 - \sum_{i=1}^{3} a^i b^i = a^0 b^0 - \mathbf{a} \cdot \mathbf{b} \,,$$

where $\mathbf{a}$ and $\mathbf{b}$ are the Euclidean 3-vectors and $\mathbf{a} \cdot \mathbf{b}$ is the scalar product of two ordinary 3-vectors using the Euclidean space metric.

6.5 Lorentz Transformations

We have seen that under Lorentz boost, the square of the interval ds^2, as given in Eq. (6.21), remains invariant. We generalize this property as a definition of Lorentz transformations. Lorentz transformations, as $x^\mu \to x'^\mu$, are those coordinate transformations which keeps the square of the infinitesimal interval between two events invariant. Lorentz boosts are then naturally a part of Lorentz transformations but there can be other kind of transformations which can keep $ds^2 = dx \cdot dx$ invariant. Suppose under a transformation $x^\mu \to x'^\mu$, we have $dx^0 = dx'^0$ and $\mathbf{dx}^2 = \mathbf{dx}'^2$ then[3] one can easily show that ds^2 remains invariant under such kind of transformations. These kind of transformations do not change the magnitude of Euclidean 3-vectors and keeps time intervals invariant which implies that such transformations are simply rotations in three-dimensional space. Rotations are orthogonal transformations and symbolically the rotated coordinates (K' coordinates) are related to the original coordinates (K frame coordinates) via

$$dx'^i = R^i{}_j dx^j , \quad dx'^0 = dx^0 , \tag{6.28}$$

where $R^i{}_j$ specifies rotation. In the above expression, we have a summation on j which runs from 1 to 3. Writing all the summations explicitly we have in this notation

$$\sum_{i=1}^{3} dx'^i dx'^j = \sum_{j,k=1}^{3} \left(\sum_{i=1}^{3} R^i{}_j R^i{}_k \right) dx^j dx^k = \sum_{j=1}^{3} dx^j dx^j ,$$

which shows that in general

$$\sum_{i=1}^{3} R^i{}_j R^i{}_k = g_{jk} , \tag{6.29}$$

where g_{jk} is the Euclidean metric whose matrix representation is

$$g = \begin{pmatrix} 1 & 0 & 0 \\ 0 & 1 & 0 \\ 0 & 0 & 1 \end{pmatrix} . \tag{6.30}$$

In matrix notation, one can write Eq. (6.29) as

$$R^\mathsf{T} R = R R^\mathsf{T} = 1 , \tag{6.31}$$

---

[3] Where $\mathbf{dx}^2 = \mathbf{dx} \cdot \mathbf{dx} = \sum_{i=1}^{3} dx^i dx^i$.

where superscript T specifies transpose operation and 1 on the right hand side represents the 3×3 unit matrix. The rotation matrix corresponding to a rotation by angle θ around the z-axis is given by

$$R = \begin{pmatrix} \cos\theta & -\sin\theta & 0 \\ \sin\theta & \cos\theta & 0 \\ 0 & 0 & 1 \end{pmatrix} . \tag{6.32}$$

A Lorentz transformation, connecting the coordinates of any event P in the frames K and K', as $x^\mu \to x'^\mu(x^\alpha)$ in the most general case can be a boost, represented by $\Lambda^\mu{}_\nu$, or a rotation, represented by $R^i{}_j$, or a product of the two operations in any order. In general, we will represent a Lorentz transformation as $L^\mu{}_\nu$. All the discussion on tensors presented before only assumed $\Lambda^\mu{}_\nu$ as Lorentz transformations but we can always use $L^\mu{}_\nu$ instead of only boost, in those discussions, and still everything presented remains valid. Under a Lorentz transformation any contravariant 4-vector transforms as

$$x'^\mu = L^\mu{}_\nu x^\nu .$$

As the square of the infinitesimal interval remains invariant under Lorentz transformations one can write

$$ds^2 = \eta_{\mu\nu} dx^\mu dx^\nu = \eta_{\alpha\beta} dx'^\alpha dx'^\beta ,$$

where we have used the fact that the metric tensor components are the same in all inertial frames. As a consequence of this and the fact that dx^μ transforms as a 4-vector under Lorentz transformations, we can write the above statement as

$$\eta_{\mu\nu} dx^\mu dx^\nu = (\eta_{\alpha\beta} L^\alpha{}_\mu L^\beta{}_\nu) dx^\mu dx^\nu .$$

Comparing both sides, we come to the conclusion that

$$\eta_{\mu\nu} = \eta_{\alpha\beta} L^\alpha{}_\mu L^\beta{}_\nu , \tag{6.33}$$

which becomes

$$\eta = L^{\mathrm{T}} \eta L . \tag{6.34}$$

If we represent the determinant of a matrix M as $|M|$ then the above matrix relation implies $|\eta| = |L^{\mathrm{T}} \eta L| = |L|^2 |\eta|$ and consequently

$$|L| = \pm 1 . \tag{6.35}$$

Lorentz transformations corresponding to $|L| = 1$ are called proper Lorentz transformations and the other case corresponds to improper Lorentz transformations. Moreover, setting $\mu = 0$ and $\nu = 0$ in Eq. (6.33), we get,

$$1 = (L^0{}_0)^2 - \sum_{i=1}^{3} (L^i{}_0)^2 \,,$$

which specifies that there can be two mutually disjoint possibilities as:

$$L^0{}_0 = \begin{cases} \geq 1 \,, & \text{Orthochronous Lorentz Transformation}\,, \\ \leq -1 \,, & \text{Non} - \text{orthochronous Lorentz Transformation}\,. \end{cases} \tag{6.36}$$

Lorentz boosts and rotations (in three dimensions) correspond to orthochronous, proper Lorentz transformations, whereas parity transformation, where the Lorentz transformation only reverses the sign of spatial coordinates as

$$(x^0, \mathbf{x}) \xrightarrow{\mathcal{P}} (x^0, -\mathbf{x})\,, \tag{6.37}$$

is an orthochronous, improper Lorentz transformation. The matrix corresponding to the parity transformation is

$$L = \begin{pmatrix} 1 & 0 & 0 & 0 \\ 0 & -1 & 0 & 0 \\ 0 & 0 & -1 & 0 \\ 0 & 0 & 0 & -1 \end{pmatrix}. \tag{6.38}$$

If one looks at the rotation transformation or boost, one can see that if someone continuously makes θ or β tend to zero $R^i{}_j$ or $\Lambda^\mu{}_\nu$ tends to the identity transformation. On the other hand, parity transformation does not depend on any continuous parameter and so one cannot continuously transform the parity transformation to an identity transformation. Parity is called a discrete Lorentz transformation as such a transformation cannot be continuously related to the identity transformation by making some parameters smaller and smaller. There can be another example of a discrete Lorentz transformation, called time reversal, which changes the sign of the time component of a 4-vector as

$$(x^0, \mathbf{x}) \xrightarrow{\mathcal{T}} (-x^0, \mathbf{x})\,. \tag{6.39}$$

The matrix corresponding to the time-reversal transformation is

$$L = \begin{pmatrix} -1 & 0 & 0 & 0 \\ 0 & 1 & 0 & 0 \\ 0 & 0 & 1 & 0 \\ 0 & 0 & 0 & 1 \end{pmatrix}, \tag{6.40}$$

showing that time-reversal transformation is a non-orthochronous, improper Lorentz transformation. In this book, we will mainly work with orthochronous, proper Lorentz transformations. The discrete transformations will, in general, not be referred to as Lorentz transformations, they will be specified by parity or time-reversal transformations.

6.5.1 Lorentz Boost with Arbitrary Velocity Direction

In the first section, we discussed the case of Lorentz boosts when $\mathbf{v}$ was parallel to the $x - x'$ axes direction. In this subsection, we will assume that the spatial axes of the coordinate systems K and K' remain parallel to each other like the previous case while the relative velocity makes an arbitrary angle with the x or x' axes direction. Figure 6.2 specifies the geometry of the frames. We define some Euclidean 3-vectors as

$$\mathbf{x}_{\|} \equiv (\mathbf{x} \cdot \widehat{\boldsymbol{\beta}})\widehat{\boldsymbol{\beta}}\,, \tag{6.41}$$

$$\mathbf{x}_{\perp} \equiv \mathbf{x} - \mathbf{x}_{\|} = \mathbf{x} - (\mathbf{x} \cdot \widehat{\boldsymbol{\beta}})\widehat{\boldsymbol{\beta}}\,, \tag{6.42}$$

where $\widehat{\boldsymbol{\beta}} = \boldsymbol{\beta}/\beta$ and $\boldsymbol{\beta}$ is defined in Eq. (6.2). We can break down any 3-vector into a part which is parallel to $\boldsymbol{\beta}$ and another part which is perpendicular to $\boldsymbol{\beta}$ such that $\mathbf{x} = \mathbf{x}_{\|} + \mathbf{x}_{\perp}$. Our strategy is to write the Lorentz boost relations in such a way that when relative velocity is parallel to $x - x'$-axes we can reproduce the results in Eq. (6.4). From Eq. (6.4), we also know that only $\mathbf{x}_{\|}$ gets transformed and $\mathbf{x}_{\perp}$ remains unaltered under a Lorentz boost. From these facts, we can infer that the time component transforms as $x'^0 = \gamma \left(x^0 - \boldsymbol{\beta} \cdot \mathbf{x}\right)$. In a similar fashion, we can say that $\mathbf{x}'_{\|} = \gamma(\mathbf{x}_{\|} - \boldsymbol{\beta}x^0)$ and $\mathbf{x}'_{\perp} = \mathbf{x}_{\perp}$. From these, one can write the expression of $\mathbf{x}' = \mathbf{x}'_{\|} + \mathbf{x}'_{\perp} = \mathbf{x}_{\perp} + \gamma(\mathbf{x}_{\|} - \boldsymbol{\beta}x^0)$. Using the expressions of $\mathbf{x}_{\|}$ and $\mathbf{x}_{\perp}$ as given in Eqs. (6.41, 6.42) one can write the general Lorentz boost transformation formulas as

$$\begin{aligned} x'^0 &= \gamma \left(x^0 - \boldsymbol{\beta} \cdot \mathbf{x}\right)\,, \\ \mathbf{x}' &= \mathbf{x} + (\gamma - 1)(\mathbf{x} \cdot \widehat{\boldsymbol{\beta}})\widehat{\boldsymbol{\beta}} - \gamma\boldsymbol{\beta}x^0\,. \end{aligned} \tag{6.43}$$

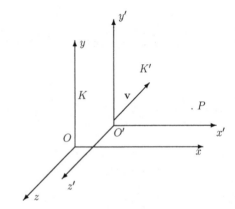

Fig. 6.2 The two inertial frames K and K' and their relative motion. Here, O and O' are the origins of the two reference frames. An arbitrary event P is specified by the dot near it. The relative velocity vector makes an arbitrary angle with the $x - x'$ direction

Before we move on to the next topic, some words on the utility of the above general result. If one assumes isotropy of space, then one is free to rotate the coordinate system in any possible way and consequently one can always rotate the K and K' coordinate systems such that $x - x'$-axes becomes parallel to the direction of the relative velocity. In such cases, only the result in Eq. (6.4) seems relevant. The results presented here become important only when there are other vector fields, like the electric or magnetic field, present in space. In the presence of vector fields, the orientation of the fields with respect to the coordinate axes becomes important and consequently arbitrary rotation of the coordinate axes does not remain a general option. In other words, vector fields do break the isotropy of space by their orientation, and in the presence of such fields, if one has to use the Lorentz boost formula, then one has to use the above result.

6.6 Why Tensors in Physics?

Suppose in frame K we can write a physical law in the tensorial form as

$$T^{\mu\nu\cdots}{}_{\alpha\beta\cdots} = 0 \, ,$$

where the dots specify more possible indices. How does the law look like in frame K'? In such a case, one has to transform the tensor $T^{\mu\nu\cdots}{}_{\alpha\beta\cdots}$ to the primed frame. The tensor transformation rules specifies that $T'^{\mu\nu\cdots}{}_{\alpha\beta\cdots} = 0$ in the primed frame, showing that the physical law looks the same in both the inertial frames. Actually, the tensor equation looks the same in all inertial frames connected to K frame via Lorentz transformations. Once we can write down a physical law in tensorial form, we know the form of the law in all inertial frames (although the tensor components change under Lorentz transformation). Such form invariant formulation of physical laws is often called covariant[4] formulation. In this book, we will like to cast all the known laws of electrodynamics in covariant form.

6.7 Brief Description of 4-Velocity, 4-Momentum and 4-Force for a Massive Particle

The position of a massive particle in spacetime is given by the 4-vector $x^\mu(\tau)$ where τ is the proper time parameter. If the particle remains at rest in any frame, then the particle is evolving only in time τ called the proper time and the invariant interval between two different temporal points is $ds^2 = c^2 d\tau^2$. So the proper time for any interval, separating two positions of the particle in spacetime, is defined as

---

[4] This covariance has nothing to do with covariant tensors.

$$d\tau = \frac{ds}{c} . \tag{6.44}$$

As ds^2 is Lorentz invariant so is proper time. Proper time does not transform under a Lorentz transformation. On the other hand, the coordinate time t changes under a Lorentz boost. For photons, which are massless localized energy packets, we have $ds^2 = 0$ for any temporal interval and consequently one cannot define the proper time for photons. In this book, we will not be dealing with quantized excitations of the electromagnetic field and consequently we will only be dealing with massive particles.

If we want to define 4-velocity of a massive particle, it becomes difficult to define it as dx^μ/dt, since in this case, the velocity does not transform as a 4-vector. On the the other hand, the quantity

$$u^\mu \equiv \frac{dx^\mu}{d\tau} , \tag{6.45}$$

do transform like a 4-vector, as dx^μ is a 4-vector and $d\tau$ is a scalar. As time is dilated in the K frame, where the particle is moving with 3-velocity $\mathbf{v}(= d\mathbf{x}/dt)$ and $dt = \gamma d\tau$, we have

$$u^0 = \frac{dx^0}{d\tau} = c\frac{dt}{d\tau} = c\gamma , \quad \text{and} \quad u^i = \frac{dx^i}{d\tau} = \gamma\frac{dx^i}{dt} = \gamma v^i .$$

So the 4-velocity of a massive particle is

$$u^\mu = \gamma(c, \mathbf{v}) , \quad \Longrightarrow \quad u_\mu = \gamma(c, -\mathbf{v}) , \tag{6.46}$$

where we have used the metric to lower the index of the contravariant 4-velocity. From the above information, we get

$$u^\mu u_\mu = (u^0)^2 - \mathbf{u} \cdot \mathbf{u} = \gamma^2(c^2 - v^2) = c^2 , \tag{6.47}$$

showing that the 4-velocity of a massive particle is normalized to c^2.

The 4-momentum of a massive particle, with mass m and 4-velocity u^μ, is

$$p^\mu \equiv mu^\mu , \tag{6.48}$$

which gives $p^0 = mu^0 = mc\gamma$ and $\mathbf{p} = m\gamma\mathbf{v}$. As u^μ is a 4-vector and m is a scalar the 4-momentum transforms like a 4-vector. One can write $p^0 = E/c$, where the energy of the particle is $E = mc^2\gamma$. From the above definition, one gets $p^\mu p_\mu = (p^0)^2 - \mathbf{p} \cdot \mathbf{p} = m^2c^2$, which yields $(E/c)^2 - \mathbf{p}^2 = m^2c^2$ and consequently

$$E^2 = \mathbf{p}^2 c^2 + m^2 c^4 . \tag{6.49}$$

For natural Euclidean 3-vectors as $\mathbf{v}$, we will write $\mathbf{v} \cdot \mathbf{v} = v^2$, but when dealing with the 3-vector part, as $\mathbf{p}$, of a 4-vector p, we will always write $\mathbf{p} \cdot \mathbf{p} = \mathbf{p}^2$ as p^2 is $p^\mu p_\mu$ in our convention.

The 4-force f^μ acting on a massive particle is defined as

$$f^\mu \equiv \frac{dp^\mu}{d\tau} = m\frac{du^\mu}{d\tau} . \tag{6.50}$$

As du^μ is a 4-vector so f^μ is also a 4-vector by definition. The 4-acceleration a^μ is defined as $a^\mu \equiv du^\mu/d\tau$. Differentiating $u^\mu u_\mu = c^2$ with respect to τ, we get

$$a^\mu u_\mu = 0, \quad \Longrightarrow \quad f^\mu p_\mu = 0. \tag{6.51}$$

The 4-momentum is orthogonal to 4-force or the 4-velocity is orthogonal to 4-acceleration. As a consequence of this $f^0 p^0 = \mathbf{f} \cdot \mathbf{p}$, we can write the 4-force as

$$f^\mu = \left(\frac{\mathbf{f} \cdot \mathbf{p}}{p^0}, \mathbf{f}\right) . \tag{6.52}$$

In this chapter, we do not discuss more on special relativity. If more details are required, we will specify them later on. Next, we discuss some preliminary parts of Fourier transform theory as required in electrodynamics.

6.8 Relevant Portions of Fourier Transform Theory

In electrodynamics we will we frequently using the theory of Fourier transforms. It is better to deal with the definitions and conventions of the Fourier transform theory, briefly in a chapter, so that readers can easily follow the text in the later chapters. For Fourier transform to exist, we require at least two continuous real variables, traditionally named x and k. One can Fourier expand $f(x)$ as

$$f(x) = \frac{1}{\sqrt{2\pi}} \int_{-\infty}^{\infty} \tilde{f}(k)e^{ikx}dk . \tag{6.53}$$

The function $\tilde{f}(k)$ is called the Fourier transform of $f(x)$. In general, the functional form of $\tilde{f}(k)$ is different from $f(x)$, but for the sake of brevity, we will use the same functional name for both the functions. Henceforth, the Fourier transform of $f(x)$ will simply be written as $f(k)$. The factor of $1/\sqrt{2\pi}$ in the Fourier transform is more a convention, the origin of such a convention is linked with the theory of Fourier series which uses 2π periodic functions. An important (generalized) function often used in physics is the Dirac delta function, defined as

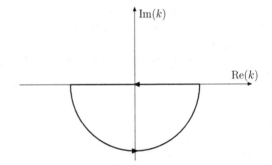

Fig. 6.3 The contour for evaluating the integral giving the Dirac delta function, for $x \neq 0$, in the complex k plane

$$\delta(x) = \frac{1}{2\pi} \int_{-\infty}^{\infty} e^{-ikx} dk \,. \tag{6.54}$$

From the above definition, one can see that when $x \neq 0$ (here x is purely real) one can construct a semicircular contour in the complex k plane in the lower half, as shown in Fig. 6.3. The radius of the semicircular region tends to infinity. The integrand e^{-ikx} has no singularities in the entire complex k plane and so the contour integral gives $\delta(x) = 0$. When $x = 0$ the integrand e^{-ikx} becomes unity and the integral in Eq. (6.54) diverges producing the familiar property of the Dirac delta function

$$\delta(x) = \begin{cases} 0, & x \neq 0, \\ \infty, & x = 0. \end{cases} \tag{6.55}$$

Both x and k are treated on the same footing and one can also write

$$\delta(k) = \frac{1}{2\pi} \int_{-\infty}^{\infty} e^{-ikx} dx \,, \tag{6.56}$$

which shows the nature of $\delta(k)$. In this case, one must be careful to note that $\delta(k)$ is not the Fourier transform of $\delta(x)$, unlike the convention for other functions. In fact, the reader can easily show that the Fourier transform of the Dirac delta function is a constant. The definition of the Dirac delta function is intimately related to Fourier transform theory. From the above definition, one can easily show that

$$\delta(x) = \delta(-x) \,, \tag{6.57}$$

which states that the Dirac delta function is an even function. The relation between Fourier transforms and the Dirac delta function becomes apparent from the following steps. From the definition of the Fourier transform, we can write

$$f(x)e^{-ikx} = \frac{1}{\sqrt{2\pi}} \int_{-\infty}^{\infty} f(k')e^{i(k'-k)x} dk' \,,$$

by multiplying both sides of Eq. (6.53) by e^{-ikx}. Integrating both sides of the above equation with respect to x yields

$$
\begin{aligned}
\int_{-\infty}^{\infty} f(x)e^{-ikx}\,dx &= \frac{1}{\sqrt{2\pi}} \int_{-\infty}^{\infty} dx \int_{-\infty}^{\infty} dk'\, f(k')e^{-ix(k-k')} \\
&= \sqrt{2\pi} \int_{-\infty}^{\infty} dk' \left[\frac{1}{2\pi} \int_{-\infty}^{\infty} e^{-ix(k-k')}\,dx \right] f(k') \\
&= \sqrt{2\pi} \int_{-\infty}^{\infty} dk'\delta(k-k') f(k') = \sqrt{2\pi} f(k),
\end{aligned}
$$

which shows that the Fourier transform of $f(x)$ is given by

$$
f(k) = \frac{1}{\sqrt{2\pi}} \int_{-\infty}^{\infty} f(x)e^{-ikx}\,dx . \tag{6.58}
$$

The above results show that the factors appearing in the Fourier transform and its inverse transform are inherently related with the way we represent the Dirac delta function. Using the definition of the Dirac delta and the Fourier transforms, one can formally derive some of the defining properties of the Dirac delta function. One can write Eq. (6.53) using Eq. (6.58) as

$$
\begin{aligned}
f(x) &= \frac{1}{\sqrt{2\pi}} \int_{-\infty}^{\infty} \left[\frac{1}{\sqrt{2\pi}} \int_{-\infty}^{\infty} f(y)e^{-iky}\,dy \right] e^{ikx}\,dk \\
&= \int_{-\infty}^{\infty} \left[\frac{1}{2\pi} \int_{-\infty}^{\infty} e^{-ik(y-x)}\,dk \right] f(y)\,dy = \int_{-\infty}^{\infty} \delta(y-x)\, f(y)\,dy \tag{6.59}
\end{aligned}
$$

If $f(x) = 1$ the above result becomes

$$
\int_{y=-\infty}^{y=\infty} \delta(y-x)\,dy = 1 ,
$$

and now shifting the integration variable as $z = y - x$, we get the familiar result

$$
\int_{z=-\infty}^{z=\infty} \delta(z)\,dz = 1 . \tag{6.60}
$$

These are all the familiar defining properties of the Dirac delta function. The Heaviside step function is closely related to the Dirac delta function and as we will require the step function later we introduce it here. The step function is defined as

$$
\Theta(x) = \begin{cases} 1, & x \geq 0, \\ 0, & x < 0. \end{cases} \tag{6.61}
$$

The function is discontinuous at $x = 0$. We have assigned the step function unit value at $x = 0$. From the defining properties of the Dirac delta function and the step function, one can write

$$\Theta(x) = \int_{-\infty}^{x} \delta(\tilde{x}) \, d\tilde{x} \, ,$$

where the above equation is ill defined only at $x = 0$, which is a point of (finite) discontinuity for the step function. In such a case, we assign $\Theta(0) = 1$. From the above equation, one can write

$$\frac{d\Theta(x)}{dx} = \delta(x) \, . \tag{6.62}$$

We will use the above relation frequently in later discussions on electromagnetic radiation by point charges.

In general, both $f(x)$ and $f(k)$ can be complex functions. So we can write

$$f^*(x) = \frac{1}{\sqrt{2\pi}} \int_{-\infty}^{\infty} f^*(k) e^{-ikx} dk \, ,$$

where the star designates complex conjugation. If in a particular case one knows that $f(x)$ is a real function then one must have

$$f^*(x) = f(x) = \frac{1}{\sqrt{2\pi}} \int_{-\infty}^{\infty} f^*(k) \, e^{-ikx} \, dk$$

$$= -\frac{1}{\sqrt{2\pi}} \int_{\infty}^{-\infty} f^*(-k) \, e^{ikx} \, dk \, ,$$

where we have replaced $-k$ for k as the integration variable in the second step. Flipping the integration limits and comparing the integrand with the integrand in Eq. (6.53), we immediately get

$$f^*(-k) = f(k) \, , \tag{6.63}$$

when $f(x)$ is real. Although the Fourier transforms can be complex it is easy to show that the Dirac delta function, as defined in Eq. (6.54), is strictly real.

When both $\mathbf{x}$ and $\mathbf{k}$ are Euclidean 3-vectors the above definitions gets modified to

$$f(\mathbf{x}) = \frac{1}{(2\pi)^{3/2}} \int_{-\infty}^{\infty} f(\mathbf{k}) \, e^{i\mathbf{k}\cdot\mathbf{x}} \, d^3k \, , \tag{6.64}$$

$$\delta^3(\mathbf{x}) \equiv \delta(x)\delta(y)\delta(z) = \frac{1}{(2\pi)^3} \int_{-\infty}^{\infty} e^{-i\mathbf{k}\cdot\mathbf{x}} \, d^3k \,, \tag{6.65}$$

$$f(\mathbf{k}) = \frac{1}{(2\pi)^{3/2}} \int_{-\infty}^{\infty} f(\mathbf{x}) \, e^{-i\mathbf{k}\cdot\mathbf{x}} \, d^3x \,. \tag{6.66}$$

It has to be noted that in our convention

$$\int_{-\infty}^{\infty} (\cdots) \, d^3k \implies \int_{-\infty}^{\infty} \int_{-\infty}^{\infty} \int_{-\infty}^{\infty} (\cdots) \, dk_x dk_y dk_z \,,$$

other triple integrals are analogously defined. For functions of time we define the Fourier transforms as

$$f(t) = \frac{1}{\sqrt{2\pi}} \int_{-\infty}^{\infty} f(\omega) e^{-i\omega t} \, d\omega \,, \tag{6.67}$$

$$f(\omega) = \frac{1}{\sqrt{2\pi}} \int_{-\infty}^{\infty} f(t) e^{i\omega t} \, dt \,. \tag{6.68}$$

Now we can use our convention to write down the Fourier transforms of functions of spacetime as

$$f(\mathbf{x}, t) = \frac{1}{(2\pi)^2} \int_{-\infty}^{\infty} d^3k \int_{-\infty}^{\infty} d\omega \, f(\mathbf{k}, \omega) \, e^{i(\mathbf{k}\cdot\mathbf{x} - \omega t)} \,, \tag{6.69}$$

and the inverse Fourier transform can easily be obtained. In terms of 4-vectors $x^\mu = (x^0, \mathbf{x})$ and $k^\mu = (\omega, \mathbf{k})$, one can write the above relation as

$$f(x) = \frac{1}{(2\pi)^2} \int_{-\infty}^{\infty} d^4k \, f(k) \, e^{-ik\cdot x} \,, \tag{6.70}$$

where now x and k are supposed to be 4-vectors. In our convention

$$\nabla \equiv \left(\frac{\partial}{\partial x^1}, \frac{\partial}{\partial x^2}, \frac{\partial}{\partial x^3} \right) = \left(\frac{\partial}{\partial x}, \frac{\partial}{\partial y}, \frac{\partial}{\partial z} \right) \,, \tag{6.71}$$

and as a result

$$\nabla f(\mathbf{x}, t) = \frac{1}{(2\pi)^2} \int_{-\infty}^{\infty} d^3k \int_{-\infty}^{\infty} d\omega \, (i\mathbf{k}) f(\mathbf{k}, \omega) \, e^{i(\mathbf{k}\cdot\mathbf{x} - \omega t)} \,, \tag{6.72}$$

$$\nabla^2 f(\mathbf{x}, t) = -\frac{1}{(2\pi)^2} \int_{-\infty}^{\infty} d^3k \int_{-\infty}^{\infty} d\omega \, \mathbf{k}^2 f(\mathbf{k}, \omega) \, e^{i(\mathbf{k}\cdot\mathbf{x} - \omega t)} \,, \tag{6.73}$$

$$\frac{\partial f(\mathbf{x}, t)}{\partial t} = \frac{1}{(2\pi)^2} \int_{-\infty}^{\infty} d^3k \int_{-\infty}^{\infty} d\omega \, (-i\omega) f(\mathbf{k}, \omega) \, e^{i(\mathbf{k}\cdot\mathbf{x} - \omega t)} \,, \tag{6.74}$$

and so on.

Sometimes an asymmetric distribution of the 2π factors become useful. In this case instead of $\frac{1}{\sqrt{2\pi}}$ in the Fourier transform we simply have $\frac{1}{2\pi}$ and the inverse Fourier transform do not have any factor before the integral. The Dirac-delta function definition remains unchanged. We use this form of Fourier transform in Chap. 10.

6.8.1   The Product of Fourier Transforms: Convolution

In electrodynamics we will require the inverse Fourier transform of the product of two Fourier transforms. Suppose $f(k)$ and $g(k)$ are the Fourier transforms of $f(x)$ and $g(x)$, and we want to know the inverse Fourier transform of the product $f(k)g(k)$. In general, it is not the product $f(x)g(x)$ but something else. The inverse Fourier transform of $f(k)g(k)$ is

$$
\frac{1}{\sqrt{2\pi}} \int_{k=-\infty}^{\infty} f(k)g(k)\, e^{ikx}\, dk
$$

$$
= \frac{1}{\sqrt{2\pi}} \int_{k=-\infty}^{\infty} \left[\frac{1}{\sqrt{2\pi}} \int_{y=-\infty}^{\infty} f(y) e^{-iky}\, dy \right] \left[\frac{1}{\sqrt{2\pi}} \int_{z=-\infty}^{\infty} g(z) e^{-ikz}\, dz \right] e^{ikx}\, dk
$$

$$
= \frac{1}{\sqrt{2\pi}} \int_{y=-\infty}^{\infty} \int_{z=-\infty}^{\infty} \left[\frac{1}{2\pi} \int_{k=-\infty}^{\infty} e^{-ik(y+z-x)}\, dk \right] f(y)g(z)\, dy dz .
$$

Using the definition of the Dirac delta function, the above step becomes

$$
\frac{1}{\sqrt{2\pi}} \int_{k=-\infty}^{\infty} f(k)g(k)\, e^{ikx}\, dk = \frac{1}{\sqrt{2\pi}} \int_{y=-\infty}^{\infty} \int_{z=-\infty}^{\infty} \delta(y+z-x) f(y)g(z)\, dy\, dz
$$

$$
= \frac{1}{\sqrt{2\pi}} \int_{y=-\infty}^{\infty} f(y)g(x-y)\, dy . \tag{6.75}
$$

The last expression gives the inverse Fourier transform of $f(k)g(k)$ and the integral with the prefactor of $1/\sqrt{2\pi}$ is generally called the convolution of the two functions $f(y)$ and $g(y)$. One can easily show that

$$
\frac{1}{\sqrt{2\pi}} \int_{y=-\infty}^{\infty} f(y)g(x-y)\, dy = \frac{1}{\sqrt{2\pi}} \int_{y=-\infty}^{\infty} f(x-y)g(y)\, dy ,
$$

and so the convolution of the two functions can be written in both possible ways.

6.9   Problems

1. Show that under a general coordinate transformation (which may not be linear and hence $\Lambda^{\mu}{}_{\nu}$ is not a constant), x^{μ} is not a contravariant vector, but dx^{μ} still remains a contravariant vector.
2. Prove the validity of Eq. (6.14).

3. Show that under a Lorentz transformation the totally antisymmetric Levi-Civita is transformed into another fourth rank tensor whose components are exactly the same as the original antisymmetric Levi-Civita.

4. Show that

$$s^2 = c^2 t^2 - x^2 - y^2 - z^2 \, ,$$

remains an invariant under a Lorentz boost and coordinate rotation.

5. Show that the Kronecker delta symbol has the same components in all inertial reference frames connected by Lorentz transformations.

6. If $T^{\mu\nu}{}_{\alpha\beta}$ is a mixed tensor and we have a mathematical quantity called A whose transformation properties we do not know a priory and we also know that B^{μ}_{α} is a mixed tensor such that $B^{\mu}{}_{\alpha} = A T^{\mu\nu}{}_{\alpha\beta}$ holds true then show that A must be a tensor and the indices which specifies it are as $A^{\nu}{}_{\beta}$.

7. If $T^{\mu}{}_{\nu\alpha} = S^{\mu}{}_{\nu\alpha}$ is a tensor equation involving two tensors $T^{\mu}{}_{\nu\alpha}$ and $S^{\mu}{}_{\nu\alpha}$ then show that the following tensor equations are also valid,

   (a) $T_{\mu\nu\alpha} = S_{\mu\nu\alpha}$.
   (b) $T^{\mu\nu}{}_{\alpha} = S^{\mu\nu}{}_{\alpha}$.
   (c) $T^{\mu\nu\alpha} = S^{\mu\nu\alpha}$.

   To derive the above results judiciously raise or lower the indices in the tensor equation with the inverse metric or the metric.

8. From the definition of the Dirac delta function in Eq. (6.54) show that $\delta(x)$ is an even function.

9. From the definition of the Dirac delta function in Eq. (6.54) show that $\delta(x)$ is a real function.

10. A general Lorentz transformation can be represented as a product of Lorentz boost and rotation. Suppose we are talking of infinitesimal rotations about the z-axis and infinitesimal boosts where the velocity vector has components along the x and y axes.

    (a) Write down the rotation matrix R corresponding to an infinitesimal rotation along the z-axis by an infinitesimal angle as $R(\delta\theta) = 1 + \delta R(\delta\theta)$. Here 1 is the unit 4×4 matrix. Write R as a 4×4 matrix and explicitly show the matrix $\delta R(\delta\theta)$.

    (b) You know the formula for the general Lorentz boost (between two inertial frames where the axes of the frames are parallel) where the relative velocity vector, $\boldsymbol{\beta} = (\beta_1, \beta_2\beta_3)$, makes an arbitrary angle with the $x - x'$ axes. Suppose we are talking about an infinitesimal Lorentz boost where $\gamma \simeq 1$ and $\beta_i \ll 1$. We write the infinitesimal velocity components as $\delta\beta_i$. Write the Lorentz boost matrix $\Lambda(\delta\beta_i) = 1 + \delta\Lambda(\delta\beta_i)$. Explicitly write the matrix $\delta\Lambda(\delta\beta_i)$ to first order in the velocity parameters.

    (c) Suppose you are given a matrix which corresponds to an infinitesimal Lorentz transformation

$$\begin{pmatrix} 1 & -\gamma_0^2 \delta\beta_1 & -\gamma_0 \delta\beta_2 & 0 \\ -\gamma_0^2 \delta\beta_1 & 1 & \frac{\gamma_0-1}{\beta}\delta\beta_2 & 0 \\ -\gamma_0 \delta\beta_2 & -\frac{\gamma_0-1}{\beta}\delta\beta_2 & 1 & 0 \\ 0 & 0 & 0 & 1 \end{pmatrix}.$$

Here, γ_0 is some parameter which may not be unity and is not related to the infinitesimal velocity parameters $\delta\beta_1$, $\delta\beta_2$. Moreover, $\delta\beta_3 = 0$. From your previous knowledge, can you represent the above Lorentz transformation matrix as a product of an infinitesimal rotation matrix and an infinitesimal Lorentz boost? If possible clearly show the infinitesimal rotation and infinitesimal boost matrices. From your result find out the rotation angle and the x and y components of infinitesimal velocity. Check whether the infinitesimal rotation and boost matrices commute.

11. If $\tilde{f}(x) = f(x - x_0)$, for some constant x_0, then show that $\tilde{f}(k) = e^{-ikx_0} f(k)$, where $f(k)$ is the Fourier transform of $f(x)$ and $\tilde{f}(k)$ is the Fourier transform of $\tilde{f}(x)$.
12. If $\tilde{f}(x) = e^{ixk_0} f(x)$, for some constant k_0, then show that $\tilde{f}(k) = f(k - k_0)$, where $f(k)$ is the Fourier transform of $f(x)$ and $\tilde{f}(k)$ is the Fourier transform of $\tilde{f}(x)$.
13. Show that if $\tilde{f}(x) = f(ax)$ then

$$\tilde{f}(k) = \frac{1}{|a|} f\left(\frac{k}{a}\right),$$

where a is any real number except zero.
14. Suppose we have a real function $f(x)$ which has finite number of distinct zeros. In such a case, try to prove that

$$\delta\left(f(x)\right) = \sum_i \frac{\delta(x - x_i)}{\left|\frac{df}{dx}\right|_{x_i}},$$

where x_i specifies the zeros of $f(x)$ and i specifies the number of zeros of the function. This result is very useful, but the proof can be tricky as it involves some formal handling of the Dirac delta function.

To prove the above relation, you may actually follow heuristic logic. Suppose $g(x)$ is another well behaved real function. If you can show that

$$\int_{-\infty}^{\infty} \delta\left(f(x)\right) g(x) dx = \sum_i \frac{g(x_i)}{\left|\frac{df}{dx}\right|_{x_i}},$$

then that can imply that our result is proved. You can try to prove the above result. To prove the above result, you may at first Taylor expand $f(x)$ up to linear order around the (various) zeros of it and then replace the Taylor expansion for the function around the zeros inside the integral. Here, we have assumed that the function $g(x)$ does not diverge at any of the zeros of $f(x)$.

If you are not satisfied with the level of rigor, then you can go in for more interesting and mathematically sound techniques to prove our original result.

15. From your understanding of Lorentz boosts in arbitrary directions show that the product of two Lorentz boosts along the x-direction and y-direction, respectively, is, in general, not a boost. If the two boosts are infinitesimal ones, as we have defined infinitesimal boosts in question 10(b), then can you figure out what kind of transformations correspond to the product of the boosts?

Chapter 7
Covariant form of Maxwell's Equations in the Absence of Bound Charges and Bound Currents

We know Maxwell's equations and have used them in previous chapters. Till now we have only written down the Maxwell equations in terms of the electric and magnetic fields, $\mathbf{E}(\mathbf{x}, t)$ and $\mathbf{B}(\mathbf{x}, t)$. The source terms were represented by the charge density and current density $\rho(\mathbf{x}, t)$ and $\mathbf{j}(\mathbf{x}, t)$. We will assume all the Euclidean 3-vectors as written as

$$\mathbf{v} \equiv (v_x, v_y, v_z) = (v^1, v^2, v^3),$$

where the usual vector components are interpreted as contravariant components. The metric in the Euclidean 3-space is simply given by

$$g_{ij} \equiv \text{Diag.}(1, 1, 1), \tag{7.1}$$

were Diag.() specifies the diagonal entries of a 3×3 matrix whose other non-diagonal entries are zero. The inner products of vectors in the Euclidean 3-space is done with respect to the metric above. Here the Latin indices runs over 1 to 3. In this chapter, we will show that the Maxwell's equations, in the absence of any material medium, can be written elegantly in a covariant way. In the present case, a material medium may imply a dielectric, conductor or a plasma medium. Maxwell's equations in presence of such material media will be dealt in the last chapter. It must be noted that a material medium always chooses a particular reference frame where it is at rest. Most of the time Maxwell's equations are generally written in that specific frame and as a consequence the covariance of the equations is not that useful a concept in these cases. On the other hand in the absence of any material medium, the covariance of the equations readily show that electromagnetism is special theory of relativity in disguise.

© Springer Nature Singapore Pte Ltd. 2021
K. Bhattacharya and S. Mukhopadhyay, *Introduction to Advanced Electrodynamics*,
https://doi.org/10.1007/978-981-16-7802-8_7

7.1 Maxwell's Equations Using only 3-Vectors

We know that the inhomogeneous Maxwell's equations are written as

$$\nabla \cdot \mathbf{E} = 4\pi\rho, \quad \nabla \times \mathbf{B} - \frac{1}{c}\frac{\partial \mathbf{E}}{\partial t} = \frac{4\pi}{c}\mathbf{J}, \tag{7.2}$$

where $\rho(\mathbf{x}, t)$ specifies free charge density and $\mathbf{J}(\mathbf{x}, t)$ specifies free current density, both of which acts as sources of the fields. The homogeneous Maxwell's equations are

$$\nabla \cdot \mathbf{B} = 0, \quad \nabla \times \mathbf{E} + \frac{1}{c}\frac{\partial \mathbf{B}}{\partial t} = 0. \tag{7.3}$$

As the magnetic induction $\mathbf{B}$ is divergence-less, we can define the magnetic vector potential $\mathbf{A}(\mathbf{x}, t)$ as

$$\mathbf{B} = \nabla \times \mathbf{A}. \tag{7.4}$$

Using this relation in Faraday's law, we can write

$$\nabla \times \mathbf{E} + \frac{1}{c}\frac{\partial}{\partial t}(\nabla \times \mathbf{A}) = 0,$$

and as partial derivatives commute the above relation can also be written as

$$\nabla \times \left[\mathbf{E} + \frac{1}{c}\frac{\partial \mathbf{A}}{\partial t}\right] = 0 \implies \mathbf{E} + \frac{1}{c}\frac{\partial \mathbf{A}}{\partial t} = -\nabla\Phi,$$

where $\Phi(\mathbf{x}, t)$ is called the electromagnetic scalar potential[1]. The above relation yields

$$\mathbf{E} = -\nabla\Phi - \frac{1}{c}\frac{\partial \mathbf{A}}{\partial t}. \tag{7.5}$$

The divergence of the second inhomogeneous Maxwell equation yields the continuity equation as

$$\frac{\partial\rho}{\partial t} + \nabla \cdot \mathbf{J} = 0. \tag{7.6}$$

---

[1] Traditionally Φ is called the electromagnetic scalar potential but we must be careful with the name. Here we must point out that Φ is a scalar in the non-relativistic sense, it does transform like a scalar under a three-dimensional spatial coordinate rotation. On the other hand in relativistic covariant notation, we will see that Φ is not a scalar at all but a component of a 4-vector which transforms nontrivially under Lorentz boosts. In this book, we will still use the traditional name of Φ as the electromagnetic scalar potential.

It must be noted that the continuity equation actually expresses conservation of electric charge and is by itself a product of the Maxwell equations. The Maxwell equations contain the information that electric charge is conserved, and one does not require to assume it as an extra input. From these basic groundwork, we can now proceed to write down the Maxwell equations in the covariant form using tensors. The fields $\mathbf{A}(\mathbf{x}, t)$ and $\Phi(\mathbf{x}, t)$ are collectively called gauge fields as one can have certain kind of transformations involving them called gauge transformations.

Before we write down Maxwell's equations in the covariant form let us try to see some properties of the above equations. The homogeneous Maxwell's equations have $\rho(\mathbf{x}, t)$ and $\mathbf{J}(\mathbf{x}, t)$ as source terms. We do not know how these terms are affected by Lorentz transformations as these terms are not parts of any 4-vector. We expect to know the transformation of $\mathbf{J}(\mathbf{x}, t)$ under a spatial rotation as it is supposed to be a 3-vector. If electromagnetism has a relativistic structure then it is fair to assume that the source terms can be combined in a 4-vector. From dimensional arguments we can assume the 4-current, which acts as the source term, can be written as

$$J^{\mu}(x) \equiv (c\rho(x), \mathbf{J}(x)), \qquad (7.7)$$

where $x \equiv (ct, \mathbf{x})$. At this moment the above assumption is too bold as it forces some specific forms of transformations on the components of J^{μ} as a 4-vector transforms in a very specific way under Lorentz transformations. Later we will show that our assumption about the constituents of J^{μ} and other 4-vectors defined below makes perfect sense as the principle of relativity will become apparent in the covariant notation written down using these 4-vectors. One immediate advantage of defining the 4-current as above is seen in rewriting the continuity equation:

$$\frac{\partial J^0}{\partial x^0} + \partial_i J^i = 0,$$

in a compact form,

$$\partial_{\mu} J^{\mu} = 0. \qquad (7.8)$$

The gauge fields $\Phi(x)$ and $\mathbf{A}(x)$ do have the same dimensions and as both the fields produce four functions it is tempting to define another 4-vector field as

$$A^{\mu}(x) \equiv (\Phi(x), \mathbf{A}(x)), \qquad (7.9)$$

which transforms as a standard 4-vector under Lorentz transformations. From our previous knowledge, Eq. (7.5), we can now write the electric and magnetic fields as

$$E^i(x) = \partial^i \Phi(x) - \partial^0 A^i(x) = \partial^i A^0(x) - \partial^0 A^i(x), \qquad (7.10)$$

where we have used the natural convention in which $\nabla \equiv (\partial_1, \partial_2, \partial_3)$, and the fact that a covariant, 3-vector component (where the index is down) can be transformed to a contravariant, 3-vector component (when the index is above) at the expense of a negative sign. For the case of the electric field, we move the indices up (or down) with respect to the metric of the four-dimensional space $\eta^{\mu\nu}$ (or $\eta_{\mu\nu}$) as time and space coordinates explicitly appear in the expression of the electric field.

The magnetic field is written in as $\mathbf{B}(x) = \nabla \times \mathbf{A}(x)$ which can be written down in component form as

$$B^i = -\varepsilon^{ijk}\partial_j A_k \,, \tag{7.11}$$

where we have a summation over indices j and k. The minus sign appears on the right-hand side as in this case also we raise (or lower) the indices with the four-dimensional metric $\eta_{\mu\nu}$ (or $\eta^{\mu\nu}$). The three-dimensional Levi-Civita symbol is defined as

$$\varepsilon^{ijk} = \begin{cases} 1 \text{ if } i,j,k \text{ are even permutations of } 1,2,3. \\ -1 \text{ if } i,j,k \text{ are odd permutations of } 1,2,3. \\ 0 \text{ if there is any repeatation of indices}. \end{cases} \tag{7.12}$$

The Levi-Civita symbol in three-dimensions forms a contravariant tensor whose components are constant and remains the same under any Lorentz transformations. From Eq. (7.10) and the expression of the magnetic field, we see that we can write both the electric field and the magnetic field in terms of the gauge field 4-vector A^μ components. Next we show that in reality the electric field and the magnetic field can be interpreted as components of a second rank, antisymmetric tensor in four dimensions.

7.2   The Field Strength Tensor and Its Dual

We have written the components of the electric and magnetic fields in the last two paragraphs. If we explicitly write them down using contravariant components only then we will have

$$E^i = \partial^i A^0 - \partial^0 A^i \,, \ B^1 = \partial^3 A^2 - \partial^2 A^3 \,, \ B^2 = \partial^1 A^3 - \partial^3 A^2 \,, \ B^3 = \partial^2 A^1 - \partial^1 A^2 \,,$$

which can be grouped together as components of a second-rank tensor as

$$F^{\mu\nu} \equiv \partial^\mu A^\nu - \partial^\nu A^\mu \,. \tag{7.13}$$

The above quantity is a second-rank tensor, which can be verified from the quotient theorem in tensor analysis. The above tensor is antisymmetric by construction and is called the field strength tensor. Many times the field strength tensor is written in the matrix notation as

$$F = \begin{pmatrix} 0 & -E^1 & -E^2 & -E^3 \\ E^1 & 0 & -B^3 & B^2 \\ E^2 & B^3 & 0 & -B^1 \\ E^3 & -B^2 & B^1 & 0 \end{pmatrix}. \tag{7.14}$$

In the above equation, one must associate the μ with the row index and ν with the column index when writing the matrix components of $F^{\mu\nu}$. We immediately see that if we group the gauge fields as 4-vector components then the electric and magnetic fields can be obtained as components of a second-rank tensor.

We can immediately define another second rank, antisymmetric tensor from $F^{\mu\nu}$ and the totally antisymmetric fourth rank Levi-Civita tensor as

$$\tilde{F}^{\mu\nu} \equiv \frac{1}{2}\varepsilon^{\mu\nu\alpha\beta} F_{\alpha\beta} , \tag{7.15}$$

called the dual field strength tensor. In the present case $F_{\alpha\beta} = \eta_{\alpha\kappa}\eta_{\beta\delta} F^{\kappa\delta}$ and from this information one can now calculate all the components of the dual field strength tensor. In Matrix form it comes out as

$$\tilde{F} = \begin{pmatrix} 0 & -B^1 & -B^2 & -B^3 \\ B^1 & 0 & E^3 & -E^2 \\ B^2 & -E^3 & 0 & E^1 \\ E^3 & E^2 & -E^1 & 0 \end{pmatrix}, \tag{7.16}$$

which is obtained from the matrix expression of Eq. (7.14) by applying the following transformations,

$$\mathbf{E} \to \mathbf{B}, \quad \text{and} \quad \mathbf{B} \to -\mathbf{E}.$$

Once one knows the form of $F^{\mu\nu}$ then applying the above transformations on the components of $F^{\mu\nu}$ will yield the dual field strength tensor $\tilde{F}^{\mu\nu}$.

7.3   Maxwell's Equations in Covariant form

We first try to write the covariant forms of the inhomogeneous Maxwell's equations. From our previous discussions, one can easily write the equation $\nabla \cdot \mathbf{E} = 4\pi\rho$ as $\partial_i F^{i0} = (4\pi/c)J^0$. As F^{i0} is antisymmetric in nature we see that the above equation can also be written as

$$\partial_\mu F^{\mu 0} = \frac{4\pi}{c} J^0 , \tag{7.17}$$

as when $\mu = 0$ the left-hand side of the above equation does not contribute. The other inhomogeneous equation can also be written down in the covariant form after some work. First we try to write down $(\nabla \times \mathbf{B})$ in component form. In our convention (as in Eq. (7.11)):

$$(\nabla \times \mathbf{B})^i = -\varepsilon^{ijk}\partial_j B_k \quad \text{where,} \quad B_k = -\varepsilon_{klm}\partial^l A^m .$$

Which leads to:

$$(\nabla \times \mathbf{B})^i = \varepsilon^{ijk}\partial_j(\varepsilon_{klm}\partial^l A^m) = \varepsilon^{ijk}\varepsilon_{klm}\partial_j\partial^l A^m = \varepsilon^{kij}\varepsilon_{klm}\partial_j\partial^l A^m .$$

To evaluate the above expression we require to know what is $\varepsilon^{kij}\varepsilon_{klm}$. From the quotient theorem of tensor analysis, we know that the result of the contraction on k must lead to a mixed tensor. Keeping this in mind we can formally write

$$\varepsilon^{kij}\varepsilon_{klm} = a\,\delta^i_l\delta^j_m + b\,\delta^i_m\delta^j_l , \tag{7.18}$$

where the values of the coefficients a and b have to be found out. Before we evaluate the values of the coefficients we require to know what is ε_{klm}. In the present case

$$\varepsilon_{klm} = \eta_{ki}\eta_{lj}\eta_{mn}\varepsilon^{ijn} ,$$

and as the metric is diagonal we immediately come to the conclusion that the components of ε_{klm} are similar in magnitude but opposite in sign to the components of ε^{klm}. Using this fact we can evaluate the values of a and b. Fixing $i = 1, j = 2$ and $l = 1, m = 2$ in Eq. (7.18) we get $a = -1$, on the other hand fixing $i = 1, j = 2$ and $l = 2, m = 1$ in Eq. (7.18) we get $b = 1$ giving us

$$\varepsilon^{kij}\varepsilon_{klm} = \delta^i_m\delta^j_l - \delta^i_l\delta^j_m . \tag{7.19}$$

Using the above result, we can write

$$(\nabla \times \mathbf{B})^i = (\delta^i_m\delta^j_l - \delta^i_l\delta^j_m)\partial_j\partial^l A^m$$
$$= \partial_j\partial^j A^i - \partial_j\partial^i A^j = \partial_j(\partial^j A^i - \partial^i A^j) = \partial_j F^{ji} . \tag{7.20}$$

Next we take up the other term in the second inhomogeneous equation and write it as

$$\frac{1}{c}\frac{\partial E^i}{\partial t} = \frac{\partial E^i}{\partial x^0} = \partial_0 F^{i0} .$$

Combining the last two results we can write the second inhomogeneous Maxwell equation as

$$\partial_j F^{ji} + \partial_0 F^{0i} = \frac{4\pi}{c}J^i \implies \partial_\mu F^{\mu i} = \frac{4\pi}{c}J^i .$$

Combining this equation with the first inhomogeneous equation as given in Eq. (7.17) we see that the two inhomogeneous equations can be combined together and written as

$$\partial_\mu F^{\mu\nu} = \frac{4\pi}{c} J^\nu. \tag{7.21}$$

In the covariant form, the two inhomogeneous Maxwell equations combine to a single tensor equation. The reader must note that in the compact covariant notation we still have four separate equations as there were in Eq. (7.2). Covariant notation only shows that the separate inhomogeneous Maxwell equations are parts of a single tensor equation.

Next we write down the homogeneous Maxwell equations as given in Eq. (7.3) in the covariant form. We know $\nabla \cdot \mathbf{B} = 0$ in the component form becomes $\partial_i B^i = 0$ and we know how to write B^i in terms of the vector potential. Using Eq. (7.11), we have the first homogeneous equation as

$$\varepsilon^{ijk} \partial_i \partial_j A_k = 0. \tag{7.22}$$

The above equation can be written in the form

$$\varepsilon_{123}(\partial^1\partial^2 A^3 - \partial^1\partial^3 A^2) + \varepsilon_{231}(\partial^2\partial^3 A^1 - \partial^2\partial^1 A^3) + \varepsilon_{312}(\partial^3\partial^1 A^2 - \partial^3\partial^2 A^1) = 0,$$

where we have used the antisymmetry of the last two indices of the three-dimensional Levi-Civita symbol. As in our convention $\varepsilon_{123} = \varepsilon_{231} = \varepsilon_{312} = -1$ the above equation becomes a single equation as

$$\partial^1 F^{23} + \partial^2 F^{31} + \partial^3 F^{12} = 0. \tag{7.23}$$

The second homogeneous equation (as given in Eq. (7.3)) in the component form can be written as

$$\varepsilon^{ijk} \partial_j E_k + \partial_0 \varepsilon^{ijk} \partial_j A_k = 0. \tag{7.24}$$

The above equation can also be written as

$$\varepsilon_{ijk} \partial^j F^{k0} + \varepsilon_{ijk} \partial^0 \partial^j A^k = 0.$$

Using $i = 1$ the above equation becomes

$$\varepsilon_{123}(\partial^2 F^{30} - \partial^3 F^{20}) + \varepsilon_{123}\partial^0(\partial^2 A^3 - \partial^3 A^2) = 0,$$

which is

$$\partial^2 F^{30} + \partial^3 F^{02} + \partial^0 F^{23} = 0. \tag{7.25}$$

Using $i = 2$ and $i = 3$ in Eq. (7.24), one obtains

$$\partial^1 F^{30} + \partial^3 F^{01} + \partial^0 F^{13} = 0, \quad \partial^1 F^{20} + \partial^2 F^{01} + \partial^0 F^{12} = 0. \tag{7.26}$$

Comparing the above expressions and the expressions in Eqs. (7.25) and (7.23), one can see that the two homogeneous equations appearing in Eq. (7.3) can be combined in one covariant form as

$$\partial^\mu F^{\nu\lambda} + \partial^\nu F^{\lambda\mu} + \partial^\lambda F^{\mu\nu} = 0 \,. \tag{7.27}$$

The above covariant equation represents two homogeneous Maxwell equations. In this discussion on the homogeneous equations, we have deduced the form of the above covariant homogeneous Maxwell equation from the original homogeneous equations. On the other hand, the above relation can very easily be proved by using the expression of $F_{\mu\nu}$ and the commuting property of the partial derivatives. The homogeneous Maxwell equation in the covariant form is more related to the structure of the antisymmetric field strength tensor.

One can write the covariant homogeneous Maxwell equation in a more compact form using the dual of $F^{\mu\nu}$ introduced in Eq. (7.15). We can proceed to calculate $\partial_\mu \tilde{F}^{\mu\nu}$ as

$$\partial_\mu \tilde{F}^{\mu\nu} = \frac{1}{2}\varepsilon^{\mu\nu\alpha\beta}\partial_\mu F_{\alpha\beta} = -\frac{1}{2}\varepsilon^{\nu\mu\alpha\beta}\partial_\mu F_{\alpha\beta} \,, \tag{7.28}$$

where we have used the antisymmetry of the four-dimensional Levi-Civita symbol. If we set $\nu = 0$ in the above equation then the right-hand side becomes

$$-\frac{1}{2}\varepsilon^{0ijk}\partial_i F_{jk} = -(\varepsilon^{0123}\partial_1 F_{23} + \varepsilon^{0231}\partial_2 F_{31} + \varepsilon^{0312}\partial_3 F_{12})$$
$$= -(\partial_1 F_{23} + \partial_2 F_{31} + \partial_3 F_{12}) = 0 \,,$$

using the result of Eq. (7.23). Similarly choosing other values of ν and using Eqs. (7.25) and (7.26) one can show that in general $(1/2)\varepsilon^{\nu\mu\alpha\beta}\partial_\mu F_{\alpha\beta} = 0$. So ultimately we can write the Maxwell equations in a compact covariant way as

$$\partial_\mu F^{\mu\nu} = \frac{4\pi}{c}J^\nu \,, \quad \partial_\mu \tilde{F}^{\mu\nu} = 0 \,, \tag{7.29}$$

where the first one represents the inhomogeneous Maxwell equations and the second one the homogeneous Maxwell equations. Due to antisymmetry of the field strength tensor $F^{\mu\nu}$, one can easily figure out from the inhomogeneous Maxwell equation

$$\partial_\nu \partial_\mu F^{\mu\nu} = \frac{4\pi}{c}\partial_\nu J^\nu = 0 \,, \tag{7.30}$$

giving back the continuity equation.

We have successfully written down the Maxwell equations in the tensor form and now we know that these forms of the Maxwell equations will remain the same under any Lorentz transformations. The Maxwell equations remain second-order linear, partial differential equations which can be solved in many cases if appropriate initial conditions or boundary conditions are specified. Although one can be satisfied in successfully writing down the covariant form of the Maxwell equations, the satisfaction comes at a cost. The dynamical field equations are now in terms of the field $A^\mu(x)$ and not the actual physical fields **E** and **B**. After solving the Maxwell equations one has to resort to Eqs. (7.4) and (7.5) to obtain the physical electric and magnetic fields. This process of solving itself brings in a new set of complexity which was not present before. The complexity is related to the fact that the field $A^\mu(x)$ is not uniquely defined, it has arbitrariness. This arbitrariness is related to gauge transformations, a topic which we will discuss later. To eradicate gauge arbitrariness one has to impose more conditions on the solutions of the Maxwell equations.

7.4 Transformation of $F^{\mu\nu}$ Under a Lorentz Transformation

We have written Maxwell's equations in the covariant form, and we know how the 4-vectors J^μ, A^μ transform under Lorentz transformations. We have imposed the transformation properties of charge density, current density and the gauge potentials by assuming that J^μ and A^μ are 4-vectors. In this book, we will see that we get back expected laws of physics when the above quantities do indeed transform as components of a 4-vector. Till now we have not seen how the electric field **E** and the magnetic field **B** transform under a Lorentz boost. To see how they transform we have to use the machinery of the covariant formulation and try to understand how a second-rank tensor $F^{\mu\nu}$ transform under a Lorentz transformation.

Suppose in the K frame of Fig. 6.1 (or the laboratory frame) we have non-zero electromagnetic fields, the electric field is **E** and the magnetic field is **B**. The components of these fields appear as components of $F^{\mu\nu}$ in frame K. As $F^{\mu\nu}$ is a second-rank contravariant tensor it transforms as (from Eq. (6.17)),

$$F'^{\mu\nu} = \frac{\partial x'^\mu}{\partial x^\alpha} \frac{\partial x'^\nu}{\partial x^\beta} F^{\alpha\beta} .$$

From the definition of $\Lambda^\mu{}_\nu$ in Eq. (6.9) we can write the above equation as

$$F'^{\mu\nu} = \Lambda^\mu{}_\alpha \Lambda^\nu{}_\beta F^{\alpha\beta} , \tag{7.31}$$

which can easily be written in the matrix form as

$$F' = \Lambda F \Lambda^{\mathrm{T}} \tag{7.32}$$

where the matrix form of Λ is given in Eq. (6.10) and the matrix form of F is given in Eq. (7.14). Using the relation $(\beta\gamma)^2 - \gamma^2 = -1$ the above matrix multiplication gives

$$\Lambda F \Lambda^{\mathrm{T}} = \begin{pmatrix} 0 & -E^1 & -\gamma(E^2 - \beta B^3) & -\gamma(E^3 + \beta B^3) \\ E^1 & 0 & -\gamma(B^3 - \beta E^2) & \gamma(B^2 + \beta E^3) \\ \gamma(E^2 - \beta B^3) & \gamma(B^3 - \beta E^2) & 0 & -B^1 \\ \gamma(E^3 + \beta B^2) & -\gamma(B^2 + \beta E^3) & B^1 & 0 \end{pmatrix}$$

$$= F'. \tag{7.33}$$

Assuming $\mathbf{E}'$ and $\mathbf{B}'$ to be the transformed electromagnetic fields, in the boosted frame K' in Fig. 6.1, one can write

$$F' = \begin{pmatrix} 0 & -E'^1 & -E'^2 & -E'^3 \\ E'^1 & 0 & -B'^3 & B'^2 \\ E'^2 & B'^3 & 0 & -B'^1 \\ E'^3 & -B'^2 & B'^1 & 0 \end{pmatrix}. \tag{7.34}$$

From the above equation, we can write the transformed electromagnetic fields in K' frame as

$$\begin{aligned} E'^1 &= E^1, & B'^1 &= B^1, \\ E'^2 &= \gamma(E^2 - \beta B^3), & B'^2 &= \gamma(B^2 + \beta E^3), \\ E'^3 &= \gamma(E^3 + \beta B^2), & B'^3 &= \gamma(B^3 - \beta E^2), \end{aligned} \tag{7.35}$$

which directly shows that the electric and magnetic fields (in the laboratory frame) get mixed up in the boosted frame. The electric field in the K' frame is made up of the old electric and magnetic fields showing that the purity of the electric and magnetic fields is not respected in special relativity.

Under a rotation of the axes, where K' is obtained by rotating the K frame, along a fixed direction, by a fixed angle θ the new fields are obtained as one obtains transformed vector components under coordinate rotation. In particular if one rotates the frame K along the z-axis by an angle θ then one will obtain

$$\begin{aligned} E'^1 &= E^1 \cos\theta - E^2 \sin\theta, & B'^1 &= B^1 \cos\theta - B^2 \sin\theta, \\ E'^2 &= E^1 \sin\theta + E^2 \cos\theta, & B'^2 &= B^1 \sin\theta + B^2 \cos\theta, \\ E'^3 &= E^3, & B'^3 &= B^3, \end{aligned} \tag{7.36}$$

and in this case the electric and magnetic fields do not mix among themselves. One can generalize the above transformation of fields under more general rotations but in no cases will the electric field mix with magnetic fields and so we do not specify more general results regarding field transformation due to rotation of the coordinate system.

We can now generalize the field transformation law under a general Lorentz boost using the conventions defined in Sect. 6.5.1. In the present case, the relative velocity of frame K' makes an arbitrary angle with the common $x - x'$ axes as shown in Fig. 6.2. The axes of the frames remain parallel to each other while K' moves with respect to an observer in K frame. We will use the general setup to write down the particular result which we have written in Eq. (7.35). In deriving Eq. (7.35) the relative velocity of frame K was parallel to the common $x - x'$ axes and we will see that the general framework can accommodate the particular result written above. Using the convention set in Sect. 6.5.1, we can write the specific result in Eq. (7.35) as

$$\mathbf{E}'_\parallel = \mathbf{E}_\parallel \,, \quad \mathbf{E}'_\perp = \gamma \left[\mathbf{E}_\perp + (\boldsymbol{\beta} \times \mathbf{B}) \right] \,, \tag{7.37}$$

$$\mathbf{B}'_\parallel = \mathbf{B}_\parallel \,, \quad \mathbf{B}'_\perp = \gamma \left[\mathbf{B}_\perp - (\boldsymbol{\beta} \times \mathbf{E}) \right] \,, \tag{7.38}$$

where $\mathbf{E}_\parallel$, $\mathbf{E}'_\parallel$ and $\mathbf{B}_\parallel$, $\mathbf{B}'_\parallel$ are components of the electric and magnetic fields parallel to the relative velocity of K' with respect to frame K, similarly $\mathbf{E}_\perp$, $\mathbf{E}'_\perp$ and $\mathbf{B}_\perp$, $\mathbf{B}'_\perp$ are components of the electric and magnetic fields perpendicular to the relative velocity. In the present case, the relative velocity $\mathbf{v}$ is along $x - x'$ axes and we have chosen the 3-vector $\boldsymbol{\beta} = (\beta, 0, 0)$ where $\beta = v/c$.

We see that the particular transformations in Eq. (7.35) have now been written down in a general form using only the parallel and perpendicular components of the fields, hence we can take Eqs. (7.37) and (7.38) to be the general relations connecting the transformed fields to the old ones under a Lorentz boost. The equations above do not depend upon any particular special direction of $\mathbf{v}$ and can reduce to the particular results in Eq. (7.35) in the appropriate limit. In the general case $\mathbf{E}_\parallel = (\mathbf{E} \cdot \hat{\boldsymbol{\beta}})\hat{\boldsymbol{\beta}}$ where the direction of $\hat{\boldsymbol{\beta}}$ can be arbitrary and

$$\gamma = \frac{1}{\sqrt{1 - \boldsymbol{\beta}^2}} \,, \tag{7.39}$$

where now $\boldsymbol{\beta}^2 = \boldsymbol{\beta} \cdot \boldsymbol{\beta} = (\beta^1)^2 + (\beta^2)^2 + (\beta^3)^2$ when we have specified the arbitrary 3-vector $\boldsymbol{\beta} = (\beta^1, \beta^2, \beta^3)$. The components are specifically $\beta^i \equiv v^i/c$ where $\mathbf{v}$ is the arbitrary relative velocity of frame K' with respect to frame K. The perpendicular component of the electric field is $\mathbf{E}_\perp = \mathbf{E} - \mathbf{E}_\parallel = \mathbf{E} - (\mathbf{E} \cdot \hat{\boldsymbol{\beta}})\hat{\boldsymbol{\beta}}$. Using these results and the transformation law in Eq. (7.37) we can write

$$\mathbf{E}' = \mathbf{E}'_\parallel + \mathbf{E}'_\perp = \mathbf{E}_\parallel + \gamma \left[\mathbf{E}_\perp + (\boldsymbol{\beta} \times \mathbf{B}) \right]$$

$$= (\mathbf{E} \cdot \hat{\boldsymbol{\beta}})\hat{\boldsymbol{\beta}} + \gamma \left[\mathbf{E} - (\mathbf{E} \cdot \hat{\boldsymbol{\beta}})\hat{\boldsymbol{\beta}} + \boldsymbol{\beta} \times \mathbf{B} \right]$$

$$= \gamma \left[\mathbf{E} + (\boldsymbol{\beta} \times \mathbf{B}) \right] + (\mathbf{E} \cdot \hat{\boldsymbol{\beta}})\hat{\boldsymbol{\beta}} - \gamma (\mathbf{E} \cdot \hat{\boldsymbol{\beta}})\hat{\boldsymbol{\beta}} \,. \tag{7.40}$$

In the above equation, we can now replace $\hat{\boldsymbol{\beta}} = \boldsymbol{\beta}/\beta$ and use the relation $(\gamma - 1)/\beta^2 = \gamma^2/(\gamma + 1)$ and rewrite it as

$$\mathbf{E}' = \gamma \left[\mathbf{E} + (\boldsymbol{\beta} \times \mathbf{B}) \right] - \frac{\gamma^2}{\gamma + 1} (\mathbf{E} \cdot \boldsymbol{\beta})\boldsymbol{\beta} \,, \tag{7.41}$$

giving us the final form of the transformation of the electric field under a general Lorentz boost. In a similar way, one can write

$$\mathbf{B}' = \mathbf{B}'_\parallel + \mathbf{B}'_\perp = \mathbf{B}_\parallel + \gamma\,[\mathbf{B}_\perp - (\boldsymbol{\beta} \times \mathbf{B})]$$

$$= (\mathbf{B} \cdot \hat{\boldsymbol{\beta}})\hat{\boldsymbol{\beta}} + \gamma\left[\mathbf{B} - (\mathbf{B} \cdot \hat{\boldsymbol{\beta}})\hat{\boldsymbol{\beta}} - \boldsymbol{\beta} \times \mathbf{E}\right]$$

$$= \gamma\,[\mathbf{B} - (\boldsymbol{\beta} \times \mathbf{E})] - \frac{\gamma^2}{\gamma+1}(\mathbf{B} \cdot \boldsymbol{\beta})\boldsymbol{\beta}\,. \qquad (7.42)$$

The above two equations give the full general formula of electromagnetic field transformations under a general Lorentz boost.

As one can see from the definition of γ that it is a function of $(v^i/c)^2$ and so if we neglect powers of $(v^i/c)^2$ in some case where $v^i \ll c$ then the above equations of transformation becomes simplified to $(\gamma \sim 1)$

$$\mathbf{E}' = \mathbf{E} + (\boldsymbol{\beta} \times \mathbf{B})\,, \quad \mathbf{B}' = \mathbf{B} - (\boldsymbol{\beta} \times \mathbf{E})\,, \qquad (7.43)$$

which can be taken to be the field transformations in the *non-relativistic* limit.

7.4.1 The Electromagnetic Field of a Point Charge Moving with Uniform Velocity

Let us discuss the classic example showing how to deduce the electric and magnetic fields of a uniformly moving charge in some inertial frame K. The picture in Fig. 7.1 shows the charge moving with uniform velocity $\mathbf{v}$ in frame K. Without any loss of generality the direction of $\mathbf{v}$ is chosen along the common $x - x'$ axes direction. Due to axial symmetry all directions orthogonal to $\mathbf{v}$ are equivalent, we can freely rotate the coordinate axes keeping the $x - x'$ direction along $\mathbf{v}$ intact. The point where the fields are calculated is shown as the point P in the figure. This point remains static in frame K and moves relative to an observer in frame K'. The distance of point P from O' at any time t' in K' is called r'. As we can rotate the coordinate axes freely perpendicular to the direction of $\mathbf{v}$, we can always choose the point P to be lying on

Fig. 7.1 The two inertial frames K and K' and their relative motion. A charge q is at rest at O' and moving with uniform velocity with respect to frame K. The fields are evaluated at point P which is r' distance from O' at time t'. Here length of the line OP in K (or in K') frame is b which is a constant

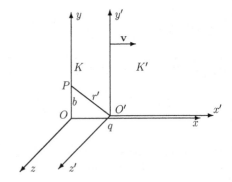

the y-axis in frame K. The distance OP is a constant called b. We orient the z' and y' axes in such a way that they always remain parallel to the z and y axes in frame K. The resulting orientation of the axes is shown in Fig. 7.1. The axes are chosen in such a way that they coincide at $t = t' = 0$.

In the present case, the 4-coordinates of P in frame K are $(ct, 0, b, 0)$. The time coordinates of point P are related via the inverse Lorentz transformations as specified in Eq. (6.4), $x'^0 = \gamma(x^0 + \beta x^1) = \gamma(ct - 0)$ as $x^1 = 0$. Here $\beta = v/c$ and $\gamma = 1/\sqrt{1 - (v/c)^2}$. The above relation gives

$$t' = \gamma t , \tag{7.44}$$

time is dilated in the K' frame as the clock in K is moving with respect to a K' observer at rest. Similarly the x' coordinate is obtained by the inverse Lorentz transformations as $x'^1 = \gamma(x^1 - \beta x^0) = -\gamma vt$. In frame K' we have $x'^1 = -vt'$ as the point P moves away uniformly in the negative x' direction. So we have $x'^1 = -vt' = -\gamma vt$ which again reproduces the result of Eq. (7.44). The 4-coordinates of point P in K' frame are then $(ct', -vt', b, 0)$. One can easily see that $r'^2 = b^2 + (x'^1)^2 = b^2 + v^2 t'^2 = b^2 + \gamma^2 v^2 t^2$.

In the K' frame the charge is at rest and so we can easily write the fields at P to be:

$$\mathbf{E}' = \frac{q}{r'^3}\mathbf{r}' , \quad \mathbf{B}' = 0 ,$$

which yields

$$E'_x = \frac{q}{r'^3}x'^1 , \quad E'_y = \frac{q}{r'^3}x'^2 , \quad E'_z = \frac{q}{r'^3}x'^3 ,$$

which ultimately gives the field components in K' at P:

$$E'_x = -\frac{qvt'}{(b^2 + v^2 t'^2)^{3/2}} , \quad E'_y = \frac{qb}{(b^2 + v^2 t'^2)^{3/2}} , \quad E'_z = 0, \tag{7.45}$$

$$B'_x = 0 , \quad B'_y = 0 , \quad B'_z = 0 . \tag{7.46}$$

As the point O', where the charge is situated, is static in K' frame we do not have any magnetic field at point P for a K' observer. One must note that the electric field components are time-dependent while the magnetic field components are all constant (in fact zero) in K' frame. This is because the time dependence of the electric field arises due to the motion of the field point P in frame K' where the charge is at rest. Although the electric field at point P is time-dependent, the field at P is an instantaneous electrostatic field.

The fields in the K frame where the charge is actually moving with uniform velocity can now be written by applying the inverse transformations obtained from Eq. (7.35). Due to the principle of relativity one can always write the inverse relations from Eq. (7.35) as

$$
\begin{aligned}
E_x &= E'_x, & B_x &= B'_x, \\
E_y &= \gamma(E'_y + \beta B'_z), & B_y &= \gamma(B'_y - \beta E'_z), \\
E_z &= \gamma(E'_z - \beta B'_y), & B_z &= \gamma(B'_z + \beta E'_y).
\end{aligned} \tag{7.47}
$$

From the above set of equations and the field components in K' frame one can write

$$
E_x = E'_x = -\frac{qv\gamma t}{(b^2 + \gamma^2 v^2 t^2)^{3/2}}, \qquad B_x = 0, \tag{7.48}
$$

$$
E_y = \gamma E'_y = \frac{q\gamma b}{(b^2 + \gamma^2 v^2 t^2)^{3/2}}, \qquad B_y = 0, \tag{7.49}
$$

$$
E_z = 0, \qquad\qquad\qquad B_z = \gamma\beta E'_y = \beta E_y. \tag{7.50}
$$

One must note that the time dependence of the fields in the K frame are due to genuine electromagnetic effects caused by the motion of the charge. Due to the motion of the charge a magnetic field appears in the K frame.

In both K and K' frame, we see that the x component of the electric fields change sign. At $t = t' = 0$ the two frames coincide and $E_x = E'_x = 0$ at that moment. Before this moment both t and t' were negative and the fields were positive, whereas after $t = 0$ the x component of the electric fields turned negative. The y components of the electric field are always positive and attains the maximum value at $t = 0$.

In the non-relativistic limit, we have $\gamma \sim 1$ and β remains much, much smaller than unity. In this limit we see that

$$
B_z = \beta E_y \sim \frac{\beta q b}{r'^3} = \frac{v}{c} q \frac{r' \sin\theta}{r'^3},
$$

where θ is the angle between $O'P$ and $O'O$ in Fig. 7.1 and $r' \sim (b^2 + v^2 t^2)^{1/2}$. In vector form the above equation can be written as

$$
\mathbf{B} \sim \frac{q}{c} \frac{\mathbf{v} \times \mathbf{r}'}{r'^3}, \tag{7.51}
$$

which is a form of the Biot-Savart law in magnetostatics. The reproduction of the Biot-Savart law shows that our covariant formulation of electrodynamics can produce well-known laws of electrodynamics known previously.

7.4.2 Covariant form of the Lorentz Force Law

Till now we have seen how electromagnetic fields are produced by charges and currents acting as sources. The theory will not be complete if we do not have the reverse interactions which predict the effect of electromagnetic fields on the sources. This effect is given by the Lorentz force law in electrodynamics. The question is can we write a tensor equation for the force which a charged particle feels in presence of electromagnetic fields? If the 4-momentum of the charged particle is p^μ then this force is $f^\mu = dp^\mu/d\tau$ as given in Eq. (6.50). Here, τ is the proper time of the inertial frame which coincides with the charged particles rest frame instantaneously. The charged particle feeling the force will in general accelerate and consequently to define proper time we require a series of inertial frames which coincide with the charged particle's (actual) frame momentarily. From our understanding of the Lorentz force law, as written in non-covariant form,

$$\frac{d\mathbf{p}}{dt} = q\left(\mathbf{E} + \frac{1}{c}\mathbf{v} \times \mathbf{B}\right), \tag{7.52}$$

we can assume that this 4-force is first order in the electromagnetic fields. In the most general way then one can write a covariant equation as

$$\frac{dp^\mu}{d\tau} = (?)u^\mu + (?)F^{\mu\nu}u_\nu + (?)\tilde{F}^{\mu\nu}u_\nu, \tag{7.53}$$

where the brackets with the question mark specify some coefficients which are made of constants such as velocity of light and/or the charge of the particle. The 4-velocity of the charged particle appears explicitly in the above equation. The brackets must contain Lorentz scalars. As we do not know what are the quantities in the brackets, the question marks are present. Immediately we can see that the first term on the right-hand side does not contain any field and so this term remains absent in the Lorentz force law. The other two terms on the right-hand side can be used to derive the covariant Lorentz force law if one neglects parity symmetry of electromagnetism. Parity operation was briefly introduced in Sect. 6.5, we will speak more over it later. As experimentally it is known that electromagnetism respects parity symmetry one is forced to drop the last term from the right-hand side of the above equation. Consequently the covariant form of the Lorentz force law must be

$$\frac{dp^\mu}{d\tau} = (?)F^{\mu\nu}u_\nu, \tag{7.54}$$

where we have to find out the quantity in the bracket and show that the above force becomes equivalent to the form as given in Eq. (7.52), in the non-relativistic limit. From the above equation, we get

$$\frac{dp^i}{d\tau} = (?)F^{i\nu}u_\nu = (?)(F^{i0}u_0 + F^{ij}u_j)$$
$$= (?)(E^i c\gamma + \epsilon^{ijk}\gamma v_j B_k),$$

where we have used $u_0 = c\gamma$ and $u_j = \gamma v_j$. Using the relation $F^{ij} = \epsilon^{ijk}B_k$ where Einstein summation convention is used, we can write the above equation as

$$\frac{dp^i}{d\tau} = (?)c\gamma \left(E^i + \frac{1}{c}\epsilon^{ijk}v_j B_k \right).$$

In our convention, for two 3-vectors $\mathbf{v}$ and $\mathbf{B}$, we have

$$(\mathbf{v} \times \mathbf{B})^i = \epsilon^{ijk}v_j B_k, \tag{7.55}$$

and consequently we can write

$$\frac{d\mathbf{p}}{d\tau} = (?)c\gamma \left(\mathbf{E} + \frac{1}{c}\mathbf{v} \times \mathbf{B} \right).$$

In the non-relativistic limit $\gamma \to 1$ and τ becomes the coordinate time t. Comparing the non-relativistic limit of the above equation with Eq. (7.52) we immediately get $(?) = q/c$, and consequently one can now write the covariant Lorentz force law as

$$\frac{dp^\mu}{d\tau} = \frac{q}{c} F^{\mu\nu} u_\nu. \tag{7.56}$$

The above 4-vector equation actually conveys a bit more than the non-relativistic Lorentz force law because the above equation also has the time component which gives

$$\frac{dp^0}{d\tau} = \frac{q}{c} F^{0i} u_i = -\frac{q}{c} F^{i0} u_i = -\frac{q\gamma}{c} E^i v_i,$$

yielding

$$\frac{dp^0}{d\tau} = \frac{q\gamma}{c}(\mathbf{E} \cdot \mathbf{v}). \tag{7.57}$$

One must note that $q\mathbf{E} \cdot \mathbf{v} = (q\mathbf{E}) \cdot \frac{d\mathbf{x}}{dt}$ gives the rate of change of the charged particle's energy in the electromagnetic field. Consequently the time component of the covariant Lorentz force law is simply related to the rate of change of the charged particle's energy in the electromagnetic field.

7.5 A Brief Discussion of Parity Symmetry

In Sect. 6.5 we discussed the parity operation. We briefly present some more information about this discrete operation in the present section. We know that under parity transformation the components of $x^\mu = (x^0, \mathbf{x})$ transforms as

$$\mathbf{x} \xrightarrow{\mathcal{P}} -\mathbf{x}, \quad x^0 \xrightarrow{\mathcal{P}} x^0, \tag{7.58}$$

and the components of 4-momentum $p^\mu = (p^0, \mathbf{p})$ transforms as

$$\mathbf{p} \xrightarrow{\mathcal{P}} -\mathbf{p}, \quad p^0 \xrightarrow{\mathcal{P}} p^0. \tag{7.59}$$

These are the basic operations by which parity transformations are defined. Next we may try to understand how does the electric field $\mathbf{E}$ transform under parity. To understand this we must note that the electric field is a normal vector and is always proportional to a unit vector in a certain direction. For a point charge at rest $\mathbf{E} \propto \frac{\mathbf{r}}{r^3}$ where $\mathbf{r}$ is the vector whose origin is on the point charge and tip is at the point where the field is evaluated, and $r = |\mathbf{r}|$. From our basic definition of parity operation, we can always say that

$$\mathbf{E} \xrightarrow{\mathcal{P}} -\mathbf{E}. \tag{7.60}$$

3-vectors which transforms under parity with a minus sign, like the electric field, are called polar 3-vectors or simply as normal 3-vectors. Actually, the above notation does not convey the whole story of parity transformation because in reality what is happening is

$$\mathbf{E}(x^0, \mathbf{x}) \xrightarrow{\mathcal{P}} -\mathbf{E}(x^0, -\mathbf{x}). \tag{7.61}$$

As the spatial vector or the 3-momentum levels always change sign under parity we always do not write down the exact dependence of the fields on the spatial coordinates or 3-momenta. The reader must accept Eq. (7.60) keeping this point in mind. Henceforth we will not write the spatial coordinate dependence of the fields anymore but will always assume something like the one presented in Eq. (7.61).

Next we want to know how does the volume current density $\mathbf{J} = \rho \mathbf{v}$ transform under parity. Here, ρ is the charge density which does not change sign under parity and $\mathbf{v}$ is the velocity of the charged lump in motion. As 3-momentum changes sign under parity we expect $\mathbf{v}$ to be also changing sign under parity. We must remember that J^0 contains ρ and consequently it does not change sign under parity, so

$$\mathbf{J} \xrightarrow{\mathcal{P}} -\mathbf{J}, \quad J^0 \xrightarrow{\mathcal{P}} J^0. \tag{7.62}$$

At the next level we can pose the question, how does the magnetic field $\mathbf{B}$ transform under parity operation. To answer this question we will use parity symmetry of classical electrodynamics. Till now we have only talked about parity operation but parity is a symmetry operation in classical electrodynamics. What we mean by symmetry operation is that if we carry out parity transformation on the laws of classical electrodynamics then the laws remain the same as before and does not transform to a new parity transformed law. Suppose we take the differential form of Ampere's law (in magnetostatics)

$$\nabla \times \mathbf{B} = \frac{4\pi}{c}\mathbf{J} \,.$$

Parity symmetry of electrodynamics predicts that the parity transformed Ampere's law must look exactly the same as the above equation. Let us now figure out how the individual terms in both sides of the above equation can change under parity such that the whole equation remains the same after the transformations. As experimentally no one has observed violation of parity symmetry in electromagnetism we can safely assume that parity is a symmetry in the present case. Under parity, $\nabla \xrightarrow{P} -\nabla$, as spatial derivatives change sign under parity, and $\mathbf{J}$ also changes sign. As a result if Ampere's law has to remain the same under parity transformation one must have

$$\mathbf{B} \xrightarrow{P} \mathbf{B} \,, \tag{7.63}$$

which shows that the magnetic field does not change sign under parity transformation. 3-vectors which do not pick up a minus sign under parity transformation are called axial 3-vectors. One can easily show that the orbital angular momentum $\mathbf{L} = \mathbf{r} \times \mathbf{p}$ is another axial 3-vector. One can now easily show that

$$A^\mu \xrightarrow{P} (\Phi, -\mathbf{A}) \,,$$

where the electromagnetic scalar potential Φ transforms as a scalar under parity.

Now we can opine on the parity transformation of $F^{\mu\nu}$ because we know how the electric field and the magnetic field transforms under parity. From what we have listed so far we can directly write

$$F^{i0} \xrightarrow{P} -F^{i0} \,, \quad F^{ij} \xrightarrow{P} F^{ij} \,, \tag{7.64}$$

and

$$\tilde{F}^{i0} \xrightarrow{P} \tilde{F}^{i0} \,, \quad \tilde{F}^{ij} \xrightarrow{P} -\tilde{F}^{ij} \,. \tag{7.65}$$

The $\tilde{F}^{\mu\nu}$ tensors transform as pseudo-tensors under parity. Now we will be able to explain why the last term on the right-hand side of Eq. (7.53) was dropped due to parity symmetry. Writing the 3-vector part of Eq. (7.53) (with the first term on the right-hand side dropped) we have

$$\frac{dp^i}{d\tau} = (?)F^{i\nu}u_\nu + (?)\tilde{F}^{i\nu}u_\nu$$
$$= (?)(F^{i0}u_0 + F^{ij}u_j) + (?)(\tilde{F}^{i0}u_0 + \tilde{F}^{ij}u_j) \,.$$

From the above equation, we see that the left-hand side transforms as a normal vector (acquires a minus sign) under parity transformation which implies the right-hand side should also acquire a minus sign under parity transformation. Here, we have assumed the terms in the brackets with question marks to be made up of charge and velocity of light both of which are normal scalars which remain invariant under parity transformation, so these brackets are untouched by parity transformations. Now let us concentrate on the individual terms as $F^{i0}u_0$, which acquires a minus sign under parity because F^{i0} does so and u_0 is unaffected by parity. Similarly, you can convince yourself that $F^{ij}u_j$ also acquires a minus sign under parity transformation. On the other hand $\tilde{F}^{i0}u_0$ and $\tilde{F}^{ij}u_j$ do not acquire negative signs under a parity transformation. So we have a problem, the right-hand side of the above equation does not totally transform as a normal vector whereas the left-hand side does. To eradicate this problem one has to drop the term containing $\tilde{F}^{\mu\nu}u_\nu$ by choosing the content of the bracket (with the question mark) multiplying it to be zero. In such a case parity transformation property of both sides of the last equation remains identical.

7.6 Problems

1. Suppose a material medium has electromagnetic fields specified by $F^{\mu\nu}$ inside it, and has a 4-velocity vector (of its center of momentum) u^μ which is normalized as $u^\mu u_\mu = 1$. Show that

   (a) the electric field in the rest frame of the medium is given by $-F^{\mu\nu}u_\mu$,
   (b) the magnetic field in the rest frame of the medium is given by $-\tilde{F}^{\mu\nu}u_\mu$.

2. Derive the form of the matrix $\Lambda F \Lambda^T$ as given in Eq. (7.33) by explicit matrix multiplication.

3. Show that $F^{ij} = \epsilon^{ijk}B_k$ where the symbols have their natural meaning.

4. Suppose we have two reference frames K and K' whose axes are parallel. The frame K' is moving with a velocity $\mathbf{v} = c\boldsymbol{\beta}$ making an arbitrary angle with the common $x - x'$-axes. In the frame K', we have a certain 4-vector S'^μ whose components are $S'^\mu = (0, \mathbf{S}')$. In the K frame, this vector is transformed to $S^\mu = (S^0, \mathbf{S})$.

   (a) From this information find out what is $u^\alpha S_\alpha$ where $u^\alpha = \gamma(c, \mathbf{v})$ and γ is conventionally defined.
   (b) Express S^0 in terms of the vector $\mathbf{S}$. The result must involve some inner product which you can guess. Find out the appropriate inner product in this case.
   (c) Suppose I let you know that S^μ is an axial 4-vector attached to a charged particle moving with 4-velocity u^α in presence of an electromagnetic field in the K frame. The electromagnetic field affects the charge via Lorentz force. The 4-vector S^μ is also affected by the electromagnetic fields and it changes in time due to this interaction with the fields.
   We want to know $ds^\mu/d\tau$ where we have the information that this rate is linearly dependent on the electromagnetic fields and S^μ itself. Can you propose a possible general Lorentz covariant form of $ds^\mu/d\tau$ in such a case.

This proposed form must have unknown coefficients, the tensor $F^{\mu\nu}$ and the vectors u^{μ}, S^{μ} and $du^{\mu}/d\tau$ suitably contracted. The form must not depend on other higher time derivatives of u^{μ}.

5. We have seen that the time variation of the electric field in the K' frame, as given in Eq. (7.46), does not give rise to any magnetic field. Does this fact contradict Ampere's law which has the Maxwell correction? Here the source current is zero.

6. Suppose we have a line, along the x-axis, with uniform charge density η per unit length. The line is infinitely long. Calculate the electric field due to this line charge at any point outside the line. Does this system admit any magnetic field? Now suppose we look at the same line charge from an inertial frame moving with uniform speed v in the direction of the positive x-axis. Calculate the electric and magnetic fields in that inertial frame. What kind of fields do you now see? From the result can you infer about the charge density and current in the line? If there is a magnetic field in the moving frame, can you connect the magnetic field with Ampere's law?

7. Find out the trajectory of a charged particle in presence of a constant electric field (and zero magnetic field) along the x-axis using the relativistic form of the Lorentz force law. Assume the particle was at $x = 0$ at $t = 0$. You can also try to find out the trajectory of the charged particle in presence of a constant magnetic field (and zero electric field) along the y-axis using the relativistic Lorentz law. Lastly, suppose you have a charged particle in a reference frame where the electric field is along the x-axis and magnetic field along the y-axis. At $t = 0$ the velocity was $\mathbf{v} = (v_{0x}, v_{0y}, v_{0z})$. Find out an inertial frame (by properly boosting your frame) where the electric field vanishes. How will the charge move in that frame? From this information can you predict the nature of motion of the charge in the laboratory frame? The last answer is complicated in general, to make life simple you can try to answer it in the non-relativistic limit. The general answer (for the relativistic case) may require a numerical solution and you may use a computer program to solve it.

Chapter 8
Gauge Invariance of Electrodynamics

Now we know the Maxwell equations in absence of any bound charges and bound currents as

$$\partial_\mu F^{\mu\nu} = \frac{4\pi}{c} J^\nu , \quad \partial_\mu \tilde{F}^{\mu\nu} = 0 ,$$

where J^μ is the free current density. The Lorenz force equation is

$$\frac{dp^\mu}{d\tau} = \frac{q}{c} F^{\mu\nu} u_\nu ,$$

where $p^\mu = mu^\mu$ is the 4-momentum of the charged particle, with charge q, in the electromagnetic field. Here m is the mass of the charged particle and u^μ is its 4-velocity. The source current $J^\mu = (c\rho, \mathbf{J})$ where ρ is the charge density and $\mathbf{J}$ is the 3-current density of the source. In the above discussion, the antisymmetric field strength tensor is $F^{\mu\nu} = \partial^\mu A^\nu - \partial^\nu A^\mu$ where the 4-potential $A^\mu = (\Phi, \mathbf{A})$. Here, Φ is the scalar electromagnetic potential and $\mathbf{A}$ is the magnetic vector potential. These equations are enough to specify the electrodynamical evolution of a system consisting of charges, currents and electromagnetic fields. Here we must point out that the above equations are not yet complete, there are some redundancy. Suppose one applies a transformation to the electromagnetic 4-potential as

$$A^\mu(x) \rightarrow A'^\mu(x) = A^\mu(x) + \partial^\mu \chi(x) , \tag{8.1}$$

where $\chi(x)$ is a normal differentiable scalar function in four dimensions, then $F^{\mu\nu}(x)$ does not change. In the above case $x \equiv (x^0, \mathbf{x})$. As a result we can see that all the relevant electrodynamic equations written down in the beginning of the chapter remains unchanged under the transformation given in Eq. (8.1). We will call such a transformation to be a gauge transformation and moreover from now on often call the 4-potential $A^\mu(x)$ to be a gauge field. It has to be noted that the gauge transformation

© Springer Nature Singapore Pte Ltd. 2021
K. Bhattacharya and S. Mukhopadhyay, *Introduction to Advanced Electrodynamics*,
https://doi.org/10.1007/978-981-16-7802-8_8

takes place at one spacetime point marked by x. One can also represent the gauge transformation in the component form (in terms of Φ and $\mathbf{A}$ which are components of A^μ) as $A'^0(x) = A^0(x) + \frac{\partial \chi(x)}{\partial (ct)}$ and $A'^i(x) = A^i(x) + \frac{\partial \chi(x)}{\partial x_i}$ which can be written as

$$\Phi'(x) = \Phi(x) + \frac{1}{c}\frac{\partial \chi(x)}{\partial t}, \quad \mathbf{A}'(x) = \mathbf{A}(x) - \nabla \chi(x). \tag{8.2}$$

As gauge transformations do not change the basic Maxwell's equations of electrodynamics or even the Lorentz force law we interpret the gauge transformations as symmetry transformations in electrodynamics. One must note that the electric field $\mathbf{E}$ and magnetic field $\mathbf{B}$ remains the same under gauge transformations as they are components of $F^{\mu\nu}$ which remains invariant under a gauge transformation. This particular symmetry makes electrodynamics more complex and subtle.

The complexity of gauge transformations can be felt in various ways, here we present one point which amply clarifies the issue. The Maxwell equations in the covariant form yields the functional form of the 4-potential as solutions when we know the source charge density and 3-current density. But due to gauge freedom the solution is arbitrary, one may get a solution $A^\mu(x)$ and another one doing the same problem may get a different solution $A'^\mu(x)$ which are apparently not identical but related by a gauge transformation as given in Eq. (8.1). It is seen that the theory in terms of A^μ becomes arbitrary, as a unique solution does not exist as all solutions can be gauge transformed into a different set of solutions. This complexity was not present in electrodynamics when the Maxwell's equations were written in terms of the electric and magnetic fields. The problem appeared when we started to talk in the language of the 4-potential. Can one then avoid this extra complexity by not using the 4-potential? The answer is no, as the covariant formulation of electrodynamics can only be written in terms of the gauge fields and not in terms of the electric and magnetic fields so a covariant formulation of electrodynamics without gauge symmetry is impossible.

Faced with such a redundancy what can be the way out? In general this redundancy implies that the Maxwell's equation $\partial_\mu F^{\mu\nu} = \frac{4\pi}{c} J^\nu$ written explicitly as

$$\partial_\mu \partial^\mu A^\nu - \partial_\mu \partial^\nu A^\mu = \frac{4\pi}{c} J^\nu, \tag{8.3}$$

in terms of the gauge potential has some extra 'spurious' degrees of freedom. This extra degree of freedom makes the solution arbitrary up to a gauge transformation. To remove this redundancy one has to apply extra conditions on A^μ such that the gauge field loses the ability of further gauge transformations. These extra conditions have to be imposed to make the gauge field solution unique. As because these extra conditions do not arise from fundamental physics (like Maxwell's theory) one can have multiple choices. The choices depend upon the context, if one is trying to do explicit relativistic calculations then one condition is applied on the other hand if relativistic formulation is not a priority then some other condition is applied. Next we will present the various gauge fixing conditions.

8.1   The Lorenz Gauge

Suppose that the gauge field $A^\mu(x)$ satisfies Eq. (8.3) as well as

$$\partial^\mu A_\mu(x) = \frac{1}{c}\frac{\partial \Phi(x)}{\partial t} + \nabla \cdot \mathbf{A}(x) = 0, \tag{8.4}$$

where $x = (t, \mathbf{x})$. The condition imposed above goes by the name Lorenz gauge condition which is explicitly Lorentz invariant.[1] If one applies a Lorentz transformation then the Lorenz gauge condition does not change its form. First of all we must be clear about one fact, although the Lorenz gauge condition is applied in an ad hoc fashion, the choice can always be made without violating any general principles. Suppose one is working with a certain A^μ which does not satisfy the above condition but only satisfies Eq. (8.3). As Eq. (8.3) permits gauge transformations one can now apply a gauge transformation as $A'^\mu = A^\mu + \partial^\mu \chi$ in such a way that $\partial^\mu A'_\mu = 0$. As a result

$$\partial^\mu A_\mu + \partial_\mu \partial^\mu \chi = 0,$$

which can always be satisfied for well behaved $A^\mu(x)$ and $\chi(x)$. Given any arbitrary A^μ which satisfies Eq. (8.3) we can always gauge transform to a new A'^μ which satisfies both Eq. (8.3) and Eq. (8.4). Once the Lorenz gauge is chosen the gauge symmetry of the problem is severely restricted but is not extinguished as we will see shortly. One can rewrite Eq. (8.3) using the commutative property of partial derivatives as

$$\partial_\mu \partial^\mu A^\nu - \partial^\nu(\partial_\mu A^\mu) = \frac{4\pi}{c}J^\nu, \tag{8.5}$$

and applying the Lorenz gauge condition the inhomogeneous Maxwell equation becomes

$$\Box A^\nu = \frac{4\pi}{c}J^\nu, \tag{8.6}$$

where the d'Alembart operator

$$\Box \equiv \partial_\mu \partial^\mu = \frac{1}{c^2}\frac{\partial^2}{\partial t^2} - \nabla^2.$$

---

[1] The reader must be careful about the names, the gauge condition goes by the name Lorenz gauge and the force felt by a charge in electromagnetic field is the Lorentz force. The names are different because they are originating from the names of two different individuals, one is Ludvig Valentin Lorenz and the other one from Hendrik Antoon Lorentz. Lorenz was a Danish physicist while Lorentz was a Dutch physicist.

In the Lorenz gauge the homogeneous Maxwell equation and the Lorentz force law remain as they were before. The Maxwell equations for Φ and $\mathbf{A}$ are obtained from the $\nu = 0$ and $\nu = i$ (where $i = 1, 2, 3$) components of Eq. (8.6) and they are:

$$\nabla^2 \Phi - \frac{1}{c^2} \frac{\partial^2 \Phi}{\partial t^2} = -4\pi\rho, \tag{8.7}$$

$$\nabla^2 \mathbf{A} - \frac{1}{c^2} \frac{\partial^2 \mathbf{A}}{\partial t^2} = -\frac{4\pi}{c} \mathbf{J}. \tag{8.8}$$

Once we find solutions for Φ and $\mathbf{A}$ they satisfy the Lorenz gauge condition. The Lorenz gauge condition as stated above does not extinguish gauge transformation property of Φ and $\mathbf{A}$, we will now show that a different gauge transformation is still allowed.

Suppose we have obtained Eq. (8.6) after applying the Lorenz gauge condition as given in Eq. (8.4). Next we apply another gauge transformation as $A'^\nu = A^\nu + \partial^\nu \Lambda$ for some well-behaved scalar function $\Lambda(x)$. In such a case Eq. (8.6) changes to

$$\Box A'^\nu + \Box \partial^\nu \Lambda = \Box A'^\nu + \partial^\nu \Box \Lambda = \frac{4\pi}{c} J^\nu,$$

which is a different equation and may contain new physics. Here we have used the commutative property of partial differential operators to take the ∂^ν to the right side of the d'Alembart operator. Apparently such a transformation must not be a gauge transformation as it changes the basic electrodynamic equation obtained in the Lorenz gauge, but if one imposes the restriction

$$\Box \Lambda(x) = 0, \tag{8.9}$$

then one again attains the inhomogeneous Maxwell equation in the Lorenz gauge. This whole exercise shows that one has the freedom of a restricted gauge transformation when working in the Lorenz gauge. The Lorenz gauge does not completely take away the arbitrariness of the gauge potential.

In the Lorenz gauge we see that both Φ and $\mathbf{A}$ satisfies similar equations, the reason being that the Lorenz gauge is explicitly Lorentz invariant and keeps the relativistic nature of the Maxwell equation intact. In calculations where one uses the relativistic formalism one is bound to use the Lorenz gauge condition. Before we end this discussion on the Lorenz gauge it is pertinent to point out that the homogeneous Maxwell's equations are not sensitive to any particular gauge choice and so we have not explicitly written down how those equations look in Lorenz gauge. The reason for this is related to the nature of the homogeneous Maxwell equations as given in Eq. (7.27). The homogeneous equations are essentially constraints arising out of the mathematical structure of $F^{\mu\nu}$, which remains true for any gauge we choose.

In many cases one does not require explicitly relativistic formalism and in those cases one may use other gauge conditions. One often used gauge condition which is not explicitly Lorentz invariant is the Coulomb gauge condition.

8.2 The Coulomb Gauge

Another widely used gauge condition, which is used in classical electrodynamics as well as in semi-classical theory of radiation, is called the Coulomb gauge. This gauge condition is sometimes called the radiation gauge and in some cases it is also called the transverse gauge. The gauge solutions in the Coulomb gauge condition must satisfy

$$\nabla \cdot \mathbf{A}(x) = 0 \,, \tag{8.10}$$

where $x = (t, \mathbf{x})$. The above gauge condition immediately shows that it is more useful in cases where explicit covariant description of electrodynamics is not required as the gauge condition is not covariant. If we make a Lorentz transformation the above gauge condition will change. If we Fourier transform $\mathbf{A}(x)$ and write it as

$$\mathbf{A}(x) = \frac{1}{(2\pi)^2} \int_{-\infty}^{\infty} d^4k \, \mathbf{A}(k) \, e^{ik \cdot x} \,,$$

where we are following the convention introduced in Sect. 6.8 of Chap. 6, then the gauge condition in the wave vector space becomes $\mathbf{k} \cdot \mathbf{A}(t, \mathbf{k}) = 0$. This shows that $\mathbf{A}(k)$ is transverse to $\mathbf{k}$, and hence the name transverse gauge. As discussed for the case of the Lorenz gauge, in the present case also it can easily be shown that, one can always choose to work in the Coulomb gauge using the gauge transformation property of Maxwell's equations.

One can write the inhomogeneous Maxwell equation as given in Eq. (8.5) in the Coulomb gauge as:

$$\partial_\mu \partial^\mu A^\nu - \partial^\nu (\partial_0 A^0) = \frac{4\pi}{c} J^\nu \,. \tag{8.11}$$

This equation can be written in terms of the gauge field components as

$$\nabla^2 \Phi = -4\pi\rho \,, \quad \nabla^2 \mathbf{A} - \frac{1}{c^2} \frac{\partial^2 \mathbf{A}}{\partial t^2} = -\frac{4\pi}{c} \mathbf{J} + \frac{1}{c} \nabla \left(\frac{\partial \Phi}{\partial t} \right) \,. \tag{8.12}$$

These are the relevant inhomogeneous Maxwell equations. The first equation looks exactly like the Coulomb's law in electrostatics, albeit here both $\Phi(t, \mathbf{x})$ and $\rho(t, \mathbf{x})$ are functions of time explicitly. The integral solution for the scalar potential turns out to be

$$\Phi(t, \mathbf{x}) = \int_{-\infty}^{\infty} \frac{\rho(t, \mathbf{x}')}{|\mathbf{x} - \mathbf{x}'|} d^3x' \,, \tag{8.13}$$

where the three dimensional integral on $\mathbf{x}'$ is over the source region. The similarity of this solution with the electrostatic case justifies the name of this gauge choice

Fig. 8.1 The source point is enclosed in the small rectangle at the tip of $\mathbf{x}'$ vector and the field point is at P. In reality the source region can be of arbitrary shape, may be localized, and be one, two or three dimensional

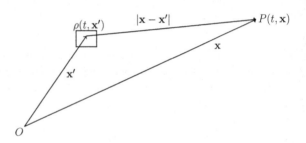

as Coulomb gauge. Figure 8.1 shows the source point and the field point P where the scalar potential is evaluated. In the source region $\rho(t, \mathbf{x}') \neq 0$. As for the scalar potential at P the source points instantaneously affect it, the non-relativistic nature of the problem is apparent in this gauge choice.

To simplify the second Maxwell equation in Coulomb gauge one writes the 3-current density as

$$\mathbf{J} = \mathbf{J}_L + \mathbf{J}_T \,, \tag{8.14}$$

where $\mathbf{J}_L$ is called the longitudinal current and $\mathbf{J}_T$ is called the transverse current which satisfies

$$\nabla \times \mathbf{J}_L = 0 \,, \quad \nabla \cdot \mathbf{J}_T = 0 \,. \tag{8.15}$$

From the way the longitudinal current is defined it can also be called irrotational current and the transverse current can also be called the solenoidal current. If one works in the Fourier space where $\mathbf{J}_{L,T}(t, \mathbf{k})$ are functions of the wave vector $\mathbf{k}$ we will have $\mathbf{J}_L(t, \mathbf{k}) \propto \mathbf{k}$ and $\mathbf{k} \cdot \mathbf{J}_T(t, \mathbf{k}) = 0$. From the definition of the longitudinal current, one can write $\mathbf{J}_L = \nabla S(t, \mathbf{x})$ where $S(t, \mathbf{x})$ is a function of t and $\mathbf{x}$. If we take

$$S = \frac{1}{4\pi} \frac{d\Phi}{dt} \,, \tag{8.16}$$

then the equation for the vector potential in Coulomb gauge becomes

$$\nabla^2 \mathbf{A} - \frac{1}{c^2} \frac{\partial^2 \mathbf{A}}{\partial t^2} = -\frac{4\pi}{c} \mathbf{J} + \frac{4\pi}{c} \mathbf{J}_L = -\frac{4\pi}{c} (\mathbf{J} - \mathbf{J}_L) \,,$$

which can also be written as

$$\nabla^2 \mathbf{A} - \frac{1}{c^2} \frac{\partial^2 \mathbf{A}}{\partial t^2} = -\frac{4\pi}{c} \mathbf{J}_T \,, \quad \text{and} \quad \nabla \left(\frac{\partial \Phi}{\partial t} \right) = 4\pi \mathbf{J}_L \,. \tag{8.17}$$

These show that in the Coulomb gauge the transverse gauge field equation has only $\mathbf{J}_T$ as source. The longitudinal part of the 3-current density does not have anything to do with the 3-vector potential in the Coulomb gauge whereas the longitudinal current sources the scalar potential in the Coulomb gauge. In the Coulomb gauge in the absence of sources, when $\rho(t, \mathbf{x}) = 0$, we simply have

$$\mathbf{E} = -\frac{1}{c}\frac{\partial \mathbf{A}}{\partial t}, \quad \mathbf{B} = \nabla \times \mathbf{A}, \tag{8.18}$$

which are simple expressions. We do not get this simplification in the Lorenz gauge.

8.3 Problems

1. Suppose in a particular situation we have a constant electric field, E_0, along the x-axis. First find out a gauge choice where $\mathbf{A} = 0$ and $\Phi = \Phi(\mathbf{x})$ which produces such a field configuration. Next figure out another gauge choice where $\mathbf{A} = \mathbf{A}(t)$ and $\Phi = 0$ which produces the same electric field along the x-axis. Find out how one can gauge transform from the first gauge configuration to the second gauge configuration.

2. Suppose we have a constant magnetic field, of magnitude B_0, along the z-axis. Find out all the distinct possible gauge potentials which can give rise to such a magnetic field. What are the gauge transformations which connect one choice to the other?

3. There can be certain kind of gauge configurations which actually do not produce any physical electric or magnetic field. These kinds of gauge configurations are called pure gauge fields. Show that a pure gauge configuration can be obtained from the gauge transformation from $A^\mu(x) = 0$.

4. In this chapter we have worked with two gauge choices, the Lorenz gauge and the Coulomb gauge. These are not all the gauge choices with which people work. Suppose we are dealing with the temporal gauge where $\Phi(x) = 0$. Write down the inhomogeneous Maxwell equations in this gauge.

5. Another gauge which is of considerable theoretical interest is called the velocity gauge. Here the gauge condition is

$$\nabla \cdot \mathbf{A}(x) + \frac{c}{v^2}\frac{\partial \Phi(x)}{\partial t} = 0,$$

where v is a constant parameter with the dimension of velocity. Using this gauge condition write down the inhomogeneous Maxwell equations. The significance of this gauge choice is related to its generality. Find out the values of v for which the velocity gauge condition reduces to the Lorenz gauge condition or the Coulomb gauge condition.

Chapter 9
Action Principle in Electrodynamics

Micheal Faraday's famous experiment on iron filing sprinkled on a piece of paper with a magnet held underneath led him to the discovery of what he called the lines of force, later renamed as field by William Thompson. When James Clerk Maxwell became aware of Faraday's work, he was particularly impressed by the idea of 'lines of force traversing all space'. Having already worked on the idea of vector field in fluid dynamics, Maxwell eventually developed the first field theory: the theory of electromagnetic (radiation) field summarized in the four Maxwell's equations. However, Maxwell's equations suffered from, what then seemed to be, a fundamental inconsistency: that of the equations not being invariant under Galilean transformation. It was left to Albert Einstein to unify Maxwell's electromagnetic field theory with 'special relativity'.

In this chapter, we will try to derive most of the things discussed till now from an underlying action principle. The advantage of the action principle is that it can be applied to the case of particles as well as fields. We will summarize the results of the action principle as applied for particles and then apply it for a particle in electromagnetic field to get the Lorentz force law. Then we generalize the action principle to the case of classical fields and apply the action principle to derive the Maxwell's equations of electrodynamics.

9.1 Classical Particle Mechanics

The laws of classical mechanics are deterministic. The basic problem of classical mechanics is to write down the equations of motion of a particle or a system of particles with given mass. The particle trajectories/orbits can be completely specified or predicted by solving these equations of motion if the initial position ($\mathbf{x}$) and velocity ($\mathbf{v}$) are known. The particle dynamics in configuration space is characterized by a function $L = L(x_j, \dot{x}_j, t)$, called the Lagrangian. The system may contain N number

© Springer Nature Singapore Pte Ltd. 2021   239
K. Bhattacharya and S. Mukhopadhyay, *Introduction to Advanced Electrodynamics*,
https://doi.org/10.1007/978-981-16-7802-8_9

of particles specified by the index j which runs from 1 to $3N$. The Lagrangian is constructed as the difference between kinetic energy T (a quadratic function of $\dot{x}_j$) and potential energy V (usually a function of position x_j except in circumstances such as motion of a charged particle in magnetic field) viz. $L = T - V$. It turns out that given the initial and final positions (x_i, t_i) and (x_f, t_f), respectively, the particle trajectory in configuration space can be uniquely determined subject to the condition that the action defined as:

$$S[L] = \int_{t_i}^{t_f} L(x_j, \dot{x}_j, t)\, dt \tag{9.1}$$

is an extremum or in other words $\delta S = 0$. Here, $S[L]$ designates that the action is a functional of the Lagrangian L. Given a Lagrangian function $L(x_j, \dot{x}_j, t)$ and a path in configuration space, we obtain the action S and so S transforms a function to a number unlike a function which transforms numbers to numbers. In this sense, the action is not a function of the Lagrangian but a functional. In the variation of the action we have assumed the two end points in configuration space are fixed such that

$$\delta x_j(t_i) = \delta x_j(t_f) = 0, \tag{9.2}$$

for $j = 1 \cdots 3N$. Assuming independent variations of δx_js, one obtains the $3N$ (if there are no constraints) second order Euler-Lagrange equations of motion

$$\frac{d}{dt}\left(\frac{\partial L}{\partial \dot{x}_j}\right) - \frac{\partial L}{\partial x_j} = 0. \tag{9.3}$$

One can apply the above equations to solve particle mechanics problems if one can construct the Lagrangian function properly.

The description of particle dynamics in configuration space using the Lagrangian formulation can be replaced by an equivalent Hamiltonian formulation where the position x_j and momentum p_j variables are treated as independent and particle dynamics is described in phase space. We now make the transition from Lagrangian description specified by independent variables $x_j, \dot{x}_j$ to Hamiltonian formulation with independent variables x_j, p_j, where p_j is called the canonical momentum defined as

$$p_j = \frac{\partial L}{\partial \dot{x}_j}. \tag{9.4}$$

In our discussion, an index up and an index down does not make any difference as for classical mechanics we use the Euclidean metric as specified in Eq. (7.1). As a result j may appear as a lower (or upper) index on both sides of the above equation. The Hamiltonian $H(x_j, p_j, t)$ is then given by

$$H = \sum_{j=1}^{3N} p_j \dot{x}_j - L(x_j, \dot{x}_j, t). \tag{9.5}$$

Assuming that the relation in Eq. (9.4) is invertible, which means we can obtain $\dot{x}_j = \dot{x}_j(x_j, p_j)$ from it, we can express the above Hamiltonian as a function of x_j and p_j. One may apply the action principle in the phase space to obtain the first order Hamilton's equations of motion

$$\dot{x}_j = \frac{\partial H}{\partial p_j}; \quad \dot{p}_j = -\frac{\partial H}{\partial x_j}. \tag{9.6}$$

Again, we have not put much emphasize on the position of the index j as we are using Euclidean metric in the present case. One may work with the Lagrangian formalism or the Hamiltonian formalism according to need. The formalisms are generally equivalent and give similar equations of motion. These formalism can be generalized to the classical fields. It is an interesting point to note that, in some cases, one can write down a Lagrangian description of field theory but the equivalent Hamiltonian description cannot be done easily. A particular example where the Hamiltonian formalism is problematic is classical electromagnetism itself. We will briefly opine on this topic later in this chapter.

9.1.1 The Lagrangian of a Relativistic Free Particle

We now construct the free particle Lagrangian in the relativistic limit. The relativistic free particle Lagrangian has to fulfill the following two criteria, the action should be invariant under Lorentz transformation and the Lagrangian should reduce to $L = \frac{1}{2}mv^2$ in the non-relativistic limit ($c \to \infty$). Here $v^2 = \sum_{i=1}^{3} \dot{x}^i \dot{x}^i$. For a single particle the only Lorentz scalar is $ds^2 = \eta_{\mu\nu} dx^\mu dx^\nu$ and so we can start with

$$S = -\alpha \int_{s_1}^{s_2} ds,$$

where α is a constant and fixes the dimension of the right-hand side. The minus sign assures us that the minimum of this action will give us the proper Euler-Lagrange equations of motion. We know that the proper time

$$d\tau = dt \sqrt{1 - \frac{v^2}{c^2}},$$

is a Lorentz invariant. Keeping this in mind we can now write

$$S = -\int_{\tau_1}^{\tau_2} \alpha c \, d\tau \, .$$

One can also write the above relation in the conventional sense as

$$S = \int_{t_1}^{t_2} L(x, \dot{x}, t) dt \, ,$$

where t is the coordinate time. From the above equations, one can now easily guess that the Lagrangian for a single relativistic particle must be

$$L = -\alpha c \sqrt{1 - \frac{v^2}{c^2}} \, .$$

In the non-relativistic limit, one can expand L as

$$L = -\alpha c \sqrt{1 - \frac{v^2}{c^2}} \simeq -\alpha c (1 - \frac{v^2}{2c^2}) = -\alpha c + \frac{\alpha v^2}{2c} .$$

The first term in the Lagrangian is a constant and hence can be excluded. In the non-relativistic limit the second term should be equal to $\frac{1}{2} m v^2$, which means $\alpha = mc$. We can now write down the Lagrangian of a relativistic free particle as

$$L = -mc^2 \sqrt{1 - \frac{v^2}{c^2}} \, . \tag{9.7}$$

The canonical momentum is given by

$$p^i = \frac{\partial L}{\partial \dot{x}^i} = \frac{m \dot{x}^i}{\sqrt{1 - \frac{v^2}{c^2}}} = m u^i \, . \tag{9.8}$$

It must be noted at this point that although we are now working in the relativistic arena, still the momentum p^i is a 3-vector component and is not a part of any 4-vector and as a result it is safe to use the Euclidean metric, as specified in Eq. (7.1), in this case. Because we are using the Euclidean metric, the index i appears as contravariant indices on both sides of the above equation. Here, u^i are defined as in Eq. (6.45).

Let us now compute the Hamiltonian. By definition, the Hamiltonian is given by

$$H = \sum_i \dot{x}^i p^i - L = \sum_i \dot{x}^i \frac{m \dot{x}^i}{\sqrt{1 - \frac{v^2}{c^2}}} + mc^2 \sqrt{1 - \frac{v^2}{c^2}} = \frac{mc^2}{\sqrt{1 - \frac{v^2}{c^2}}} \, . \tag{9.9}$$

Thus, in relativistic mechanics, the energy of a free particle remains non-zero even when $v = 0$, this energy mc^2 is called the rest energy of the particle.

9.1.2 Lorentz Force Law from Action Principle

The Lorentz force law was given in Eq. (7.56) in Chap. 7. Presently, we want to derive the force law from the principle of least action. The action should be made up of two parts: one describing the free relativistic charged particle and the other describing the interaction of the particle with electromagnetic field. We have already written down the free particle Lagrangian. For the interaction part, we use the vector field in the form of 4-potential $A^\mu(x)$ which describes the electromagnetic field (electric field and the magnetic field are after all derived from the 4-potential) and construct a Lorentz invariant scalar out of it. We take a small segment of the trajectory dx^μ. The product $A_\mu dx^\mu$ is a Lorenz invariant scalar. We construct the action describing the interaction of the particle with the field by integrating it over the trajectory in space time as $-\frac{q}{c} \int A_\mu dx^\mu$, q being the charge on the particle. The Lagrangian for the interaction part can be extracted by writing

$$\int_{t_1}^{t_2} L_{\text{int}}(x, \dot{x}) dt = -\frac{q}{c} \int_{x_1}^{x_2} A_\mu dx^\mu ,$$

where

$$-\frac{q}{c} A_\mu dx^\mu = -\frac{q}{c} A^0 c\, dt + \frac{q}{c} \mathbf{A} \cdot d\mathbf{x} = \left(-q\Phi + \frac{q}{c} \mathbf{A} \cdot \mathbf{v} \right) dt . \qquad (9.10)$$

Using the last equation, the total Lagrangian for the relativistic charged particle moving in the presence of electromagnetic field is then given by

$$L(\mathbf{x}, \mathbf{v}, t) = -mc^2 \sqrt{1 - \frac{v^2}{c^2}} - q\Phi(t, \mathbf{x}) + \frac{q}{c} \mathbf{A}(t, \mathbf{x}) \cdot \mathbf{v} . \qquad (9.11)$$

The Lagrangian is obviously not in manifestly covariant form. But it is instructive to derive the Euler-Lagrange equation using the Lagrangian. Moreover, we can identify the velocity dependent potential energy term in the form of $\frac{q}{c} \mathbf{A} \cdot \mathbf{v}$ which gives rise to force on the charged particle in a magnetic field. The canonical momentum is given by

$$P_i = \frac{\partial L}{\partial \dot{x}_i} = mu_i + \frac{q}{c} A_i = p_i + \frac{q}{c} A_i . \qquad (9.12)$$

Here, $\mathbf{P}$ is the canonical momentum conjugate to $\mathbf{x}$ and $\mathbf{p} \equiv m\mathbf{u}$ is the kinematic momentum.

In the present case, we can write

$$\gamma^2 = 1 + \frac{\gamma^2 v^2}{c^2} = \left(1 + \frac{\mathbf{p}^2}{m^2 c^2} \right) ,$$

where γ is defined for relativistic particle motion as in Eq. (6.2) in Chap. 6. As $\mathbf{p} = m\gamma\mathbf{v} = \mathbf{P} - (q/c)\mathbf{A}$, we can write

$$\mathbf{v} = \frac{c\mathbf{P} - q\mathbf{A}}{\sqrt{\mathbf{p}^2 + m^2c^2}}.$$

Using these equations, we can now write the Hamiltonian for the charged particle in electromagnetic field as

$$H(\mathbf{x}, \mathbf{P}) = \mathbf{P} \cdot \mathbf{v} - L(\mathbf{x}, \mathbf{v})$$
$$= c\frac{\mathbf{p} \cdot \left(\mathbf{p} + \frac{q}{c}\mathbf{A}\right)}{\sqrt{\mathbf{p}^2 + m^2c^2}} + \frac{mc^2}{\sqrt{\frac{\mathbf{p}^2}{m^2c^2} + 1}} - \frac{q}{c}\mathbf{A} \cdot \frac{c\mathbf{P} - q\mathbf{A}}{\sqrt{\mathbf{p}^2 + m^2c^2}} + q\Phi.$$

One can now easily simplify the right-hand side of the above equation and get

$$H(\mathbf{x}, \mathbf{P}, t) = \sqrt{\left(P - \frac{q}{c}\mathbf{A}(t, \mathbf{x})\right)^2 c^2 + m^2c^4} + q\Phi(t, \mathbf{x}), \qquad (9.13)$$

which is the proper Hamiltonian of a charged particle in the presence of electromagnetic fields. In the above equations, we have consistently used the three-dimensional Euclidean metric for all the scalar products as we are dealing with pure 3-vectors for most of the cases.

Now we can derive the Lorentz force law in the Lagrangian formalism. For this, we first require to find out $(\partial L/\partial x_i)$ where the Lagrangian is of the charged particle in an electromagnetic field as given in Eq. (9.11). One can easily see that

$$\frac{\partial L}{\partial x_i} = \frac{q}{c}[\nabla(\mathbf{A} \cdot \mathbf{v})]_i - q[\nabla\Phi]_i,$$

where again we are using the 3-dimensional Euclidean metric and upper and lower indices change positions without a sign change. From vector analysis, we know that

$$\nabla(\mathbf{A} \cdot \mathbf{v}) = (\mathbf{A} \cdot \nabla)\mathbf{v} + (\mathbf{v} \cdot \nabla)\mathbf{A} + \mathbf{v} \times (\nabla \times \mathbf{A}) + \mathbf{A} \times (\nabla \times \mathbf{v}), \qquad (9.14)$$

where $\mathbf{A}$ and $\mathbf{v}$ are 3-vectors which depend on both space and time. In our particular case $\mathbf{v} = \mathbf{v}(t)$, as because it is the velocity vector of a particle and consequently all the terms involving spatial derivatives of $\mathbf{v}$ will be zero. Consequently, the Euler-Lagrange equation gives

$$\frac{d\mathbf{P}}{dt} = \nabla L = \frac{q}{c}[(\mathbf{v} \cdot \nabla)\mathbf{A} + \mathbf{v} \times (\nabla \times \mathbf{A})] - q\nabla\Phi. \qquad (9.15)$$

In our case

$$\frac{d\mathbf{P}}{dt} = \frac{d\mathbf{p}}{dt} + \frac{q}{c}\frac{d\mathbf{A}(t, \mathbf{x})}{dt},$$

and as $\mathbf{A}(t, \mathbf{x})$ is a field depending upon both space and time we must have

$$\frac{d\mathbf{A}(t, \mathbf{x})}{dt} = \frac{\partial \mathbf{A}(t, \mathbf{x})}{\partial t} + (\mathbf{v} \cdot \nabla)\mathbf{A}.$$

Using these relations, we can write Eq. (9.15) as

$$\frac{d\mathbf{p}}{dt} = \frac{q}{c}[\mathbf{v} \times (\nabla \times \mathbf{A})] - q\nabla\Phi - \frac{q}{c}\frac{\partial \mathbf{A}}{\partial t}.$$

From the relations in Eqs. (7.4) and (7.5), which relates the gauge fields to the electric and magnetic fields in Chap. 7, we can now write

$$\frac{d\mathbf{p}}{dt} = q\left[\mathbf{E} + \frac{1}{c}\mathbf{v} \times \mathbf{B}\right],$$

which coincides with the form of the Lorentz force law as discussed in Sect. 7.4.2 in Chap. 7. As $dt = \gamma d\tau$ the above relation can also be written as

$$\frac{d\mathbf{p}}{d\tau} = q\gamma\left[\mathbf{E} + \frac{1}{c}\mathbf{v} \times \mathbf{B}\right],$$

which is the relativistic generalization of the Lorentz force law as noted down in Chap. 7. It must be noted that as we started with a non-covariant Lagrangian of a charged particle in an electromagnetic field, we do not get the explicitly covariant form of the Lorentz equation as discussed previously. One can also obtain the Lorentz force law from the Hamiltonian equations

$$\frac{dx_i}{dt} = \frac{\partial H}{\partial P_i}; \quad \frac{dP_i}{dt} = -\frac{\partial H}{\partial x_i}. \tag{9.16}$$

We do not present the derivation of the Lorentz force law in the Hamiltonian picture, the reader can do this derivation as an exercise.

9.2   Lagrangian Formalism for Electromagnetic Fields

In this section, we will mostly use a fully covariant formalism as we are now interested in deriving the covariant Maxwell equations from an action principle. To derive the Maxwell field equations, at first, we must have to extend the action principle to the case of fields and then use an appropriate Lagrangian (density) for the electromagnetic fields. Before we fully embark on this business, we will like to introduce the canonical formalism for 4-vector fields.

9.2.1 Introduction to the Lagrangian Formalism

The degrees of freedom which specify the electromagnetic fields in the fully covariant formalism of electrodynamics is the 4-vector gauge field $A^\mu(x)$ where $x \equiv (t, \mathbf{x})$. As like particle mechanics, one may think that the Lagrangian of the electromagnetic field may be written as $L(A^\mu, \dot{A}^\mu)$, where the dot represents derivative with respect to coordinate time t. It turns out that the above choice in a fully covariant theory is not permissible as the Lagrangian cannot depend only on time derivatives, relativity demands the Lagrangian must also depend on spatial derivatives of A^μ. Consequently, the Lagrangian of electromagnetic fields can be $L(A^\mu, \partial^\nu A^\mu)$ which depends upon both A^μ and $\partial^\nu A^\mu$. The Lagrangian cannot depend upon higher derivatives of the gauge fields as $\partial^\kappa \partial^\nu A^\mu$ or $\partial^\sigma \partial^\kappa \partial^\nu A^\mu$ because in that case the equations of motion coming out from the action will be higher than second order in time or spatial derivatives, whereas the covariant Maxwell's equations are second order equations. We can now write

$$ S = \int_{t_1}^{t_2} L(A^\mu, \partial^\nu A^\mu) c \, dt , \qquad (9.17) $$

where t_1 and t_2 are two coordinate time instants and the factor of c has been deliberately introduced here for later convenience. To start with, we demand the action of the electromagnetic field must be

1. Lorentz invariant,
2. it must be gauge invariant,
3. and of course by extremizing the action we must get the covariant Maxwell equations.

Shortly we will see how these points are taken care of while we develop the canonical formalism for the 4-vector fields. For a continuous system, one can define a Lagrangian density $\mathcal{L}(A^\mu, \partial^\nu A^\mu)$ as

$$ L(A^\mu, \partial^\nu A^\mu) \equiv \int_V d^3x \mathcal{L}(A^\mu, \partial^\nu A^\mu) , \qquad (9.18) $$

where V is the volume over which A^μ and $\partial^\nu A^\mu$ are non-zero. Here, $\mathcal{L}(A^\mu, \partial^\nu A^\mu)$ is called the Lagrangian density, which by assumption is a Lorentz invariant quantity. The Lagrangian density is constructed in such a way that it is always Lorentz invariant. This naturally imply the Lagrangian L by itself is not Lorentz invariant as d^3x transforms nontrivially under a Lorentz transformation. Now we can write down the action in Eq. (9.17) as

$$ S = \int_{t_1}^{t_2} c \, dt \int_V d^3x \, \mathcal{L}(A^\mu, \partial^\nu A^\mu) = \int_\Omega d^4x \, \mathcal{L}(A^\mu, \partial^\nu A^\mu) , \qquad (9.19) $$

where Ω is the 4-volume over which A^μ and $\partial^\nu A^\mu$ are non-zero. We know that d^4x is Lorentz invariant and $\mathcal{L}(A^\mu, \partial^\nu A^\mu)$ is by construction Lorentz invariant, as a result we see that S is indeed Lorentz invariant.

To obtain the field equations, one must vary the action and find out the condition for the extremum $\delta S = 0$. To vary the action, we vary the field and its derivative, around the configuration which extrimizes the action, as

$$A^\mu(x) \to A^\mu(x) + \delta A^\mu(x), \quad \partial_\nu A^\mu(x) \to \partial_\nu A^\mu(x) + \delta(\partial_\nu A^\mu(x)).$$

Here, δ represents field variations and the variation of field derivatives at a particular spacetime point x and consequently $\delta(\partial_\nu A^\mu(x)) = \partial_\nu(\delta A^\mu(x))$. The variation of the action gives

$$\delta S[\mathcal{L}] = \int_\Omega d^4x \left[\frac{\partial \mathcal{L}}{\partial A^\mu} \delta A^\mu + \frac{\partial \mathcal{L}}{\partial(\partial_\mu A^\nu)} \delta(\partial_\mu A^\nu) \right], \tag{9.20}$$

where the variations are done in such a way that

$$\delta A^\mu(x) = 0, \quad \delta(\partial_\mu A^\nu) = 0, \quad \text{on } \partial\Omega, \tag{9.21}$$

where $\partial\Omega$ specifies the closed three-dimensional hypersurface boundary of the four-dimensional spacetime volume Ω. Noting that

$$\partial_\mu \left[\frac{\partial \mathcal{L}}{\partial(\partial_\mu A^\nu)} \delta A^\nu \right] = \partial_\mu \left[\frac{\partial \mathcal{L}}{\partial(\partial_\mu A^\nu)} \right] \delta A^\nu + \frac{\partial \mathcal{L}}{\partial(\partial_\mu A^\nu)} \partial_\mu(\delta A^\nu),$$

we can write

$$\delta S[\mathcal{L}] = \int_\Omega d^4x \left\{ \frac{\partial \mathcal{L}}{\partial A^\nu} - \partial_\mu \left[\frac{\partial \mathcal{L}}{\partial(\partial_\mu A^\nu)} \right] \right\} \delta A^\nu$$
$$+ \int_\Omega d^4x \, \partial_\mu \left[\frac{\partial \mathcal{L}}{\partial(\partial_\mu A^\nu)} \delta A^\nu \right]. \tag{9.22}$$

By the application of Gauss theorem in four-dimensional Minkowski spacetime, which states

$$\int_\Omega d^4x \, \partial_\mu V^\mu = \oint_{\partial\Omega} V^\mu dS_\mu, \tag{9.23}$$

where V^μ is an arbitrary 4-vector field and dS_μ is a 4-vector equal to the area of an infinitesimal element of the closed hypersurface $\partial\Omega$. dS_μ is along the outward normal to the closed hypersurface $\partial\Omega$. Using the Gauss law, we can write

$$\int_\Omega d^4x \, \partial_\mu \left[\frac{\partial \mathcal{L}}{\partial(\partial_\mu A^\nu)} \delta A^\nu \right] = \int_{\partial\Omega} \left[\frac{\partial \mathcal{L}}{\partial(\partial_\mu A^\nu)} \delta A^\nu \right] dS_\mu.$$

As δA^ν vanishes on $\partial \Omega$ we have

$$\delta S[\mathcal{L}] = \int_\Omega d^4x \left\{ \frac{\partial \mathcal{L}}{\partial A^\nu} - \partial_\mu \left[\frac{\partial \mathcal{L}}{\partial(\partial_\mu A^\nu)} \right] \right\} \delta A^\nu . \tag{9.24}$$

As δA^ν are supposed to be independent variations around the field configuration which extrimizes the action, the above integral can vanish only when

$$\partial_\mu \left[\frac{\partial \mathcal{L}}{\partial(\partial_\mu A^\nu)} \right] - \frac{\partial \mathcal{L}}{\partial A^\nu} = 0 , \tag{9.25}$$

giving us the Euler-Lagrange equation for the 4-vector electromagnetic field.

9.2.2   The Maxwell Equations

Now we know the Euler-Lagrange equation for the electromagnetic field, but we do not know the Lagrangian density $\mathcal{L}(A^\mu, \partial^\nu A^\mu)$ for the electromagnetic field. To find its form, we must note that it must be

1. Lorentz invariant and parity symmetric.
2. gauge symmetric,
3. must include interaction between fields and charges and currents,
4. lastly, it must produce the known form of covariant Maxwell's equations as we know.

Moreover, we can also assume that the pure field Lagrangian density (which does not contain the interaction of field and source) must contain terms which are quadratic in the field derivatives, as it happens for free particles. The free particle Lagrangian is purely quadratic in position derivatives. We see that all these properties are satisfied by the following form of the Lagrangian density,

$$\mathcal{L} = -\frac{1}{16\pi} F^{\mu\nu} F_{\mu\nu} - \frac{1}{c} J^\mu A_\mu , \tag{9.26}$$

where the factors $-(1/16\pi)$ and $-(1/c)$ originate from the fourth point. As we are now using Lorentz covariant formalism, we are not writing the summations explicitly, Einstein summation convention is used explicitly. The above expression has a a double sum over μ and ν on the first term on the right hand side where as a sum on μ alone on the last term. The Lagrangian is manifestly Lorentz invariant and contains interaction terms between fields and sources. We cannot write a term as $\tilde{F}^{\mu\nu} F_{\mu\nu}$ as this term violates parity. From the discussion of parity symmetry in Chap. 7, one can easily show that if the Lagrangian density contains both $F^{\mu\nu} F_{\mu\nu}$ and $\tilde{F}^{\mu\nu} F_{\mu\nu}$ separately then parity symmetry is broken. One may try to add a term

$\tilde{F}^{\mu\nu}\tilde{F}_{\mu\nu}$ to the Lagrangian density, but it turns out that this term is really not an independent term as $\tilde{F}^{\mu\nu}\tilde{F}_{\mu\nu} = -F^{\mu\nu}F_{\mu\nu}$.

Looking at the Lagrangian density of the electromagnetic field one may see that the $F^{\mu\nu}F_{\mu\nu}$ is manifestly gauge invariant but the interaction term $J^{\mu}A_{\mu}$ is apparently not gauge invariant. Under a gauge transformation $A^{\mu}(x) \rightarrow A'^{\mu}(x) = A^{\mu}(x) + \partial^{\mu}\chi(x)$ it is seen that $J^{\mu}A_{\mu} \rightarrow J^{\mu}(A_{\mu} + \partial_{\mu}\chi)$ which shows that under a gauge transformation the action changes as

$$\int_{\Omega} d^4x \mathcal{L}(A^{\mu}, \partial^{\nu}A^{\mu}) \rightarrow \int_{\Omega} d^4x \mathcal{L}(A^{\mu}, \partial^{\nu}A^{\mu}) + \int_{\Omega} d^4x J^{\mu}\partial_{\mu}\chi .$$

We will show below how the action of the electromagnetic field is really gauge invariant if the electromagnetic current is conserved and the current vanishes on the boundary of spacetime volume Ω. To prove this, we first note that $\partial_{\mu}(\chi J^{\mu}) = \chi\partial_{\mu}J^{\mu} + J^{\mu}\partial_{\mu}\chi$, which implies if we have $\partial_{\mu}J^{\mu} = 0$ then $\partial_{\mu}(\chi J^{\mu}) = J^{\mu}\partial_{\mu}\chi$. Assuming electromagnetic current conservation the apparently gauge non-invariant part of the action can now be written as

$$\int_{\Omega} d^4x J^{\mu}\partial_{\mu}\chi = \int_{\Omega} d^4x \partial_{\mu}(\chi J^{\mu}) = \int_{\partial\Omega} (\chi J^{\mu}) dS_{\mu} ,$$

where we have used Gauss theorem in the last step. As we assume that $J^{\mu} = 0$ on $\partial\Omega$, we see that the above term actually vanishes making the action of the electromagnetic field gauge invariant.

Now we show that the Lagrangian density as proposed do indeed produce the Maxwell equations of Electrodynamics. We can write the Euler-Lagrange equation for the electromagnetic field as

$$\partial^{\rho}\left[\frac{\partial\mathcal{L}}{\partial(\partial^{\rho}A^{\sigma})}\right] - \frac{\partial\mathcal{L}}{\partial A^{\sigma}} = 0,$$

where the Lagrangian density can be also written as

$$\mathcal{L} = -\frac{1}{8\pi}[(\partial^{\mu}A^{\nu})(\partial_{\mu}A_{\nu}) - (\partial^{\mu}A^{\nu})(\partial_{\nu}A_{\mu})] - \frac{1}{c}J^{\mu}A_{\mu} , \qquad (9.27)$$

showing that the noninteracting part of the Lagrangian is really quadratic in the field derivatives. From the above expression, one can write

$$\begin{aligned}
\frac{\partial\mathcal{L}}{\partial(\partial^{\rho}A^{\sigma})} &= -\frac{1}{8\pi}\left[\frac{\partial(\partial^{\mu}A^{\nu})}{\partial(\partial^{\rho}A^{\sigma})}(\partial_{\mu}A_{\nu}) + (\partial^{\mu}A^{\nu})\frac{\partial(\partial_{\mu}A_{\nu})}{\partial(\partial^{\rho}A^{\sigma})}\right.\\
&\quad \left. -\frac{\partial(\partial^{\mu}A^{\nu})}{\partial(\partial^{\rho}A^{\sigma})}(\partial_{\nu}A_{\mu}) - (\partial^{\mu}A^{\nu})\frac{\partial(\partial_{\nu}A_{\mu})}{\partial(\partial^{\rho}A^{\sigma})}\right]\\
&= -\frac{1}{8\pi}\left[\frac{\partial(\partial^{\mu}A^{\nu})}{\partial(\partial^{\rho}A^{\sigma})}(\partial_{\mu}A_{\nu}) + \eta_{\mu\kappa}\eta_{\nu\tau}(\partial^{\mu}A^{\nu})\frac{\partial(\partial^{\kappa}A^{\tau})}{\partial(\partial^{\rho}A^{\sigma})}\right.
\end{aligned}$$

$$-\frac{\partial(\partial^\mu A^\nu)}{\partial(\partial^\rho A^\sigma)}(\partial_\nu A_\mu) - \eta_{\mu\tau}\eta_{\nu\kappa}(\partial^\mu A^\nu)\frac{\partial(\partial^\kappa A^\tau)}{\partial(\partial^\rho A^\sigma)}\Bigg] \; .$$

Using the relation

$$\frac{\partial(\partial^\mu A^\nu)}{\partial(\partial^\rho A^\sigma)} = \delta^\mu_\rho \delta^\nu_\sigma \; ,$$

we can write the last equation as

$$
\begin{aligned}
\frac{\partial \mathcal{L}}{\partial(\partial^\rho A^\sigma)} &= -\frac{1}{8\pi}\Big[\delta^\mu_\rho\delta^\nu_\sigma(\partial_\mu A_\nu) + \eta_{\mu\kappa}\eta_{\nu\tau}(\partial^\mu A^\nu)\delta^\kappa_\rho\delta^\tau_\sigma - \delta^\mu_\rho\delta^\nu_\sigma(\partial_\nu A_\mu) \\
&\quad - \eta_{\mu\tau}\eta_{\nu\kappa}(\partial^\mu A^\nu)\delta^\kappa_\rho\delta^\tau_\sigma\Big] \\
&= -\frac{1}{4\pi}(\partial_\rho A_\sigma - \partial_\sigma A_\rho) = -\frac{1}{4\pi}F_{\rho\sigma} \; .
\end{aligned}
\tag{9.28}
$$

One can easily see that

$$\frac{\partial \mathcal{L}}{\partial A^\sigma} = -\frac{1}{c}J_\sigma \; . \tag{9.29}$$

The Euler-Lagrange equations now produce

$$\partial^\rho F_{\rho\sigma} = \frac{4\pi}{c}J_\sigma \; ,$$

which is the inhomogeneous Maxwell equation as presented in Eq. (7.29) in Chap. 7. The homogeneous Maxwell equation do not come out by extremizing the action as these equations are always valid as long as we are using the definitions of the gauge fields as specified in Eqs. (7.4) and (7.5). As we are using the gauge fields A^μ as defined in Chap. 7, the homogeneous Maxwell equation remains as it was specified in Eq. (7.29).

9.2.3 Difficulty with the Hamiltonian Formulation

Like particle mechanics we can construct the Hamiltonian H of the electromagnetic field $A^\mu(x)$. The Hamiltonian depends upon $A^\mu(x)$ and its conjugate momentum $\Pi^\mu(x)$. In particle mechanics, the conjugate momentum is defined as $p = \partial L/\partial\dot{x}$, similarly, in continuous systems, we define the conjugate momentum as

$$\Pi^\mu = \frac{\partial \mathcal{L}}{\partial(\partial_0 A_\mu)} \; . \tag{9.30}$$

As in the Lagrangian formulation of continuous systems we can define a Hamiltonian density as

$$H(A^\mu, \partial_i A^\mu, \Pi^\mu) = \int_V d^3x \, \mathcal{H}(A^\mu, \partial_i A^\mu, \Pi^\mu), \qquad (9.31)$$

where $\mathcal{H}(A^\mu, \partial_i A^\mu, \Pi^\mu)$ is the Hamiltonian density of the electromagnetic field. As in particle mechanics, we can also write

$$\mathcal{H}(A^\mu, \partial_i A^\mu, \Pi^\mu) = \Pi^\mu(\partial_0 A_\mu) - \mathcal{L}(A^\mu, \partial^\nu A^\mu), \qquad (9.32)$$

which gives the relation between the Hamiltonian density and the Lagrangian density. In general, we expect that each field component A^μ has its corresponding conjugate momentum. The difficulty of the Hamiltonian formalism starts here as we can see that as $\mathcal{L}(A^\mu, \partial^\nu A^\mu)$ is made up of $F^{\mu\nu} F_{\mu\nu}$ (only relevant term as far as field momentum is concerned) the Lagrangian density is independent of the term $\partial_0 A_0$ as this term never appears in the field strength tensor. As a result of this, it is seen that although $A^0(x) \neq 0$ in general we must have

$$\Pi^0 = \frac{\partial \mathcal{L}}{\partial(\partial_0 A_0)} = 0.$$

This is a difficulty which we never faced in particle mechanics, all degrees of freedom had non-zero conjugate momentum. Here, we face a difficulty as $A^0(x)$ has its conjugate momentum to be zero. It becomes really difficult to quantize such a system as the quantization procedure requires commutation relations between fields and their conjugate momentum and as $\Pi^0 = 0$ the quantization of the zeroth component of the electromagnetic field becomes problematic. These kind of problems were seriously studied by P. A. M. Dirac in his analysis of constrained systems. We do not present more about Dirac's method here as it is out of scope of this book.

9.3 Continuous Symmetries and Noether's Theorem

The big advantage of the Lagrangian formalism for the fields is that it sets up a platform on which more abstract but useful concepts can be built. One such concept is related to continuous symmetries which we will discuss in this section. At first, we will present our discussion on symmetries in pure electromagnetic fields without sources and in the end of this section, we will include the sources and generalize our results.

9.3.1 Pure Electromagnetic Field and Continuous Symmetries

In this subsection, we will assume that the Lagrangian density to be

$$\mathcal{L} = -\frac{1}{16\pi} F^{\mu\nu} F_{\mu\nu}, \tag{9.33}$$

for the pure electromagnetic field $A^{\mu}(x)$. Let us introduce the general infinitesimal continuous transformation as

$$x^{\mu} \rightarrow x'^{\mu} = x^{\mu} + \delta x^{\mu}, \tag{9.34}$$

where $|\delta x^{\mu}| \ll x^{\mu}$. Once these kind of transformations are applied, the fields will also transform. The overall infinitesimal change in the field due to the above transformation can be written in terms of the change of the field at the same spacetime point and something else. Let us first define the infinitesimal change in the field at the same spacetime point. This change is given as

$$\delta_0 A^{\mu}(x) \equiv A'^{\mu}(x) - A^{\mu}(x). \tag{9.35}$$

These change is more like the field variation which give us the Euler-Lagrange equation of the field. The most general infinitesimal change induced on the fields due to the continuous transformation in Eq. (9.34) is given as

$$\delta A^{\mu}(x) \equiv A'^{\mu}(x') - A^{\mu}(x). \tag{9.36}$$

From the above definition, we see that

$$\delta A^{\mu}(x) = A'^{\mu}(x + \delta x) - A^{\mu}(x) = A'^{\mu}(x) + \delta x^{\nu}[\partial_{\nu} A'^{\mu}(x)] + \cdots - A^{\mu}(x),$$

where the dots represent higher order infinitesimals which are neglected in this calculation. As consequence, we can write

$$\delta A^{\mu}(x) = [A'^{\mu}(x) - A^{\mu}(x)] + \delta x^{\nu}[\partial_{\nu} A'^{\mu}(x)] = \delta_0 A^{\mu}(x) + \delta x^{\nu} \partial_{\nu}[A^{\mu}(x) + \delta_0 A^{\mu}(x)].$$

To first order in infinitesimals, we can now write

$$\delta A^{\mu}(x) = \delta_0 A^{\mu}(x) + \delta x^{\nu} \partial_{\nu} A^{\mu}(x), \tag{9.37}$$

showing that the general infinitesimal change of the field due to the continuous transformation in Eq. (9.34) contains a term which is related to the change in the field at the same spacetime point and a second term specifying change in the field due to spacetime transformation. An example of such a change can be given when we think of infinitesimal coordinate rotation by angle $\delta\theta$ about the z-axis. The infinitesimal coordinate transformations in this case is given by

$$x'^1 = x^1 - \delta\theta \, x^2, \quad x'^2 = x^2 + \delta\theta \, x^1, \quad x'^3 = x^3, \quad x'^0 = x^0,$$

which implies

$$\delta x^1 = -\delta\theta \, x^2, \quad \delta x^2 = \delta\theta \, x^1, \quad \delta x^3 = 0, \quad \delta x^0 = 0. \tag{9.38}$$

Under such a transformation, the 3-vector part of $A^\mu(x)$ rotates at the same spacetime point as

$$\delta_0 A^1(x) = -\delta\theta \, A^2(x), \quad \delta_0 A^2(x) = \delta\theta \, A^1(x), \quad \delta_0 A^3(x) = 0, \quad \delta_0 A^0(x) = 0. \tag{9.39}$$

Under spatial rotation these are the only possible changes in the 3-vector part of $A^\mu(x)$ and in this particular case $\delta A^\mu(x) = \delta_0 A^\mu(x)$. On the other hand, in the case of infinitesimal spacetime, translation by a constant 4-vector ϵ^μ, we have

$$x'^\mu = x^\mu + \epsilon^\mu. \tag{9.40}$$

Here $\delta x^\mu = \epsilon^\mu$. Under such a spacetime transformation in general we have

$$A'^\mu(x') = A^\mu(x), \tag{9.41}$$

which states that under spacetime translation the transformed 4-vector at the changed position is equal to the old 4-vector in the old position. In such a case, we must have

$$\delta A^\mu(x) = 0, \tag{9.42}$$

implying $\delta_0 A^\mu(x) = -\delta x^\nu \partial_\nu A^\mu(x) = -\epsilon^\nu \partial_\nu A^\mu(x)$. In this case, we see that under spacetime translation the change of the 4-vector at the same spacetime point compensates for the other change on A^μ due to translation.

9.3.2 Noether's Theorem

Noether's theorem states that corresponding to any continuous symmetry of the field action there corresponds a conserved charge. This is a very important theorem in classical field theory and we will spend some time in explaining the meaning of this theorem. We will prove this theorem as the proof will yield important technical components which are regularly used in field theory. Suppose under the continuous infinitesimal spacetime transformations, as specified in Eq. (9.34), the action of the free electromagnetic field

$$S = \int_\Omega d^4x\, \mathcal{L}(A^\mu, \partial^\nu A^\mu),$$

where $\mathcal{L}(A^\mu, \partial^\nu A^\mu)$ is as given in Eq. (9.33), remains invariant. In such cases, the spacetime transformation is called a symmetry of the theory. One must note that although under a continuous symmetry transformation, as given in Eq. (9.34), the action remains the same the fields $A^\mu(x)$ can transform to a new configuration.

For Noether's theorem, we only deal with continuous infinitesimal spacetime transformations and do not employ the finite transformations. There is a reason to it. Given any finite spacetime transformations as rotation, Lorentz boost or spacetime translation we can find out their infinitesimal forms as the infinitesimal rotation about the z-axis, as given in Eq. (9.38), can be obtained from the finite rotation formulas by assuming the angle of rotation to be very small. In the opposite way, if we know the infinitesimal spacetime transformations, we can build up the finite spacetime transformations if we want. We will not show the validity of the last statement here, but any elementary group theory book which deals with Lie groups shows the proof of the last statement in details. The important point is if we have shown that an infinitesimal continuous transformation to be a symmetry of the action then it naturally implies the finite transformation, corresponding to the infinitesimal one, is also a symmetry of the action.

The other point is related to the meaning of the conserved charge. In this case the word charge is used in a more general sense and has nothing to do with electric charge except its constancy. By conserved charge, we mean any quantity which is independent of spatial coordinates and remains constant in time. The conserved charge does not change in spacetime although the field configurations can change in spacetime. The beauty of Noether's theorem is that it connects spacetime symmetries with such conserved quantities.

Under continuous infinitesimal spacetime transformations in Eq. (9.34), the change in the action is like

$$\begin{aligned}
\delta S &= \delta\left[\int_\Omega d^4x\, \mathcal{L}(A^\mu, \partial^\nu A^\mu)\right] \\
&= \int_\Omega d^4x\, \delta\mathcal{L}(A^\mu, \partial^\nu A^\mu) + \int \delta(d^4x)\, \mathcal{L}(A^\mu, \partial^\nu A^\mu).
\end{aligned} \tag{9.43}$$

Here

$$\int \delta(d^4x)\, \mathcal{L}(A^\mu, \partial^\nu A^\mu) \equiv \left(\int_{\Omega'} d^4x' - \int_\Omega d^4x\right) \mathcal{L}(A^\mu, \partial^\nu A^\mu). \tag{9.44}$$

In our case $\mathcal{L}(A^\mu, \partial^\nu A^\mu)$ is a Lorentz scalar and we can write

$$\delta\mathcal{L} = \delta_0\mathcal{L} + \delta x^\mu \partial_\mu \mathcal{L}, \tag{9.45}$$

which is exactly similar in spirit to the corresponding relation in Eq. (9.37). Following the derivation of Eq. (9.37), one can easily prove the above formula. In our case, $\delta\mathcal{L} = \mathcal{L}'(A'^\mu, \partial'^\nu A'^\mu) - \mathcal{L}(A^\mu, \partial^\nu A^\mu)$. As $\delta_0\mathcal{L}$ involves change in the Lagrangian density at the same spacetime point, like the variations of $\mathcal{L}(A^\mu, \partial^\nu A^\mu)$ in the action while deriving the Euler-Lagrange equation for electromagnetic field, we must have

$$\delta_0\mathcal{L} = \frac{\partial\mathcal{L}}{\partial A^\mu}\delta_0 A^\mu + \frac{\partial\mathcal{L}}{\partial(\partial_\alpha A_\mu)}\delta_0(\partial_\alpha A_\mu), \tag{9.46}$$

where δ_0 commutes with ∂_μ. Following the steps used to derive the Euler-Lagrange equation for the field, we can write

$$\delta_0\mathcal{L} = \left\{\frac{\partial\mathcal{L}}{\partial A^\mu} - \partial_\alpha\left[\frac{\partial\mathcal{L}}{\partial(\partial_\alpha A_\mu)}\right]\right\}\delta_0 A_\mu + \partial_\alpha\left[\frac{\partial\mathcal{L}}{\partial(\partial_\alpha A_\mu)}(\delta_0 A_\mu)\right].$$

Assuming the field to be satisfying, the Euler-Lagrange equation, as given in Eq. (9.25), we can write

$$\delta\mathcal{L} = \partial_\alpha\left[\frac{\partial\mathcal{L}}{\partial(\partial_\alpha A_\mu)}(\delta_0 A_\mu)\right] + \delta x^\mu \partial_\mu\mathcal{L}. \tag{9.47}$$

We will now find out the infinitesimal change in the integration measure for the transformation $x'^\mu = x^\mu + \delta x^\mu$. The integration measure changes as

$$d^4x' = J\, d^4x, \tag{9.48}$$

where

$$J = \frac{\partial(x'^0, x'^1, x'^2, x'^3)}{\partial(x^0, x^1, x^2, x^3)}, \tag{9.49}$$

is the Jacobian of the transformation in Eq. (9.34). The Jacobian is given by the determinant

$$J = \begin{vmatrix} \frac{\partial x'^0}{\partial x^0} & \frac{\partial x'^1}{\partial x^0} & \frac{\partial x'^2}{\partial x^0} & \frac{\partial x'^3}{\partial x^0} \\[6pt] \frac{\partial x'^0}{\partial x^1} & \frac{\partial x'^1}{\partial x^1} & \frac{\partial x'^2}{\partial x^1} & \frac{\partial x'^3}{\partial x^1} \\[6pt] \frac{\partial x'^0}{\partial x^2} & \frac{\partial x'^1}{\partial x^2} & \frac{\partial x'^2}{\partial x^2} & \frac{\partial x'^3}{\partial x^2} \\[6pt] \frac{\partial x'^0}{\partial x^3} & \frac{\partial x'^1}{\partial x^3} & \frac{\partial x'^2}{\partial x^3} & \frac{\partial x'^3}{\partial x^3} \end{vmatrix}. \tag{9.50}$$

We do not require to expand the determinant, we will be able to extract relevant information, we require, from the above equation. We first note that

$$\frac{\partial x'^\mu}{\partial x^\nu} = \delta^\mu_\nu + \partial_\nu\delta x^\mu,$$

which shows that all the derivatives $(\partial x'^\mu/\partial x^\nu)$ are first order infinitesimals when $\mu \neq \nu$. Here, one must note that for rotations and Lorentz transformations δx^μs are linear functions of x^μ, where the coefficients multiplying the x^μs are some infinitesimal parameter, and consequently $\partial_\nu \delta x^\mu$ are first-order infinitesimals when δx^μs are infinitesimals. Looking at the determinant we see that one term in its expansion contains the product of all the diagonal terms and has a positive sign, this term is

$$\left(\frac{\partial x'^0}{\partial x^0}\right)\left(\frac{\partial x'^1}{\partial x^1}\right)\left(\frac{\partial x'^2}{\partial x^2}\right)\left(\frac{\partial x'^3}{\partial x^3}\right) = (1+\partial_0\delta x^0)(1+\partial_1\delta x^1)(1+\partial_2\delta x^2)(1+\partial_3\delta x^3)$$

$$= 1 + \partial_0\delta x^0 + \partial_1\delta x^1 + \partial_2\delta x^2 + \partial_3\delta x^3, \qquad (9.51)$$

up to first order of infinitesimals. Except this term all other terms originating from the Laplace expansion of the above determinant will contain at least two non-diagonal terms. It is easy to verify those other terms will contain higher order infinitesimals and hence can be neglected. Consequently, the Jacobian of the continuous infinitesimal spacetime transformation is given by

$$J = (1 + \partial_\mu \delta x^\mu), \qquad (9.52)$$

as a result of which we can write Eq. (9.44) as

$$\left(\int_\Omega (1+\partial_\mu\delta x^\mu)d^4x - \int_\Omega d^4x\right)\mathcal{L}(A^\mu, \partial^\nu A^\mu) = \int_\Omega (\partial_\mu\delta x^\mu)d^4x\, \mathcal{L}(A^\mu, \partial^\nu A^\mu). \qquad (9.53)$$

The overall change in the action due to the continuous infinitesimal change, as written in Eq. (9.43), can now be expressed as

$$\delta S = \int_\Omega d^4x \left\{ (\partial_\alpha\delta x^\alpha)\mathcal{L} + \partial_\alpha\left[\frac{\partial\mathcal{L}}{\partial(\partial_\alpha A_\mu)}(\delta_0 A_\mu)\right] + \delta x^\alpha \partial_\alpha\mathcal{L} \right\}$$

$$= \int_\Omega d^4x\, \partial_\alpha\left[\mathcal{L}\delta x^\alpha + \frac{\partial\mathcal{L}}{\partial(\partial_\alpha A_\mu)}(\delta_0 A_\mu)\right].$$

This relation can be modified to give a very important result. We write the above equation as

$$\delta S = \int_\Omega d^4x\, \partial_\alpha\left[\mathcal{L}\delta x^\alpha + \frac{\partial\mathcal{L}}{\partial(\partial_\alpha A_\mu)}(\delta A_\mu - \delta x^\nu \partial_\nu A_\mu)\right]$$

$$= \int_\Omega d^4x\, \partial_\alpha\left[\frac{\partial\mathcal{L}}{\partial(\partial_\alpha A_\mu)}\delta A_\mu + \mathcal{L}\,\eta^{\alpha\nu}\delta x_\nu - \frac{\partial\mathcal{L}}{\partial(\partial_\alpha A_\mu)}(\partial^\nu A_\mu)\delta x_\nu\right].$$

Defining the energy-momentum tensor for the free electromagnetic field as

$$T^{\alpha\nu} \equiv \frac{\partial \mathcal{L}}{\partial(\partial_\alpha A_\mu)}(\partial^\nu A_\mu) - \eta^{\alpha\nu}\mathcal{L}, \tag{9.54}$$

we can now write

$$\delta S = \int_\Omega d^4x\, \partial_\alpha \left[\frac{\partial \mathcal{L}}{\partial(\partial_\alpha A_\mu)}\delta A_\mu - T^{\alpha\nu}\delta x_\nu\right]. \tag{9.55}$$

Next we define the Noether current corresponding to the continuous infinitesimal spacetime transformation in Eq. (9.34) as

$$J^\alpha \equiv \frac{\partial \mathcal{L}}{\partial(\partial_\alpha A_\mu)}\delta A_\mu - T^{\alpha\nu}\delta x_\nu. \tag{9.56}$$

With respect to this 4-current, the infinitesimal change in action due to the continuous infinitesimal spacetime transformation is

$$\delta S = \int_\Omega d^4x\, \partial_\alpha J^\alpha. \tag{9.57}$$

Now comes the crucial point, assuming the continuous infinitesimal spacetime transformation in Eq. (9.34) to be a symmetry transformation, we must have

$$\delta S = \int_\Omega d^4x\, \partial_\alpha J^\alpha = 0,$$

corresponding to it. The above integral in general vanishes when the Noether 4-current is conserved

$$\partial_\alpha J^\alpha = 0. \tag{9.58}$$

The above discussion shows that every continuous spacetime symmetry of the field action has a corresponding conserved Noether 4-current. Corresponding to any continuous symmetry transformation of the action, we can now define a charge from the Noether current as

$$\mathcal{Q}(t) \equiv \int_V d^3x\, J^0(t,\mathbf{x}), \tag{9.59}$$

where V is the volume of 3-space where $J^0(t,\mathbf{x}) \neq 0$. If we want to see how this charge changes with time, then we must calculate the time derivative of it as

$$\frac{d\mathcal{Q}(t)}{dt} = c\int_V d^3x\, \partial_0 J^0 = -c\int_V d^3x\, (\nabla \cdot \mathbf{J}),$$

where we have used the fact that the Noether current is conserved. Using Gauss' law, we get

$$\frac{d\mathcal{Q}(t)}{dt} = -c \oint_{\partial V} \mathbf{J} \cdot d\mathbf{a},$$

where ∂V is the closed boundary surface of the volume V and $d\mathbf{a}$ is the vector representing an infinitesimal area on ∂V. It is outward oriented. In general, $\mathbf{J} = 0$ on ∂V and in that case $\mathcal{Q}(t)$ is conserved giving us the conserved Noether charge corresponding to the continuous symmetry transformation in Eq. (9.34). Now we apply Noether's theorem for various symmetry transformations.

9.3.2.1 Spacetime Translation

We discussed about infinitesimal spacetime translation, as given in Eq. (9.40), in the end of Sect. 9.3.1. For such transformations, it was noted that $\delta A^\mu(x) = 0$. If we assume that such a spacetime translation is a symmetry of the free electromagnetic field action then the Noether conserved current corresponding to this symmetry is

$$J^\mu = -T^{\mu\nu}\delta x_\nu = -T^{\mu\nu}\epsilon_\nu, \tag{9.60}$$

where ϵ_ν is a constant 4-vector. The current conservation condition

$$\partial_\mu J^\mu = 0 \quad \Longrightarrow \quad \partial_\mu T^{\mu\nu} = 0, \tag{9.61}$$

because of the constancy of ϵ_ν. This shows that for free electromagnetic fields the 4-divergence of the energy-momentum tensor vanishes as long as spacetime translation symmetry remains valid. The vanishing of the 4-divergence of the energy-momentum tensor helps us to construct a conserved 4-vector for free electromagnetic fields as

$$\mathcal{M}^\mu \equiv \int_V d^3x\, T^{0\mu}(t, \mathbf{x}). \tag{9.62}$$

From this definition, we see that

$$\frac{d\mathcal{M}^\mu}{dt} = c \int_V d^3x\, \partial_0 T^{0\mu} = -c \int_V d^3x\, \partial_i T^{i\mu}. \tag{9.63}$$

The last integral can be converted to an integral over the boundary ∂V by using the general Gauss' law for tensors which states

$$\int_V d^3x\, \partial_\mu T^{\mu\nu\kappa\cdots} = \oint_{\partial V} T^{\mu\nu\kappa\cdots} da_\mu, \tag{9.64}$$

where da_μ is a 4-vector whose magnitude is the infinitesimal area on the closed surface ∂V and which is outward oriented. If we assume $T^{i\mu}$ to be zero on ∂V, then we immediately get

$$\frac{d\mathcal{M}^\mu}{dt} = 0, \tag{9.65}$$

specifying that $\mathcal{M}^\mu$ is a constant 4-vector. Here $\mathcal{M}^\mu$ is related to the constant 4-momentum of the field and it's components are

$$\mathcal{M}^0 = \int_V d^3x \, T^{00}(t, \mathbf{x}) \equiv E_{\text{field}}, \tag{9.66}$$

giving the constant total energy of the free electromagnetic field configuration and

$$\mathcal{M}^i = \int_V d^3x \, T^{0i}(t, \mathbf{x}) \equiv c P^i_{\text{field}}, \tag{9.67}$$

where $\mathcal{M}^i/c$ gives the total constant 3-momentum of the free field configuration. The constant c comes from purely dimensional reasons.

9.3.2.2 Lorentz Transformations

Rotations and boosts constitute Lorentz transformations. Any infinitesimal Lorentz transformation can be specified as

$$x'^\mu = x^\mu + \omega^{\mu\nu} x_\nu, \tag{9.68}$$

where $\omega^{\mu\nu}$ is a second rank tensor whose components are constants and independent of space and time. Using the fact that $x^\mu x_\mu = x'^\mu x'_\mu$ under a Lorentz transformation, one can find out that $\omega^{\mu\nu} = -\omega^{\nu\mu}$. Comparing the infinitesimal changes of the coordinates under a spatial rotation about z-axis as given in Eq. (9.38), we see that for such a Lorentz transformation $\omega^{12} = \delta\theta$ and $\omega^{21} = -\delta\theta$. All the other $\omega^{\mu\nu}$ components for such a rotation are zero. For Lorentz boosts, we have to be a bit more careful. First, we note that the relevant part of the Lorentz boost formula as given in Eq. (6.4) in Chap. 6 can also be written as

$$\begin{pmatrix} x'^0 \\ x'^1 \end{pmatrix} = \begin{pmatrix} \cosh\phi & -\sinh\phi \\ -\sinh\phi & \cosh\phi \end{pmatrix} \begin{pmatrix} x^0 \\ x^1 \end{pmatrix}, \tag{9.69}$$

where ϕ is called rapidity and is given by the formula

$$\phi = \tanh^{-1}\beta. \tag{9.70}$$

For infinitesimal Lorentz boosts $\gamma \sim 1$, $\beta \ll 1$ and as a result $\phi \sim \beta$. For such a case, we get

$$\delta x^0 = -\phi x^1, \quad \delta x^1 = -\phi x^0, \tag{9.71}$$

showing that for infinitesimal Lorentz boosts $\omega^{01} = \phi$ and $\omega^{10} = -\phi$. In this section, we will not discuss more about the boost symmetry of the action, we will be more interested on rotations.

The 4-vector fields also transform because of the infinitesimal Lorentz transformations. The transformed fields are conventionally written as

$$A'^{\mu}(x') = L^{\mu}{}_{\nu}(\omega) A^{\nu}(x), \tag{9.72}$$

where the transformation matrix corresponding to a continuous infinitesimal Lorentz transformation is given by

$$L^{\mu}{}_{\nu}(\omega) = \delta^{\mu}_{\nu} + \frac{1}{2}\omega_{\alpha\beta}(\Sigma^{\alpha\beta})^{\mu}{}_{\nu}. \tag{9.73}$$

Here, $\omega_{\alpha\beta}$ represents the infinitesimal parameters specifying a Lorentz transformation and

$$(\Sigma^{\mu\nu})_{\alpha\beta} = \delta^{\mu}_{\alpha}\delta^{\nu}_{\beta} - \delta^{\mu}_{\beta}\delta^{\nu}_{\alpha}, \tag{9.74}$$

represent the generator for a Lorentz transformation for 4-vector gauge fields. From these considerations, we see that

$$\delta A^{\mu}(x) = A'^{\mu}(x') - A^{\mu}(x) = \frac{1}{2}\omega_{\alpha\beta}(\Sigma^{\alpha\beta})^{\mu}{}_{\nu}A^{\nu}(x). \tag{9.75}$$

For an infinitesimal rotation about z-axis, where $\omega^{12} = \delta\theta$ and $\omega^{21} = -\delta\theta$ and the spin matrix form as given above one can easily reproduce the $\delta A^{\mu}(x)$ as given in Eq. (9.39) from the above formula.

The Noether current, as given in Eq. (9.56), corresponding to the continuous infinitesimal Lorentz transformations can be written as

$$J^{\alpha} = \frac{\partial \mathcal{L}}{\partial(\partial_{\alpha}A_{\mu})}\left[\frac{1}{2}\omega_{\kappa\beta}(\Sigma^{\kappa\beta})_{\mu\nu}A^{\nu}(x)\right] - T^{\alpha\nu}\delta x_{\nu}.$$

From Eq. (9.28), we know

$$\frac{\partial \mathcal{L}}{\partial(\partial_{\alpha}A_{\mu})} = -\frac{1}{4\pi}F^{\alpha\mu},$$

and for Lorentz transformations $\delta x_{\nu} = \omega_{\nu\beta}x^{\beta}$. Using these facts, we can write

$$J^\alpha = -\frac{1}{8\pi}F^{\alpha\mu}\omega_{\nu\beta}(\Sigma^{\nu\beta})_{\mu\kappa}A^\kappa(x) - T^{\alpha\nu}x^\beta\omega_{\nu\beta}$$

$$= -\frac{1}{8\pi}F^{\alpha\mu}\omega_{\nu\beta}(\Sigma^{\nu\beta})_{\mu\kappa}A^\kappa(x) - \frac{1}{2}(T^{\alpha\nu}x^\beta\omega_{\nu\beta} + T^{\alpha\beta}x^\nu\omega_{\beta\nu}),$$

where we have deliberately expanded the term $T^{\alpha\nu}x^\beta\omega_{\nu\beta}$ as half of the sum of two equal terms by interchanging the dummy indices ν and β. Now one can write the Noether current for Lorentz transformations as:

$$J^\alpha = \frac{1}{2}\left[-\frac{1}{4\pi}F^{\alpha\mu}(\Sigma^{\nu\beta})_{\mu\kappa}A^\kappa(x) - T^{\alpha\nu}x^\beta + T^{\alpha\beta}x^\nu\right]\omega_{\nu\beta}, \qquad (9.76)$$

where $\omega_{\nu\beta}$ are constant infinitesimal parameters. As $\partial_\alpha J^\alpha = 0$, we can also write

$$\partial_\alpha M^{\alpha\nu\beta} = 0, \qquad (9.77)$$

where

$$M^{\alpha\nu\beta} = -\frac{1}{4\pi}F^{\alpha\mu}(\Sigma^{\nu\beta})_{\mu\kappa}A^\kappa(x) + (T^{\alpha\beta}x^\nu - T^{\alpha\nu}x^\beta). \qquad (9.78)$$

We can now proceed to define the conserved quantities, but before we do that we will have to redefine the energy-momentum tensor as it lacks some important properties which are expected from it. After revising the form of the energy-momentum tensor, we will come back to the above equation and see its consequences.

9.4   Drawbacks of the Energy-Momentum Tensor $T^{\mu\nu}$

For the free electromagnetic field, we can write the energy-momentum tensor $T^{\alpha\beta}$, whose form is given in Eq. (9.54), as

$$T^{\alpha\beta} = \frac{\partial\mathcal{L}}{\partial(\partial_\alpha A^\lambda)}(\partial^\beta A^\lambda) - \eta^{\alpha\beta}\mathcal{L},$$

where the free electromagnetic field Lagrangian density is given in Eq. (9.33). The energy-momentum tensor can also be written as

$$T^{\alpha\beta} = -\frac{1}{4\pi}\eta^{\alpha\mu}F_{\mu\lambda}(\partial^\beta A^\lambda) - \eta^{\alpha\beta}\mathcal{L}$$

$$= -\frac{1}{4\pi}\eta^{\alpha\mu}F_{\mu\lambda}(\partial^\beta A^\lambda) + \frac{\eta^{\alpha\beta}}{16\pi}F^{\mu\nu}F_{\mu\nu}. \qquad (9.79)$$

This expression of the energy-momentum tensor for the free electromagnetic field has some serious drawbacks. The drawbacks will become clear once we state what we demand from an ideal expression of the energy-momentum tensor for any field. Ideally, the energy-momentum tensor for the free electromagnetic field should be

1. invariant under a gauge transformation,
2. symmetric, and lastly
3. traceless.

We have seen that the total energy and 3-momentum of the free electromagnetic field is obtained by integrating the appropriate $T^{\alpha\beta}$ components. As the total energy and 3-momentum of the electromagnetic field does not depend on which gauge we are using to write the 4-potentials, it is natural to expect that $T^{\alpha\beta}$ components should also be gauge invariant. From the expression of $T^{\alpha\beta}$ in Eq. (9.79), we see that it is not gauge invariant.

Why do we require that $T^{\alpha\beta}$ be symmetric? This requirement can be justified in the following way. In Einstein's general theory of relativity, the energy-momentum tensor appears in the field equation. The field equation of general relativity demands that $T^{\alpha\beta}$ must be a second rank symmetric tensor. In general, most of the other energy-momentum tensors are symmetric, as the energy-momentum tensor for a perfect fluid or the energy-momentum tensor for a scalar field. All these considerations demand a symmetric $T^{\alpha\beta}$ for the electromagnetic field. From the expression of $T^{\alpha\beta}$ in Eq. (9.79), we see that it is not a symmetric tensor.

Why should $T^{\alpha\beta}$ be traceless? This point is too technical to be discussed in this book. We just state the reason for the traceless nature of the energy-momentum tensor without giving any justification. As the free electromagnetic field does not have any constant parameter with mass dimension the trace of $T^{\alpha\beta}$ should vanish. From the expression of $T^{\alpha\beta}$ in Eq. (9.79) one can calculate that $T^{\alpha}_{\alpha} \neq 0$.

All the points discussed above show that the energy-momentum tensor we obtained for the free electromagnetic field from Noether's analysis is not up to the mark. To make the point more stronger, we may actually calculate a component of $T^{\alpha\beta}$ explicitly. To do so we first note that

$$\mathcal{L} = -\frac{1}{16\pi} F^{\mu\nu} F_{\mu\nu} = \frac{1}{8\pi}(\mathbf{E}^2 - \mathbf{B}^2) \,, \tag{9.80}$$

where $\mathbf{E}$ and $\mathbf{B}$ are the electric and magnetic fields comprising the free electromagnetic field. When we write $\mathbf{E}^2$ or $\mathbf{B}^2$, we mean the scalar product of these 3-vectors using the Euclidean metric as discussed in the beginning of Chap. 7. In the following calculations, we will start with the fully covariant expressions and then switch over to scalar products and component manipulations using the Euclidean metric as given in Eq. (7.1). Now, we can calculate T^{00} as

$$T^{00} = -\frac{1}{4\pi}\eta^{0\mu} F_{\mu\lambda}(\partial^0 A^\lambda) - \frac{\eta^{00}}{8\pi}(\mathbf{E}^2 - \mathbf{B}^2)$$
$$= -\frac{1}{4\pi}E^i(\partial^0 A^i) - \frac{1}{8\pi}(\mathbf{E}^2 - \mathbf{B}^2) \,.$$

Using the fact that $\partial^0 A^i = \partial^i \Phi - E^i$, where Φ is the electromagnetic scalar potential, we can write the above equation as

$$T^{00} = \frac{1}{8\pi}(\mathbf{E}^2 + \mathbf{B}^2) + \frac{1}{4\pi}\nabla \cdot (\Phi\mathbf{E}),\tag{9.81}$$

where we have used the fact that $\nabla \cdot \mathbf{E} = 0$ for free electromagnetic field. In our calculations, $\nabla = (\partial_1, \partial_2, \partial_3)$. The reader must be aware of the fact that the energy density for free electromagnetic field is given by only the first term on the right-hand side of the above equation, the second term should not be there. Fortunately, the unwanted second term drops out when we calculate the total energy of the free electromagnetic field as

$$\int_V d^3x\, T^{00} = \frac{1}{8\pi}\int_V d^3x\,(\mathbf{E}^2 + \mathbf{B}^2) + \frac{1}{4\pi}\oint_{\partial V}(\Phi\mathbf{E})\cdot d\mathbf{a},$$

applying Gauss theorem. The second term on the right hand side drops out if $\Phi\mathbf{E}$ vanishes on the closed boundary ∂V. One can similarly calculate the other components of $T^{\alpha\beta}$ and see that all the other components will also have extra terms which are unwanted in the energy-momentum tensor. The existence of these unwanted terms in the expression of the energy density itself shows a particular drawback of the form of $T^{\alpha\beta}$ with which we are working.

9.4.1 Constructing the Proper Energy-Momentum Tensor

The method we are going to describe in this subsection was first used by F. J. Belinfante and consequently this way of constructing the energy-momentum tensor is called the Belinfante construction. To proceed with the Belinfante construction, we at first summarize the most important defining properties of the energy-momentum tensor $T^{\alpha\beta}$. One of the most important features of the energy-momentum tensor is that the spatial integral of $T^{0\beta}$ gives the conserved energy and conserved 3-momentum of the free field configuration. The second important property of $T^{\alpha\beta}$ is that its 4-divergence is zero, $\partial_\alpha T^{\alpha\beta} = 0$. Both of these properties were discussed in the last section. Any second rank tensor which satisfies primarily these conditions can be called an energy-momentum tensor for the field. Based on these facts we will proceed to construct a new second rank, gauge invariant, symmetric and traceless tensor $\Theta^{\alpha\beta}$ which satisfies the defining properties of an energy-momentum tensor.

We can write $T^{\alpha\beta}$ as given in Eq. (9.79) as

$$T^{\alpha\beta} = -\frac{1}{4\pi}\eta^{\alpha\mu}F_{\mu\lambda}(-F^{\lambda\beta} + \partial^\lambda A^\beta) + \frac{\eta^{\alpha\beta}}{16\pi}F^{\mu\nu}F_{\mu\nu}$$
$$= \Theta^{\alpha\beta} + T_{\mathrm{D}}^{\alpha\beta},\tag{9.82}$$

where the new tensors are defined as

$$\Theta^{\alpha\beta} \equiv \frac{1}{4\pi} \left[\eta^{\alpha\mu} F_{\mu\lambda} F^{\lambda\beta} + \frac{\eta^{\alpha\beta}}{4} F^{\mu\nu} F_{\mu\nu} \right] , \qquad (9.83)$$

and

$$T_{\mathrm{D}}^{\alpha\beta} \equiv -\frac{1}{4\pi} \eta^{\alpha\mu} F_{\mu\lambda} (\partial^{\lambda} A^{\beta}) . \qquad (9.84)$$

We already know from the last section that $\partial_{\alpha} T^{\alpha\beta} = 0$, but we do not know what is $\partial_{\alpha}\Theta^{\alpha\beta}$. We will be able to answer this question later. Before we proceed to find out the nature of $\Theta^{\alpha\beta}$, we have to know something about the other tensor $T_{\mathrm{D}}^{\alpha\beta}$. In general, we have

$$\partial^{\lambda} (F_{\mu\lambda} A^{\beta}) = A^{\beta} (\partial^{\lambda} F_{\mu\lambda}) + F_{\mu\lambda} (\partial^{\lambda} A^{\beta}) ,$$

where $\partial^{\lambda} F_{\mu\lambda} = 0$ for the free electromagnetic field. Using this fact, we can now write

$$T_{\mathrm{D}}^{\alpha\beta} = \frac{1}{4\pi} \partial_{\lambda} (F^{\lambda\alpha} A^{\beta}) . \qquad (9.85)$$

Using the antisymmetry of $F^{\lambda\alpha}$, the reader can easily verify that $\partial_{\alpha} T_{\mathrm{D}}^{\alpha\beta} = 0$. Next, one can calculate the 3-volume integral of $T_{\mathrm{D}}^{0\beta}$ as

$$\int_{V} d^{3}x \, T_{\mathrm{D}}^{0\beta} = \frac{1}{4\pi} \int_{V} d^{3}x \, \partial_{i} (F^{i0} A^{\beta}) = \frac{1}{4\pi} \oint_{\partial V} da_{i} \, (E^{i} A^{\beta}) ,$$

where we have used Gauss theorem. From the above equation, one can see that the volume integral of $T_{\mathrm{D}}^{0\beta}$ vanishes when the electric and magnetic fields vanish on the closed boundary ∂V. As we always assume that the fields vanish on the boundary, we can safely use

$$\int_{V} d^{3}x \, T_{\mathrm{D}}^{0\beta} = 0 . \qquad (9.86)$$

Writing $\Theta^{\alpha\beta} = T^{\alpha\beta} - T_{\mathrm{D}}^{\alpha\beta}$ we see that

$$\int_{V} d^{3}x \, \Theta^{0\beta} = \int_{V} d^{3}x \, T^{0\beta} = \mathcal{M}^{\beta} , \quad \partial_{\alpha}\Theta^{\alpha\beta} = 0 , \qquad (9.87)$$

where $\mathcal{M}^{\beta}$ is the conserved 4-momentum of the free electromagnetic field introduced in the last section. It is seen that $\Theta^{\alpha\beta}$ satisfies all the relevant defining properties of $T^{\alpha\beta}$. We can henceforth work with $\Theta^{\alpha\beta}$ as the energy-momentum tensor for free electromagnetic fields. We will show below that this redefined energy-momentum tensor has all the nice properties which we demanded from an ideal energy-momentum tensor.

From the expression of $\Theta^{\alpha\beta}$ in Eq. (9.83), we see that it is gauge invariant. The reader can easily show that $\Theta^{\alpha\beta} = \Theta^{\beta\alpha}$. Lastly

$$
\begin{aligned}
\Theta^{\alpha}_{\alpha} &= \frac{1}{4\pi}\left[\eta^{\alpha\mu}F_{\mu\lambda}F^{\lambda}_{\ \alpha} + \frac{1}{4}\delta^{\alpha}_{\beta}F^{\mu\nu}F_{\mu\nu}\right] \\
&= \frac{1}{4\pi}\left[F^{\alpha}_{\ \lambda}F^{\lambda}_{\ \alpha} + F^{\mu\nu}F_{\mu\nu}\right] \\
&= \frac{1}{4\pi}\left[F^{\alpha\lambda}F_{\lambda\alpha} + F^{\mu\nu}F_{\mu\nu}\right] = 0\,,
\end{aligned} \tag{9.88}
$$

showing that $\Theta^{\alpha\beta}$ is traceless. These shows that $\Theta^{\alpha\beta}$ is a gauge invariant, symmetric and traceless energy-momentum tensor for the free electromagnetic field.

Now we can calculate the components of $\Theta^{\alpha\beta}$. We will see that the components of $\Theta^{\alpha\beta}$ are physically transparent and does not contain any unwanted terms. Θ^{00} is

$$
\begin{aligned}
\Theta^{00} &= \frac{1}{4\pi}\left[\eta^{0\mu}F_{\mu\lambda}F^{\lambda 0} + \frac{1}{4}F^{\mu\nu}F_{\mu\nu}\right] \\
&= \frac{1}{4\pi}\left[F^{0}_{\ \lambda}F^{\lambda 0} + \frac{1}{4}(\mathbf{B}^2 - \mathbf{E}^2)\right]\,.
\end{aligned}
$$

As $F^{0}_{\ \lambda}F^{\lambda 0} = \mathbf{E}^2$, the above expression gives

$$
\Theta^{00} = \frac{1}{8\pi}(\mathbf{E}^2 + \mathbf{B}^2)\,, \tag{9.89}
$$

giving the proper energy density of the free electromagnetic field. Similarly, we have

$$
\Theta^{i0} = \Theta^{0i} = \frac{1}{4\pi}\eta^{0\mu}F_{\mu\lambda}F^{\lambda i} = -\frac{1}{4\pi}F_{j0}F^{ji} = \frac{1}{4\pi}(\mathbf{E}\times\mathbf{B})^{i}\,, \tag{9.90}
$$

giving the linear 3-momentum density of the free electromagnetic field. In this equation, the index i at the right most position can be lowered without the expense of any minus sign, whereas to lower the index i on Θ^{i0}, Θ^{0i} terms, one has to introduce a minus sign as these are proper 4-tensors.

Next, we calculate Θ^{ij}. We will see that these components of the energy-momentum tensor are related to the Maxwell's stress tensor components. In the present case,

$$
\Theta^{ji} = \Theta^{ij} = \frac{1}{4\pi}\left[\eta^{i\mu}F_{\mu\lambda}F^{\lambda j} - \frac{1}{4}\delta^{ij}F^{\mu\nu}F_{\mu\nu}\right]\,,
$$

where we have used $\eta^{ij} = -\delta^{ij}$. We can write

$$\eta^{i\mu} F_{\mu\lambda} F^{\lambda j} = -F_{i\lambda} F^{\lambda j} = -[F_{i0} F^{0j} + F_{ik} F^{kj}]$$

$$= -[E^i E^j + \sum_{k=1}^{3} F^{ik} F^{kj}]. \tag{9.91}$$

The reader can easily work out that

$$\sum_{k=1}^{3} F^{ik} F^{kj} = \delta^{ij} \mathbf{B}^2 - B^i B^j, \tag{9.92}$$

using which one can finally write

$$\Theta^{ji} = \Theta^{ij} = -\frac{1}{4\pi} \left[E^i E^j + B^i B^j - \frac{1}{2} \delta^{ij} (\mathbf{E}^2 + \mathbf{B}^2) \right]. \tag{9.93}$$

This differs from T^{ij} of Chap. 3 by a sign. The sign change is because we are using the four dimensional metric here. If we have an infinitesimal surface area da oriented along the direction of unit vector n_j along the jth direction then $\sum_{j=1}^{3} \Theta^{ij} da\, n_j$ specifies the ith component of electromagnetic stress on the surface.

9.4.2 Energy-Momentum Conservation

We will now see that $\Theta^{\alpha\beta}$ can yield energy-momentum conservation because $\partial_\alpha \Theta^{\alpha 0} = 0$ or $\partial_0 \Theta^{00} + \partial_i \Theta^{i0} = 0$. Naming the free field energy density as u, where

$$u \equiv \frac{1}{8\pi} (\mathbf{E}^2 + \mathbf{B}^2), \tag{9.94}$$

we can write the conservation equation as

$$\frac{\partial u}{\partial t} + \nabla \cdot \mathbf{S} = 0, \tag{9.95}$$

where

$$\mathbf{S} \equiv \frac{c}{4\pi} \mathbf{E} \times \mathbf{B} \tag{9.96}$$

is the Poynting vector. We know $\mathbf{S}$ represents energy flowing out per unit time through an unit area held normal to the direction of energy flow and consequently Eq. (9.95) represent energy conservation for free electromagnetic field.

We can also write the conservation equation for $\Theta^{\alpha\beta}$ as $\partial_0 \Theta^{0i} + \partial_j \Theta^{ji} = 0$. Defining the 3-momentum density of the free electromagnetic field as

$$\mathbf{g} \equiv \frac{1}{4\pi c}(\mathbf{E} \times \mathbf{B}),\qquad (9.97)$$

the conservation equation becomes

$$\frac{\partial g^i}{\partial t} + \partial_j \Theta^{ji} = 0.\qquad (9.98)$$

Integrating the above equation over any volume V, we get

$$\int_V d^3x\, \frac{\partial g^i}{\partial t} = -\int_V d^3x\, \partial_j \Theta^{ij}.$$

The total 3-momentum of the free electromagnetic field is given by

$$P^i_{\text{field}} = \int_V d^3x\, g^i,\qquad (9.99)$$

which prompts us to write the integral form of the conservation equation as

$$\frac{dP^i_{\text{field}}}{dt} = -\oint_{\partial V} \Theta^{ij} n_j\, da.\qquad (9.100)$$

where we have applied Gauss theorem. The summation on j has been suppressed here. The above equation says that the rate of change of the free field 3-momentum is given by the force transmitted across the boundary of V. In the present case, V can be any volume containing finite field configuration (and need not be the whole volume) over which the field configuration is non-zero.

The above calculations show us that $\Theta^{\mu\nu}$ can indeed be taken as the real energy-momentum tensor for the free electromagnetic field. Next, we go back to the discussion initiated at the end of the last section and try to reconsider Eq. (9.78) in the light of new developments.

9.4.3 Conserved Tensors Corresponding to Lorentz Transformations

We stopped our discussion on the tensor $M^{\alpha\nu\beta}$ corresponding to Lorentz transformations after Eq. (9.78) because we wanted to use the correct energy-momentum tensor $\Theta^{\mu\nu}$ instead of $T^{\mu\nu}$. The corrected expression for $M^{\alpha\nu\beta}$ is

$$M^{\alpha\nu\beta} = -\frac{1}{4\pi}F^{\alpha\mu}(\Sigma^{\nu\beta})_{\mu\kappa}A^{\kappa}(x) + (\Theta^{\alpha\beta}x^{\nu} - \Theta^{\alpha\nu}x^{\beta}).\qquad (9.101)$$

From the condition $\partial_\alpha M^{\alpha\nu\beta} = 0$ and the application of Gauss law, one can easily show that

$$\frac{dM^{\nu\beta}}{dt} = 0, \tag{9.102}$$

where the conserved tensor $M^{\nu\beta}$ is

$$M^{\nu\beta} \equiv \int_V d^3x \, M^{0\nu\beta}, \tag{9.103}$$

and V is the entire 3-volume in which the electromagnetic field is non-zero. We assume the fields vanish on the closed boundary ∂V. The expression of $M^{0\nu\beta}$ is

$$M^{0\nu\beta} = -\frac{1}{4\pi} F^{0i} (\Sigma^{\nu\beta})_{ik} A^\kappa(x) + (\Theta^{0\beta} x^\nu - \Theta^{0\nu} x^\beta)$$

$$= \frac{1}{4\pi} E^i (\Sigma^{\nu\beta})_{i\kappa} A^\kappa(x) + (\Theta^{0\beta} x^\nu - \Theta^{0\nu} x^\beta). \tag{9.104}$$

For angular momentum conservation, we require the form of M^{0jm} which is given as

$$M^{0jm} = \frac{1}{4\pi} E^i (\Sigma^{jm})_{ik} A^k(x) + (\Theta^{0m} x^j - \Theta^{0j} x^m), \tag{9.105}$$

where we have used our knowledge about the form of $(\Sigma^{\nu\beta})_{\mu\kappa}$ as given in Eq. (9.74) and all the indices j, m, i, k run from 1 to 3. Using the definition of Θ^{0m} and $(\Sigma^{jm})_{ik}$, we can write the above equation as

$$M^{0jm} = \frac{1}{4\pi} (E^j A^m - E^m A^j) + \frac{1}{4\pi} \left[x^j (\mathbf{E} \times \mathbf{B})^m - x^m (\mathbf{E} \times \mathbf{B})^j \right]. \tag{9.106}$$

It is to be noted that M^{0jm} is antisymmetric in the indices m and j and consequently we can define a 3-vector as

$$cM_i \equiv \frac{1}{2} \epsilon_{ijm} M^{0jm}, \tag{9.107}$$

where the factor c is for purely dimensional reason. From this definition, we can write

$$\mathbf{M} = \frac{1}{4\pi c} \int_V d^3x \, (\mathbf{E} \times \mathbf{A}) + \frac{1}{4\pi c} \int_V d^3x \, [\mathbf{x} \times (\mathbf{E} \times \mathbf{B})]. \tag{9.108}$$

The second term on the right-hand side is

$$\int_V d^3x \, (\mathbf{x} \times \mathbf{g}) \,,$$

where $\mathbf{g}$ is the linear momentum density of the free electromagnetic field. From the structure of the second term, we can easily understand that it represents the normal angular momentum of the field configuration. On the other hand, the integrand in the first term on the right-hand side of Eq. (9.108) is independent of position coordinates and cannot be the normal angular momentum of the field configuration. The first term on the right-hand side of Eq. (9.108) has the dimension of angular momentum, and for this reason, we are compelled to think about this term as the spin angular momentum of the electromagnetic field. Radiation can be circularly polarized and these polarization states can have spin angular momentum.

The power of Noether's theorem becomes manifest now, starting from Lorentz symmetry we come to the conclusion that free electromagnetic field can have two kinds of angular momentum, the first one is the standard angular momentum in fields and the other one is related to its polarization states.

9.5 Energy-Momentum Conservation in the Presence of Sources

Till now we were working with pure electromagnetic field and there was no source for the fields. The source for electromagnetic field comes through the 4-current J^μ. In the present section, we will like to introduce the source term and try to see how energy-momentum conservation conditions are modified in the presence of sources. For the field, we still have the energy-momentum tensor $\Theta^{\alpha\beta}$ as given in Eq. (9.83). But now $\partial_\alpha \Theta^{\alpha\beta} \neq 0$ as the field can exchange energy momentum with the source. In the present case, we can write

$$\partial_\alpha \Theta^{\alpha\beta} = \frac{1}{4\pi} \left[\partial^\mu (F_{\mu\lambda} F^{\lambda\beta}) + \frac{1}{4} \partial^\beta (F^{\mu\nu} F_{\mu\nu}) \right]$$

$$= \frac{1}{4\pi} \left[(\partial^\mu F_{\mu\lambda}) F^{\lambda\beta} + F_{\mu\lambda} (\partial^\mu F^{\lambda\beta}) + \frac{1}{2} F_{\mu\lambda} (\partial^\beta F^{\mu\lambda}) \right] . \quad (9.109)$$

In the presence of sources $\partial^\mu F_{\mu\lambda} \neq 0$ but is given by $(4\pi/c)J_\lambda$, and consequently we have

$$\partial_\alpha \Theta^{\alpha\beta} = \frac{1}{c} J_\lambda F^{\lambda\beta} + \frac{1}{4\pi} \left[F_{\mu\lambda} (\partial^\mu F^{\lambda\beta}) + \frac{1}{2} F_{\mu\lambda} (\partial^\beta F^{\mu\lambda}) \right] . \quad (9.110)$$

The above equation can also be written as

$$\partial_\alpha \Theta^{\alpha\beta} + \frac{1}{c} J_\lambda F^{\beta\lambda} = \frac{1}{8\pi} F_{\mu\lambda} (2\partial^\mu F^{\lambda\beta} + \partial^\beta F^{\mu\lambda})$$

$$= \frac{1}{8\pi} F_{\mu\lambda} \left[\partial^\mu F^{\lambda\beta} + (\partial^\mu F^{\lambda\beta} + \partial^\beta F^{\mu\lambda}) \right]. \qquad (9.111)$$

Using the homogeneous Maxwell equation $\partial^\mu F^{\lambda\beta} + \partial^\lambda F^{\beta\mu} + \partial^\beta F^{\mu\lambda} = 0$, we can write the above equation as

$$\partial_\alpha \Theta^{\alpha\beta} + \frac{1}{c} J_\lambda F^{\beta\lambda} = \frac{1}{8\pi} F_{\mu\lambda} (\partial^\mu F^{\lambda\beta} + \partial^\lambda F^{\mu\beta}). \qquad (9.112)$$

One can see that $\partial^\mu F^{\lambda\beta} + \partial^\lambda F^{\mu\beta}$ is symmetric in the indices μ and λ, whereas $F_{\mu\lambda}$ is antisymmetric in those indices, as a result of which we can now write the energy-momentum conservation equation in the presence of sources as

$$\partial_\alpha \Theta^{\alpha\beta} = -\frac{1}{c} J_\lambda F^{\beta\lambda}. \qquad (9.113)$$

Setting $\beta = 0$ in the above equation, we get

$$\partial_\alpha \Theta^{\alpha 0} = \frac{1}{c} J_i F^{i0} = \frac{1}{c} E^i J_i = -\frac{1}{c} \mathbf{J} \cdot \mathbf{E},$$

which can be written down explicitly as

$$\frac{\partial u}{\partial t} + \nabla \cdot \mathbf{S} = -\mathbf{J} \cdot \mathbf{E}, \qquad (9.114)$$

where u is the field energy density and $\mathbf{S}$ is Poynting vector. This equation specifies energy conservation of electromagnetic field in presence of sources.

On the other hand, when β is any spatial index as i, then we have

$$\partial_\alpha \Theta^{\alpha i} = -\frac{1}{c} J_\lambda F^{i\lambda} = -\frac{1}{c} (F^{i0} J_0 + F^{ij} J_j)$$

$$= -\frac{1}{c} \left[E^i (c\rho) + (\mathbf{J} \times \mathbf{B})^i \right].$$

From the calculations at the end of the last section, we can now write the above equation as

$$\frac{\partial g^i}{\partial t} + \partial_j \Theta^{ji} = -\left[E^i \rho + \frac{1}{c} (\mathbf{J} \times \mathbf{B})^i \right], \qquad (9.115)$$

giving us the 3-momentum density balance of the field and the sources. We see that the tensor $\Theta^{\alpha\beta}$ and source current J^α are enough to specify the energy-momentum conservation conditions when the electromagnetic field interacts with sources. The

Lagrangian formalism of electrodynamics gives us a basic framework based on which we could discuss all these important and practical topics which we will use as we proceed.

9.6 Problems

1. Show that in special relativity the proper time is maximized over a straight line path when a particle moves along that straight line with constant velocity.
2. Derive the Lorentz force equations from the Hamilton's equations of motion using the Hamiltonian given in Eq. (9.13).
3. The Lagrangian of a charged particle in presence of an electromagnetic filed is given in Eq. (9.11) and it depends on the gauge fields Φ and $\mathbf{A}$. Under a gauge transformation

$$\Phi'(x) = \Phi(x) + \frac{1}{c}\frac{\partial \chi(x)}{\partial t}, \quad \mathbf{A}'(x) = \mathbf{A}(x) - \nabla\chi(x),$$

the Lagrangian changes from L to L'. Here, $x \equiv (t, \mathbf{x})$. Write down L' explicitly and show that

$$L' = L - \frac{q}{c}\frac{d\chi}{dt}.$$

Does the equations of motion of a charged particle in an electromagnetic field change under a gauge transformation?
4. From the discussion of parity symmetry in Chap. 7 show that if the Lagrangian density of the electromagnetic field contains both $F^{\mu\nu}F_{\mu\nu}$ and $\tilde{F}^{\mu\nu}F_{\mu\nu}$ as

$$\mathcal{L} = \alpha F^{\mu\nu}F_{\mu\nu} + \beta \tilde{F}^{\mu\nu}F_{\mu\nu} - \frac{1}{c}J^{\mu}A_{\mu},$$

where α and β are Lorentz scalars, then parity symmetry is broken.
5. Show that $\tilde{F}^{\mu\nu}\tilde{F}_{\mu\nu} = -F^{\mu\nu}F_{\mu\nu}$.
6. For the Lorentz transformation in Eq. (9.68), we know $x^{\mu}x_{\mu} = x'^{\mu}x'_{\mu}$. Using this fact show that $\omega^{\mu\nu} = -\omega^{\nu\mu}$.
7. For an infinitesimal rotation about z-axis, where $\omega^{12} = \delta\theta$ and $\omega^{21} = -\delta\theta$ (all other ω components are zero) and the spin matrix is given in Eq. (9.74), show that the expression in Eq. (9.75) reproduces the results given in Eq. (9.39).
8. Show that

$$\mathcal{L} = -\frac{1}{16\pi}F^{\mu\nu}F_{\mu\nu} = \frac{1}{8\pi}(\mathbf{E}^2 - \mathbf{B}^2).$$

9. Calculate T^{0i} and T^{i0} from the expression of $T^{\alpha\beta}$ in Eq. (9.79) in terms of electric, magnetic and gauge fields.

10. Show that $\Theta^{\alpha\beta} = \Theta^{\beta\alpha}$, where $\Theta^{\alpha\beta}$ is given in Eq. (9.83).
11. Prove the relation given in Eq. (9.92).
12. Suppose you are given the free field (in absence of sources) Lagrangian density of the electromagnetic field. You know the gauge field can undergo symmetry transformations as the gauge transformations as $A^\mu(x) \rightarrow A'^\mu(x) = A^\mu(x) + \partial^\mu \chi(x)$, where $\chi(x)$ is a well behaved infinitesimal function of x. In this case, there are no spacetime transformations but infinitesimal gauge transformations.

   (a) What is the energy-momentum tensor corresponding to such a gauge transformation?
   (b) What is the Noether current corresponding to such a gauge transformation? Is the current conserved? Justify your answer.
   (c) What is the conserved charge in the present case? What will be the value of the conserved charge if $\chi(x)$ is a constant?

Chapter 10
Electromagnetic Field Produced by an Arbitrarily Moving Point Charge

Till now we have written down the Maxwell equations in covariant form and we also know the Lagrangian formalism yielding the covariant form of the Maxwell equations but we have not solved the Maxwell equations. In this chapter, we will solve the Maxwell equations for a very particular case, the case where the electromagnetic field originates from an arbitrarily moving charged particle. We will see that in classical electrodynamics an accelerated point charge can produce electromagnetic radiation.

To solve the Maxwell equations, we require to use some gauge. We have seen in Chap. 8 that the solution of the inhomogeneous Maxwell equation is arbitrary and can only be written uniquely once we have eliminated the gauge degree of freedom. As because we are interested in a manifestly covariant treatment of radiation we use the Lorenz gauge. We know that in the Lorenz gauge the inhomogeneous Maxwell equation is

$$\Box A^\nu = \frac{4\pi}{c} J^\nu ,$$

as given in Eq. (8.6), which can also be written as

$$\left(\frac{1}{c^2} \frac{\partial^2}{\partial t^2} - \nabla^2 \right) A^\nu(x) = \frac{4\pi}{c} J^\nu(x) . \tag{10.1}$$

This is a linear, second-order inhomogeneous partial differential equation which can be solved by the method of Green function. The solutions depend upon the kind of conditions we impose to solve it. The Green function $G(x, x')$ depends upon two spacetime points out of which x specifies the spacetime point where the electromagnetic field is to be evaluated and x' specifies a point in the spacetime region where the source is distributed. We will often call x to be the field point and x' to be the source point. The Green function, we will employ to solve the inhomogeneous Maxwell equation, satisfies

© Springer Nature Singapore Pte Ltd. 2021
K. Bhattacharya and S. Mukhopadhyay, *Introduction to Advanced Electrodynamics*,
https://doi.org/10.1007/978-981-16-7802-8_10

$$\Box_x G(x, x') = \delta^4(x - x'),   \tag{10.2}$$

where $\delta^4(x - x') \equiv \delta(x_0 - x_0')\delta^3(\mathbf{x} - \mathbf{x}')$ is the four-dimensional Dirac delta function and the subscript x on the d'Alembertian operator specifies that the spacetime differentiations must be carried out at the field point x. Using the Green function method, the general solution of the inhomogeneous Maxwell equation can be written as

$$A^\alpha(x) = A_{\text{free}}^\alpha(x) + \frac{4\pi}{c} \int_\Omega d^4x' G(x, x') J^\alpha(x'),   \tag{10.3}$$

where $A_{\text{free}}^\alpha(x)$ is the solution of the wave equation

$$\left(\frac{1}{c^2}\frac{\partial^2}{\partial t^2} - \nabla^2\right) A_{\text{free}}^\alpha(x) = 0.   \tag{10.4}$$

One can easily verify that the expression in Eq. (10.3) is the solution of the inhomogeneous Maxwell equation in the Lorenz gauge. To see it we apply the d'Alembertian operator on the gauge field as

$$\Box A^\alpha(x) = \Box A_{\text{free}}^\alpha(x) + \frac{4\pi}{c} \int_\Omega d^4x' [\Box_x G(x, x')] J^\alpha(x')$$

$$= \frac{4\pi}{c} \int_\Omega d^4x' \delta^4(x - x') J^\alpha(x') = \frac{4\pi}{c} J^\alpha(x).$$

If we know the spacetime dependence of $A_{\text{free}}^\alpha(x)$, the nature of the source current $J^\alpha(x)$ and the functional form of $G(x, x')$ then we can in principle find out the solution of the inhomogeneous Maxwell equation. In the present chapter we will work with the 4-current J^α produced by an arbitrarily moving point charge.

10.1   Evaluation of the Green Function

The Green function was derived earlier while studying electromagnetic radiation. In this chapter, we employ a different method to derive the Green function keeping an eye on the explicit Lorentz invariant form of it. As spacetime is assumed to be homogeneous the Green function $G(x, x')$ becomes translation invariant and hence depends only upon the separation of the two points as

$$G(x, x') = G(x - x').   \tag{10.5}$$

For the sake of brevity, we introduce $y^\alpha \equiv x^\alpha - x'^\alpha$ and introduce the Fourier transform of $G(y)$ as

$$G(y) = \int_{-\infty}^{\infty} \frac{d^4k}{(2\pi)^4} \, G(k) e^{-ik \cdot y} , \tag{10.6}$$

where $k^\mu = (k^0, \mathbf{k})$ is a 4-vector with dimension of inverse length. It is the four-dimensional generalization of the wave vector. Here, we are following the asymmetric convention of Fourier transforms introduced in in Chap. 6. Here, the integrals represent four integrals where the limits of integration are as shown. The four-dimensional Dirac delta function is written as

$$\delta^4(y) = \int_{-\infty}^{\infty} \frac{d^4k}{(2\pi)^4} e^{-ik \cdot y} . \tag{10.7}$$

Using the Fourier expansion of $G(y)$ and the above definition of the Dirac delta function, we can write Eq. (10.2) as

$$\int_{-\infty}^{\infty} \frac{d^4k}{(2\pi)^4} \, G(k) (\partial_0^2 - \nabla^2)_x e^{-ik \cdot (x - x')} = \int_{-\infty}^{\infty} \frac{d^4k}{(2\pi)^4} e^{-ik \cdot y} ,$$

which becomes

$$\int_{-\infty}^{\infty} \frac{d^4k}{(2\pi)^4} \, [k^2 G(k) + 1] e^{-ik \cdot y} = 0 . \tag{10.8}$$

The above integral vanishes in general only when

$$G(k) = -\frac{1}{k^2} , \tag{10.9}$$

and as a consequence we can write

$$G(y) = \int_{-\infty}^{\infty} \frac{d^4k}{(2\pi)^4} \frac{e^{-ik \cdot y}}{k^2 - k_0^2} . \tag{10.10}$$

Henceforth, we will call $\mathbf{k}^2 = |\mathbf{k}|^2 = \kappa^2$ and write the above integral as

$$G(y) = \int_{-\infty}^{\infty} \frac{d^3k}{(2\pi)^3} \, e^{i\mathbf{k} \cdot \mathbf{y}} \int_{-\infty}^{\infty} \frac{dk_0}{(2\pi)} \frac{e^{-ik_0 y_0}}{\kappa^2 - k_0^2} . \tag{10.11}$$

We will first like to do the k_0 integral. It can be easily observed that the k_0 integral is difficult to do in the real plane as the integrand has poles at $k_0 = \pm\kappa$. To do the k_0 integral we use the complex k_0 plane. The k_0 integral is written as

$$I_0 \equiv \frac{1}{2\pi} \int_{-\infty}^{\infty} dk_0 \, \frac{e^{-ik_0 y_0}}{\kappa^2 - k_0^2} . \tag{10.12}$$

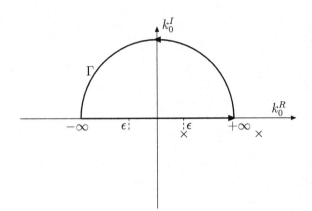

Fig. 10.1 The contour Γ in the complex k_0 plane used for evaluating I when $y_0 < 0$. The two poles on the real axis are shifted below by ϵ amount. In reality the diameter extends from $-\infty$ to ∞. The real contour on the upper half plane has infinite radius

In the complex k_0 plane, we have $k_0 = k_0^R + i k_0^I$ where k_0^R and k_0^I are the real and imaginary parts of k_0. We will see that the above integral result depends upon the sign of y_0. Suppose we have $y_0 = c(t - t') < 0$ or $t < t'$. In such a case we define a new variable $\tilde{y}_0 \equiv -y_0$ such that $\tilde{y}_0 > 0$. The relevant integral to be evaluated is

$$I = \frac{1}{2\pi} \oint_\Gamma dk_0 \, \frac{e^{ik_0 \tilde{y}_0}}{\kappa^2 - k_0^2}, \tag{10.13}$$

where Γ is the closed contour of integration in the complex k_0 plane. The aim is to find out I_0 by evaluating I. To evaluate I we note that $e^{ik_0 \tilde{y}_0} = e^{ik_0^R \tilde{y}_0} e^{-k_0^I \tilde{y}_0}$ tends to zero as $k_0^I > 0$ and increases without bound. We can choose the contour of integration as shown in Fig. 10.1. The direction of the arrows shows the direction to traverse the contour for the integration. As the poles are lying on the contour at $k_0 = \pm \kappa$, we will be unable to do the contour integral unless and until we change the contour or displace the poles. In this discussion, we will choose to displace the poles by an infinitesimal amount ϵ in the negative imaginary direction. This implies that we will be working with a modified integrand with poles at $k_0 = \kappa - i\epsilon$ and $k_0 = -\kappa - i\epsilon$ where $\epsilon > 0$. At the end of the integration, we should ideally take the limit $\epsilon \to 0$ and replace the poles on the real axis. If the result of such a limit is unambiguous and nonsingular then we will be able to successfully evaluate I_0. Choosing the contour Γ on the upper half of the complex k_0 plane as shown in Fig. 10.1 we can write

$$I = \frac{1}{2\pi} \int_{-\infty}^{\infty} dk_0 \, \frac{e^{-ik_0 y_0}}{\kappa^2 - (k_0 + i\epsilon)^2} + \frac{1}{2\pi} \int_{\text{semicirc}} dk_0 \, \frac{e^{ik_0 \tilde{y}_0}}{\kappa^2 - (k_0 + i\epsilon)^2}, \tag{10.14}$$

where the second term on the right-hand side represents the integration on the semi-circle on the upper complex plane. It has to be noted that the integrand in Eq. (10.13) has been modified. The poles of the above integrand are obtained by solving

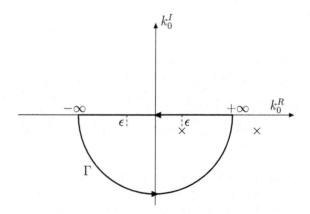

Fig. 10.2 The contour Γ in the complex k_0 plane used for evaluating I when $y_0 > 0$. The two poles on the real axis are shifted below by ϵ amount. In reality, the diameter extends from $-\infty$ to ∞. The real contour on the lower half plane has infinite radius

$$\kappa^2 - (k_0 + i\epsilon)^2 = (\kappa + k_0 + i\epsilon)(\kappa - k_0 - i\epsilon),$$

showing that the integrand has simple poles at $k_0 = \kappa - i\epsilon$ and $k_0 = -\kappa - i\epsilon$, as we want them to be. Although in the above drawing the semicircle has a finite radius but in reality the radius is infinity. The drawing shows the essence of the contour. As the radius of the semicircle tends to infinity k_0^I also tends to infinity and as a result the integrand on the second term in the right-hand side of the above equation vanishes. As the contour Γ does not contain any poles by Cauchy's residue theorem, we must have $I = 0$ giving

$$I = \frac{1}{2\pi} \int_{-\infty}^{\infty} dk_0 \, \frac{e^{-ik_0 y_0}}{\kappa^2 - (k_0 + i\epsilon)^2} = 0.$$

We see that the integration result is independent of ϵ and consequently in the $\epsilon \to 0$ limit we obtain the value of I_0 from the above equation as

$$I_0 = \frac{1}{2\pi} \int_{-\infty}^{\infty} dk_0 \, \frac{e^{-ik_0 y_0}}{\kappa^2 - k_0^2} = 0, \quad \text{when } t < t'. \tag{10.15}$$

This result shows that $G(y) = 0$ if $y_0 < 0$. In this case, we do not require to do the other integrals over k^1, k^2 and k^3 in Eq. (10.11). The result depends crucially on the sign of ϵ. If ϵ was negative the poles should have been on the upper half of the complex k_0 plane and then $I \neq 0$ because of the residues from the poles.

Now we want to evaluate I_0, as given in Eq. (10.12), for the case when $y_0 > 0$ or $t > t'$. In this case, we see that $e^{-ik_0 y_0} = e^{-ik_0^R y_0} e^{k_0^I y_0}$ tends to zero as $k_0^I < 0$ and decreases without bound. Consequently, we have to choose the semicircular contour in the lower half of the complex k_0 plane so that the integral contribution from the purely semicircular arc vanishes. Choosing the contour Γ as shown in Fig. 10.2 we can write the relevant integral over k_0 in the complex plane as

$$I = \frac{1}{2\pi} \oint_\Gamma dk_0 \, \frac{e^{-ik_0 y_0}}{\kappa^2 - (k_0 + i\epsilon)^2} , \tag{10.16}$$

where the contour contains the poles poles at $k_0 = \kappa - i\epsilon$ and $k_0 = -\kappa - i\epsilon$ when $t > t'$. This integral can now be written as

$$I = \frac{1}{2\pi} \int_{+\infty}^{-\infty} dk_0 \, \frac{e^{-ik_0 y_0}}{\kappa^2 - (k_0 + i\epsilon)^2} + \frac{1}{2\pi} \int_{\text{semicirc}} dk_0 \, \frac{e^{-ik_0 y_0}}{\kappa^2 - (k_0 + i\epsilon)^2} , \tag{10.17}$$

where the direction of integration is clear from the arrows on the contour. As the semicircular arc below has infinite radius so the integrand in the second term of the right-hand side of the above equation vanishes. From Cauchy's residue theorem, we can now write

$$I = \frac{1}{2\pi} \int_{+\infty}^{-\infty} dk_0 \, \frac{e^{-ik_0 y_0}}{\kappa^2 - (k_0 + i\epsilon)^2} = i \left[a(\kappa - i\epsilon) + a(-\kappa - i\epsilon) \right] , \tag{10.18}$$

where $a(\pm\kappa - i\epsilon)$ are the residues of the integrand at $k_0 = \pm\kappa - i\epsilon$. The residue of the integrand at $k_0 = \kappa - i\epsilon$ is obtained as

$$
\begin{aligned}
a(\kappa - i\epsilon) &= \lim_{k_0 \to \kappa - i\epsilon} (k_0 - \kappa + i\epsilon) \frac{e^{-ik_0 y_0}}{(\kappa + k_0 + i\epsilon)(\kappa - k_0 - i\epsilon)} \\
&= -\frac{e^{-i(\kappa - i\epsilon)y_0}}{2\kappa} .
\end{aligned} \tag{10.19}
$$

Similarly, one can find out

$$a(-\kappa - i\epsilon) = \frac{e^{i(\kappa + i\epsilon)y_0}}{2\kappa} . \tag{10.20}$$

The above results show

$$I = -\frac{1}{2\pi} \int_{-\infty}^{+\infty} dk_0 \, \frac{e^{-ik_0 y_0}}{\kappa^2 - (k_0 + i\epsilon)^2} = -\frac{1}{\kappa} \left[\frac{e^{i(\kappa + i\epsilon)y_0} - e^{-i(\kappa - i\epsilon)y_0}}{2i} \right] .$$

Now we can take the limit $\epsilon \to 0$. It is seen that we do not encounter any difficulty in this limiting process and the result is

$$I_0 = \frac{1}{2\pi} \int_{-\infty}^{+\infty} dk_0 \, \frac{e^{-ik_0 y_0}}{\kappa^2 - (k_0 + i\epsilon)^2} = \frac{\sin(\kappa y_0)}{\kappa} , \quad \text{when } t > t'. \tag{10.21}$$

We have succeeded in integrating k_0 from the integral in Eq. (10.11), the other three integrations remain. From the expression of $G(y)$ in Eq. (10.11) and the forms of I_0 for $y > 0$ and $y < 0$ from Eqs. (10.15), (10.21) we can write

Fig. 10.3 The fixed vector **y** is along the z-axis of the coordinate system while the vector **k** can vary around **y** if θ and ϕ changes continuously. Here, the azimuthal angle ϕ for **k** is not shown

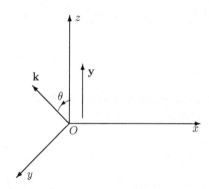

$$G(y) = \frac{\Theta(y_0)}{(2\pi)^3} \int_{-\infty}^{\infty} d^3k \, e^{i\mathbf{k}\cdot\mathbf{y}} \, \frac{\sin(\kappa y_0)}{\kappa}, \qquad (10.22)$$

where the step function $\Theta(y_0)$ is as defined in Eq. (6.61) in Chap. 6. To do the next part of the integration we define $R \equiv |\mathbf{y}| = |\mathbf{x} - \mathbf{x}'|$ and choose the coordinate system in such a way that the vector **y** is along the z-axis. The geometry of the system is shown in Fig. 10.3. In the present integral **y** is a fixed vector, whereas **k** can vary with θ. Using isotropy of space we could choose the z-axis of the coordinate system in any direction. In the present case, we have chosen the z-axis to be along the direction of **y**. In our case $d^3k = \kappa^2 d\kappa \, d\phi \, \sin\theta d\theta$ where ϕ is the azimuthal angle subtended by **k**. With all these considerations we can now write

$$G(y) = \frac{\Theta(y_0)}{(2\pi)^3} \int_0^{\infty} \kappa^2 d\kappa \int_0^{2\pi} d\phi \int_0^{\pi} d\theta \, \sin\theta \, e^{i\kappa R\cos\theta} \, \frac{\sin(\kappa y_0)}{\kappa}. \quad (10.23)$$

The ϕ integral is trivial and the θ integral is a standard one yielding,

$$
\begin{aligned}
G(y) &= \frac{\Theta(y_0)}{i(2\pi)^2 R} \int_0^{\infty} d\kappa \, \sin(\kappa y_0)\,(e^{i\kappa R} - e^{-i\kappa R}) \\
&= \frac{\Theta(y_0)}{i(2\pi)^2 R} \int_0^{\infty} d\kappa \, e^{i\kappa R} \sin(\kappa y_0) - \int_0^{-\infty} d\kappa \, e^{i\kappa R} \sin(\kappa y_0),
\end{aligned}
$$

to get the second integral on the right-hand side we have replaced the integration variable κ with $-\kappa$. We can now write

$$
\begin{aligned}
G(y) &= \frac{\Theta(y_0)}{i(2\pi)^2 R} \int_0^{\infty} d\kappa \, e^{i\kappa R} \sin(\kappa y_0) + \int_{-\infty}^{0} d\kappa \, e^{i\kappa R} \sin(\kappa y_0) \\
&= \frac{\Theta(y_0)}{i(2\pi)^2 R} \int_{-\infty}^{\infty} d\kappa \, e^{i\kappa R} \sin(\kappa y_0).
\end{aligned}
$$

In the above integral κ is just an integration variable which can have negative values. We can also write the above integral as

$$G(y) = \frac{\Theta(y_0)}{i(2\pi)^2 R} \int_{-\infty}^{\infty} d\kappa \, e^{i\kappa R} \left[\frac{e^{i\kappa y_0} - e^{-i\kappa y_0}}{2i} \right]$$

$$= -\frac{\Theta(y_0)}{2(2\pi)^2 R} \int_{-\infty}^{\infty} d\kappa \left[e^{i\kappa(y_0+R)} - e^{-i\kappa(y_0-R)} \right].$$

From the definition of the Dirac delta function in Eq. (6.56) in Chap. 6 and its evenness we can write

$$\int_{-\infty}^{\infty} d\kappa \, e^{i\kappa(y_0+R)} = 2\pi\delta(y_0 + R), \quad \int_{-\infty}^{\infty} d\kappa \, e^{-i\kappa(y_0-R)} = 2\pi\delta(y_0 - R).$$

Due to the presence of the step function and the fact that R cannot be negative, we finally have

$$G_R(x - x') = \frac{\Theta(x_0 - x_0')}{4\pi R} \delta(x_0 - x_0' - R). \tag{10.24}$$

The form of the Green function above shows that it only contributes when $x_0 = x_0' + R$ or when $t = t' + (R/c)$. As $t > t'$ the above form of the Green function is called the retarded Green function.

10.1.1   A Different Pole Prescription and the Advanced Green Function

In the previous discussion, we pushed the singularities of the integrand in the $G(y)$ expression in the lower k_0 complex half plane. The resulting Green function is the retarded one. The poles can be moved from the real axis of the complex k_0 plane in multiple ways. Instead of pushing them down one can push them slightly up, by ϵ amount, in the imaginary direction. In this case, the relevant k_0 integration becomes

$$I = \frac{1}{2\pi} \oint_\Gamma dk_0 \, \frac{e^{-ik_0 y_0}}{\kappa^2 - (k_0 - i\epsilon)^2}, \tag{10.25}$$

where $\epsilon > 0$. The calculations are similar to the ones presented in the previous case. The main difference with the last calculation is that now $G(y) \neq 0$ when $y_0 < 0$. The contour in the upper half of the complex k_0 plane contains the poles, as shown in Fig. 10.4, and consequently the non-zero contribution to the Green function comes from the contour in the upper half plane. The contour on the lower half plane, for $y_0 > 0$, does not contribute to the Green function and as a result we can write the final result as

$$G_A(x - x') = \frac{\Theta(x_0' - x_0)}{4\pi R} \delta(x_0 - x_0' + R). \tag{10.26}$$

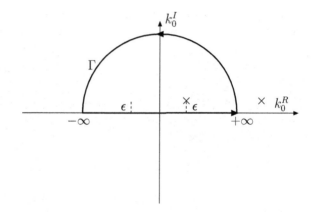

Fig. 10.4 The contour Γ in the complex k_0 plane used for evaluating I when $y_0 < 0$. The two poles on the real axis are shifted above by ϵ amount. In reality, the diameter extends from $-\infty$ to ∞. The real contour on the upper half plane has infinite radius

This form of the Green function shows that it only contributes when $x_0 = x_0' - R$ or when $t = t' - (R/c)$. As $t < t'$ the above form of the Green function is called the advanced Green function. In the present case, the response of the signal from the source point emanated at t' has reached the field point at t where $t < t'$. The signal has reached the field point before it was emitted by the source and hence the name advanced Green function.

These two choices of pole placements give rise to the retarded and the advanced response functions. One still has the freedom to displace the poles in different manners, as shifting one up and shifting the other down. This pole shift will produce a different Green function whose causal property will not be useful in classical electrodynamics. The advanced Green function itself is a bit odd, but it is a valid solution of the Green function equation as given in Eq. (10.2). The reader must now have noted that the placement of the poles on the complex k_0 plane dictates the causal structure of the theory.

10.1.2 Green Functions Under Lorentz Transformations

We have the Green functions required to solve the inhomogeneous Maxwell equation, but we do not know how these functions transform under a Lorentz transformation. In this subsection, we will see how the functions transform under Lorentz transformations. The Green functions contain the Dirac delta function. To proceed further we state an important property of the Dirac delta functions. If there exists a real function $f(x)$, of a single real variable x, which has n simple zeros at distinctly different points x_i where $i = 1, 2, \cdot, \cdot, \cdot, n$ then

$$\delta(f(x)) = \sum_{i=1}^{n} \frac{\delta(x - x_i)}{\left|\frac{df}{dx}\right|_{x_i}}, \tag{10.27}$$

where $|df/dx|_{x_i}$ is the absolute value of df/dx at $x = x_i$. Here, x is a real variable and does not represent 4-vectors. From the above result, we can easily see that

$$\delta(x^2 - a^2) = \delta[(x + a)(x - a)] = \frac{1}{2a}[\delta(x + a) + \delta(x - a)], \quad (10.28)$$

where a is a real, positive constant. Now we discuss the case where x stands for 4-vectors and apply the above result to rewrite the Green functions. As $x^\alpha - x'^\alpha = (y^0, \mathbf{R})$, we can write

$$\delta\left((x - x')^2\right) = \delta\left((y_0^2 - R^2)\right)$$
$$= \frac{1}{2R}\left[\delta(x_0 - x_0' + R) + \delta(x_0 - x_0' - R)\right].$$

From the above expressions, it is clear that the retarded and advanced Green functions can be written as

$$G_R(x - x') = \frac{1}{2\pi}\Theta(x_0 - x_0')\delta((x - x')^2), \quad (10.29)$$

$$G_A(x - x') = \frac{1}{2\pi}\Theta(x_0' - x_0)\delta((x - x')^2). \quad (10.30)$$

The Dirac delta function, in the Green functions, is Lorentz invariant as it is a function of the Lorentz invariant quantity $(x - x')^2$. The function $\delta(x - x')^2$ peaks at the same value in all the coordinate systems related via Lorentz transformation. Under a Lorentz transformation, the 4-vector $y^\mu = x^\mu - x'^\mu$ which satisfies $y^2 = 0$ transforms as $y'^\mu = L^\mu{}_\nu y^\nu$. The reader must be careful about the primes above the 4-vectors, x' specifies the source point whereas y'^α specifies the Lorentz transformed 4-vector corresponding to y^α. In the present case $y'^2 = 0$ and consequently we have

$$y^0 = \pm[(y^1)^2 + (y^2)^2 + (y^3)^2]^{1/2}, \quad y'^0 = \pm[(y'^1)^2 + (y'^2)^2 + (y'^3)^2]^{1/2},$$
$$(10.31)$$

where the plus signs represent light signals advancing in time and the minus sign stands for light signals going back in time. The plus sign corresponds to retarded signals, whereas the minus sign stands for advanced signals. From the above equation and the the transformation law $y'^0 = L^0{}_\nu y^\nu = \gamma(y^0 \pm \beta y^1)$, where $\beta < 1$, it can be seen that if $y_0 > 0$ for retarded signals then $y_0' > 0$ also. On the other hand if $y_0 < 0$ for the advanced signals then we must also have $y_0' < 0$. The signs remain same as $|y^0| \geq |y^1|$. As a consequence the sign functions $\Theta(x_0 - x_0')$ or $\Theta(x_0' - x_0)$ must be Lorentz invariant. These points show clearly that the retarded and the advanced Green's functions are actually Lorentz invariant.

In Fig. 10.5, a lightcone diagram has been presented. The present spacetime point at the origin has coordinates x^μ. Spacetime points x'^μ which can affect the field at x^μ via the retarded Green function lies on the past lightcone surface. Similarly, source

Fig. 10.5 Lightcone where
the origin specified by x^μ is
the present field point. x'^μ on
the past lightcone surface can
only affect the field point via
the retarded Green function

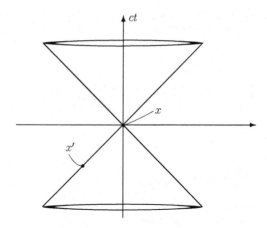

points which affect x^μ via the advanced Green's function can only lie on the future
lightcone surface.

The gauge field solutions can now be written as

$$A^\alpha(x) = A^\alpha_{\text{in}}(x) + \frac{4\pi}{c} \int_\Omega d^4x' G_R(x, x') J^\alpha(x'), \qquad (10.32)$$

where $A^\alpha_{\text{in}}(x)$ is the solution of the electromagnetic wave equation when the source
vanishes. In a similar way, we can also write

$$A^\alpha(x) = A^\alpha_{\text{out}}(x) + \frac{4\pi}{c} \int_\Omega d^4x' G_A(x, x') J^\alpha(x'), \qquad (10.33)$$

where $A^\alpha_{\text{out}}(x)$ is the solution of the electromagnetic wave equation without the source
term. In general, people work with the retarded solution as given in Eq. (10.32) for
obvious reasons.

10.2 The 4-Current for a Point Charge Moving Arbitrarily

Till now we have not discussed the source of the field $A^\alpha(x)$. Out of the various
possible sources we will consider a simple but important case, the case of an arbitrarily moving point charge. Suppose a point charge q is moving arbitrarily and its
position at coordinate time t is $\mathbf{r}(t)$ and its velocity at that instant is $\mathbf{v}(t) = d\mathbf{r}/dt$.
The trajectory of the charged particle is shown in Fig. 10.6. The charge density and
the 3-current density related to the charged particle are given as

$$\rho(x) = q\delta^3(\mathbf{x} - \mathbf{r}(t)), \quad \mathbf{J}(x) = q\mathbf{v}(t)\delta^3(\mathbf{x} - \mathbf{r}(t)). \qquad (10.34)$$

Fig. 10.6 The trajectory of an arbitrarily moving point charge q whose position at coordinate time t is $\mathbf{r}(t)$ and its velocity at that instant is $\mathbf{v}(t)$

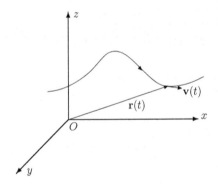

The above components can be combined to write a 4-current density $J^\alpha(x)$. First of all we introduce the position 4-vector of the charged particle as $r^\alpha(\tau) = (r^0(\tau), \mathbf{r}(\tau))$ where τ is the proper time of the charged particle. The charged particle is undergoing an arbitrary motion and so the proper time is defined in an instantaneous coordinate system which is moving with a uniform velocity $\mathbf{v}$ which coincides with the particle's instantaneous velocity at time t. There are a plethora of such kind of coordinate systems where the particle is at rest momentarily. These multitudes of instantaneous rest frames of the particles which are moving with uniform velocity are connected with the laboratory frame (the x, y, z, t coordinates system) via Lorentz transformations. Now we can write the 4-current corresponding to the charged particle's motion as

$$J^\alpha(x) = qc \int_{-\infty}^{\infty} d\tau \, u^\alpha(\tau) \delta^4(x - r(\tau)) , \qquad (10.35)$$

where $u^\alpha = dx^\alpha/d\tau$. Here $\delta^4(x - r(\tau)) = \delta(x_0 - r_0)\delta^3(\mathbf{x} - \mathbf{r}(t))$. Using the facts $x_0 = ct$, $u^0(\tau) = c\gamma$, $dt = \gamma d\tau$ where $\gamma = (1 - \beta^2)^{-1/2}$ and $\beta = v/c$, we can see from the above expression[1] that

$$J^0(x) = qc \int_{-\infty}^{\infty} \left(\frac{dt}{\gamma}\right) (c\gamma)\frac{1}{c}\delta\left(t - \frac{r_0}{c}\right) \delta^3(\mathbf{x} - \mathbf{r}(t)) = qc\delta^3(\mathbf{x} - \mathbf{r}(t)) ,$$

where $t = r_0/c$. This is the result we expected. In a similar fashion

$$\mathbf{J}(x) = qc \int_{-\infty}^{\infty} \left(\frac{dt}{\gamma}\right) (\gamma\mathbf{v})\frac{1}{c}\delta\left(t - \frac{r_0}{c}\right) \delta^3(\mathbf{x} - \mathbf{r}(t)) = q\mathbf{v}\delta^3(\mathbf{x} - \mathbf{r}(t)) ,$$

where $t = r_0/c$. This is the result we have in Eq. (10.34). We see that the expression of the 4-current $J^\alpha(x)$ in Eq. (10.35) correctly produces the source current for an arbitrarily moving point charge. From Eq. (10.3), we see that as we now know the

---

[1] In the present case r simply specifies the four components $(r^0, \mathbf{r})$ and not the magnitude of $\mathbf{r}$. The reader must be careful about this point.

form of the Green function and the form of the 4-current we can in principle write down the gauge fields produced by an arbitrarily moving point charge. In the next section, we will attempt to find out the electromagnetic fields produced by such an arbitrarily moving point charge.

10.3   Liénard-Wiechert Potentials

In this section, we want to find out the gauge field $A^\alpha(x)$ produced by an arbitrarily moving massive point charge. To do so we will be using the Lorenz gauge condition $\partial_\alpha A^\alpha(x) = 0$. To find out the form of $A^\alpha(x)$ the retarded Green function will be used and we will assume that the radiation field $A^\alpha_{\text{in}}(x)$ is absent in the present case. All the field contributions at x come from the source current $J^\alpha(x)$ and not from any previous radiation field. The resulting gauge field $A^\alpha(x)$ is known as the Liénard-Wiechert potential. We have to find out

$$A^\alpha(x) = \frac{4\pi}{c} \int_\Omega d^4x' G_R(x, x') J^\alpha(x'), \qquad (10.36)$$

where the expression of the retarded Green function $G_R(x, x')$ is given in Eq. (10.29) and the expression for the 4-current corresponding to an arbitrarily moving point charge is given in Eq. (10.35). The trajectory of the point charge is given by $r^\alpha(\tau) = (r^0(\tau), \mathbf{r}(\tau))$ where τ is the proper time of the charged particle. In the present case

$$u^\alpha(\tau) = \frac{dr^\alpha}{d\tau} = (\gamma c, \gamma v), \qquad (10.37)$$

gives the 4-velocity of the point charge. Now we can write the gauge field as

$$A^\alpha(x) = \frac{4\pi}{c} \int_\Omega d^4x' \left[\frac{1}{2\pi} \Theta(x_0 - x'_0) \delta((x - x')^2) \right] \left[qc \int_{-\infty}^{\infty} d\tau\, u^\alpha(\tau) \delta^4(x' - r(\tau)) \right]$$

$$= 2q \int_{-\infty}^{\infty} d\tau\, u^\alpha(\tau) \left[\int_\Omega d^4x'\, \Theta(x_0 - x'_0) \delta^4(x' - r(\tau)) \delta((x - x')^2) \right]. \qquad (10.38)$$

Here, q is the magnitude of charge carried by the moving point particle. Before we proceed we want to make one point clear. The proper time τ corresponds to the source coordinate time t' and not t. In our whole discussion, on the fields produced by arbitrarily moving point charge, coordinate time t corresponds to the field point and not the source point. We have used this convention because we will never use the proper time of the field point and consequently the only useful proper time, without a prime, is used for the source point. Integrating the four-dimensional integral using the four-dimensional Dirac delta function, we have

Fig. 10.7 Here, the origin with coordinates x^{μ} is the field point and the lightcone is drawn with respect to an observer at the origin. The trajectory of the point charge is shown with respect to the lightcone. The trajectory intersects the past lightcone at one unique point called $r^{\alpha}(\tau_0)$ at some proper time τ_0 corresponding to some coordinate time t'

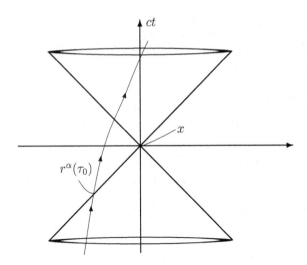

$$A^{\alpha}(x) = 2q \int_{-\infty}^{\infty} d\tau \, u^{\alpha}(\tau) \, \Theta(x_0 - r_0(\tau)) \, \delta\left((x - r(\tau))^2\right), \qquad (10.39)$$

which contributes when $x_0 > r_0(\tau)$ and $[x - r(\tau_0)]^2 = 0$ for a unique value of $\tau = \tau_0$. In Fig. 10.7, the trajectory of the point charged particle with respect to the observer's lightcone is presented. The observation or field point is at the origin of the lightcone and has coordinates x^{μ}. As trajectories of real particles cannot bend downwards in a spacetime diagram, the trajectory must cut the past lightcone of the observer only once and pierce the future lightcone only once. Because electromagnetic signals propagate at the speed of light, the moving charged particle can only affect the observation point at the origin when its trajectory touches the past lightcone of the field point observer. To take care of the Dirac delta function in Eq. (10.39) we must have $[x - r(\tau)]^2 = 0$, and henceforth we will interpret it as an equation for τ. Defining

$$f(\tau) \equiv [x - r(\tau)]^2, \qquad (10.40)$$

we can write the relevant equation as $f(\tau) = 0$ where τ_0 is the root of this equation. From Eq. (10.27), we can write

$$\delta\left(f(\tau)\right) = \frac{\delta(\tau - \tau_0)}{\left|\frac{df}{d\tau}\right|_{\tau_0}}. \qquad (10.41)$$

As $f(\tau) = (x_{\alpha} - r_{\alpha})(x^{\alpha} - r^{\alpha})$ we have

$$\frac{df}{d\tau} = 2(x_{\alpha} - r_{\alpha}) \left(\frac{dx^{\alpha}}{d\tau} - \frac{dr^{\alpha}}{d\tau}\right).$$

As τ measures the proper time of the source and x^α is the field point which is independent of τ we have $dx^\alpha/d\tau = 0$. As a result of this we can write

$$\frac{df}{d\tau} = -2(x_\alpha - r_\alpha)u^\alpha = -2u(\tau) \cdot [x - r(\tau)], \qquad (10.42)$$

and using this relation we can write Eq. (10.39) as

$$A^\alpha(x) = 2q \int_{-\infty}^{\infty} d\tau \, u^\alpha(\tau) \, \Theta(x_0 - r_0(\tau)) \frac{\delta(\tau - \tau_0)}{2u(\tau_0) \cdot [x - r(\tau_0)]}. \qquad (10.43)$$

This equation can easily be integrated using the Dirac delta function, the result is

$$A^\alpha(x) = \left. \frac{qu^\alpha(\tau)}{u \cdot [x - r(\tau)]} \right|_{\tau=\tau_0}, \qquad \text{when} \ \ x_0 > r_0(\tau_0), \qquad (10.44)$$

and τ_0 is obtained by solving $[x - r(\tau)]^2 = 0$. We can write the condition $[x - r(\tau_0)]^2 = [x_0 - r_0(\tau_0)]^2 - [\mathbf{x} - \mathbf{r}(\tau_0)]^2 = 0$ as

$$[x_0 - r_0(\tau_0)]^2 = R^2, \qquad (10.45)$$

where

$$\mathbf{R}(\tau_0) \equiv \mathbf{x} - \mathbf{r}(\tau_0), \qquad (10.46)$$

and $R = |\mathbf{R}|$. As $x_0 > r_0(\tau_0)$ we have

$$x_0 - r_0(\tau_0) = R(\tau_0). \qquad (10.47)$$

Using these relations, we have

$$\begin{aligned} u \cdot [x - r(\tau_0)] &= u_0[x_0 - r_0(\tau_0)] - \mathbf{u} \cdot [\mathbf{x} - \mathbf{r}(\tau_0)] \\ &= \gamma c R - \gamma \mathbf{v} \cdot \mathbf{R} \\ &= \gamma c R (1 - \boldsymbol{\beta} \cdot \hat{\mathbf{n}}), \end{aligned} \qquad (10.48)$$

where the unit vector along $\mathbf{R}$ is called $\hat{\mathbf{n}}$ defined as

$$\hat{\mathbf{n}} \equiv \frac{\mathbf{R}}{R}, \qquad (10.49)$$

and $\boldsymbol{\beta} = \mathbf{v}/c$. We can now write the resulting gauge fields as

$$\Phi(x) = \left. \frac{q}{(1 - \boldsymbol{\beta} \cdot \hat{\mathbf{n}})R} \right|_{\tau=\tau_0}, \qquad \mathbf{A}(x) = \left. \frac{q\boldsymbol{\beta}}{(1 - \boldsymbol{\beta} \cdot \hat{\mathbf{n}})R} \right|_{\tau=\tau_0}, \qquad (10.50)$$

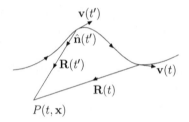

Fig. 10.8 Here, P is the field point at time t and position **x**. The diagram shows the retarded position of the moving charge which affects P at time t. The velocity, the distance of the charge from the field point and the unit vector $\hat{\mathbf{n}}$ at the retarded time t' are clearly pointed out. The position of the charge at time t is also shown

where the right-hand side of the above fields have to be calculated at $\tau = \tau_0$. As $x_0 = ct$ and $r_0(\tau_0) = ct'$, Eq. (10.47) predicts

$$t = t' + \frac{R}{c}, \qquad (10.51)$$

showing the retarded nature of the potentials. The potentials at time t are produced by the motion of the point charge R/c seconds back in time. Electromagnetic signal emitted from the point charge at coordinate time t' reaches the field point after a time period of R/c where $R(\tau_0)$ is the separation of the field point and source point at t'. Here, $\boldsymbol{\beta} = \mathbf{v}/c$ contains the velocity of the charge at coordinate time t' and $\hat{\mathbf{n}}$ points towards the field point from the source point at t'. The retarded nature of the problem is shown in Fig. 10.8. The expression of the electromagnetic potentials in Eq. (10.50) or the gauge field expression in Eq. (10.44) are traditionally called the Liénard-Wiechert potentials. In evaluating the potentials one generally do not explicitly use the proper time of the arbitrarily moving point source τ_0 but instead work with the coordinate time t' which corresponds to τ_0 and as a result the potentials are written as

$$\Phi(x) = \left[\frac{q}{(1 - \boldsymbol{\beta} \cdot \hat{\mathbf{n}})R} \right]_{\text{ret}}, \quad \mathbf{A}(x) = \left[\frac{q\boldsymbol{\beta}}{(1 - \boldsymbol{\beta} \cdot \hat{\mathbf{n}})R} \right]_{\text{ret}}, \qquad (10.52)$$

where the square brackets with the retarded subscript specifies that the quantities inside the brackets must be computed at the retarded time t' corresponding to the time at the field point t.

10.4   The Field Strength Tensor $F^{\mu\nu}$ Obtained from the Liénard-Wiechert Potentials

We know the gauge field or potential produced by an arbitrarily moving point charge, it's formal expression is given in Eq. (10.44). From this expression, we can proceed to find the electric and magnetic fields via the field strength tensor $F^{\alpha\beta}$. In this section we will not proceed in this path, rather we will calculate $F^{\alpha\beta}$ from the expression of $A^{\alpha}(x)$ as given in Eq. (10.39). The difficulty one faces if one tries to find out $F^{\alpha\beta}$ directly from the expression of the gauge field in Eq. (10.44) is related to the action of partial derivatives $\partial^{\alpha} = \partial/\partial x_{\alpha}$ on the objects on the right hand side of Eq. (10.44). The objects on the right-hand side of Eq. (10.44) must be calculated in the retarded time, whereas the partial derivative ∂^{α} involves differentiation at t, which is the coordinate time at the field point. To tackle this difficulty we use a method where the time derivative of the gauge potentials can be taken without bothering about retarded time. To figure out the form of $F^{\alpha\beta}$ we first try to calculate $\partial^{\alpha} A^{\beta}(x)$ using the expression of the gauge field as given in Eq. (10.39). We can write

$$\partial^{\alpha} A^{\beta}(x) = 2q \int_{-\infty}^{\infty} d\tau\, u^{\beta}(\tau) \left[\partial^{\alpha}\Theta(x_0 - r_0(\tau))\right] \delta\left((x - r(\tau))^2\right)$$
$$+ 2q \int_{-\infty}^{\infty} d\tau\, u^{\beta}(\tau)\, \Theta(x_0 - r_0(\tau)) \left[\partial^{\alpha}\delta\left((x - r(\tau))^2\right)\right],\quad (10.53)$$

where ∂^{α} only acts on functions dependent on the field point x^{α}. As $u^{\beta}(\tau)$ and $r(\tau)$ defines the movement of the source ∂^{α} has no effect on them. The term $\partial^{\alpha}\Theta(x_0 - r_0(\tau))$ only contributes when $\alpha = 0$ and using the result in Eq. (6.62) in Chap. 6 we can write $\partial^0\Theta(x_0 - r_0(\tau)) = \delta(x_0 - r_0(\tau))$. Using this result, we have

$$\partial^{\alpha} A^{\beta}(x) = 2q \int_{-\infty}^{\infty} d\tau\, u^{\beta}(\tau)\, \delta(x_0 - r_0(\tau))\,\delta\left(-(\mathbf{x} - \mathbf{r}(\tau))^2\right)$$
$$+ 2q \int_{-\infty}^{\infty} d\tau\, u^{\beta}(\tau)\, \Theta(x_0 - r_0(\tau)) \left[\partial^{\alpha}\delta\left((x - r(\tau))^2\right)\right],\quad (10.54)$$

which can be also be written as

$$\partial^{\alpha} A^{\beta}(x) = 2q \int_{-\infty}^{\infty} d\tau\, u^{\beta}(\tau)\, \delta(R)\,\delta(-R^2)$$
$$+ 2q \int_{-\infty}^{\infty} d\tau\, u^{\beta}(\tau)\, \Theta(x_0 - r_0(\tau)) \left[\partial^{\alpha}\delta\left((x - r(\tau))^2\right)\right],\quad (10.55)$$

using Eq. (10.46) and Eq. (10.47). As we are generally interested in cases where the source point and the field point are separate spacetime points, cases where $R \neq 0$, we have

$$\partial^\alpha A^\beta(x) = 2q \int_{-\infty}^{\infty} d\tau \, u^\beta(\tau) \, \Theta(x_0 - r_0(\tau)) \, \partial^\alpha \delta\left(f(\tau)\right) , \qquad (10.56)$$

where we have used the expression of $f(\tau)$ as defined in Eq. (10.40). The partial derivative of the Dirac delta function can be written as

$$\partial^\alpha \delta\left(f(\tau)\right) = \left[\partial^\alpha f(\tau)\right] \frac{d}{df} \delta\left(f(\tau)\right) = \left[\partial^\alpha f(\tau)\right] \left[\frac{d}{d\tau} \delta\left(f(\tau)\right)\right] \frac{d\tau}{df}$$

$$= \left[\partial^\alpha f(\tau)\right] \frac{\frac{d}{d\tau} \delta\left(f(\tau)\right)}{\frac{df}{d\tau}} , \qquad (10.57)$$

where $df/d\tau$ is given in Eq. (10.42). We have

$$\partial^\alpha f = 2(x - r)^\alpha ,$$

using which we can now write

$$\partial^\alpha \delta\left(f(\tau)\right) = -\frac{(x - r(\tau))^\alpha}{u(\tau) \cdot [x - r(\tau)]} \frac{d}{d\tau} \delta\left(f(\tau)\right) . \qquad (10.58)$$

Using the above result, we have

$$\partial^\alpha A^\beta(x) = -2q \int_{-\infty}^{\infty} d\tau \, u^\beta(\tau) \, \Theta(x_0 - r_0(\tau)) \frac{(x - r(\tau))^\alpha}{u(\tau) \cdot [x - r(\tau)]} \frac{d}{d\tau} \delta\left(f(\tau)\right) .$$

$$(10.59)$$

To do the above integral we require to know what is $d\delta(f)/d\tau$. We will actually not calculate $d\delta(f)/d\tau$ but devise a method where we do not require to calculate the derivative of a Dirac delta function. To do so we notice that the derivative of the Dirac delta is appearing inside an integral. This fact will help us to devise a method where the integral can be done without knowing the derivative of the Dirac delta function. Let us think of a function $g(x)$ which depends on a real variable x. The problem is how do we calculate the integral

$$\int_{-\infty}^{\infty} dx \, g(x) \left[\frac{d}{dx} \delta(x)\right] .$$

We can apply the method of integration by parts and obtain

$$\int_{-\infty}^{\infty} dx \, g(x) \left[\frac{d}{dx} \delta(x)\right] = [g(x)\delta(x)]_{-\infty}^{\infty} - \int_{-\infty}^{\infty} \left(\frac{dg}{dx}\right) \delta(x) \, dx .$$

The first term on the right-hand side is obviously zero, as a consequence

$$\int_{-\infty}^{\infty} dx\, g(x) \left[\frac{d}{dx} \delta(x) \right] = - \int_{-\infty}^{\infty} \left(\frac{dg}{dx} \right) \delta(x)\, dx,$$

(10.60)

which shows how the derivative of a Dirac delta function can be handled inside an integral. Using the above result, we can now write

$$\partial^{\alpha} A^{\beta}(x) = 2q \int_{-\infty}^{\infty} d\tau\, \delta\left(f(\tau) \right) \frac{d}{d\tau} \left\{ \frac{u^{\beta}(\tau)(x - r(\tau))^{\alpha}}{u(\tau) \cdot [x - r(\tau)]} \Theta(x_0 - r_0(\tau)) \right\}.$$

(10.61)

The derivative term in the above integrand can be calculated as

$$\frac{d}{d\tau} \left\{ \frac{u^{\beta}(x - r)^{\alpha}}{u \cdot (x - r)} \Theta(x_0 - r_0) \right\} = \frac{u^{\beta}(x - r)^{\alpha}}{u \cdot (x - r)} \frac{d}{d\tau} \Theta(x_0 - r_0)$$
$$+ \Theta(x_0 - r_0) \frac{d}{d\tau} \left[\frac{u^{\beta}(x - r)^{\alpha}}{u \cdot (x - r)} \right], \quad (10.62)$$

which can be simplified by noticing

$$\frac{d}{d\tau} \Theta(x_0 - r_0) = -\delta(x_0 - r_0) \frac{dr_0}{d\tau} = -\delta(R) \frac{dr_0}{d\tau}.$$

Thus when the derivative term in Eq. (10.62) with the above simplification is inserted inside the integrand in Eq. (10.61) the term containing $\delta(R)$ can only exist when $R = 0$. As we are only interested in situations where $R \neq 0$ we can omit that integral altogether and write

$$\partial^{\alpha} A^{\beta}(x) = 2q \int_{-\infty}^{\infty} d\tau\, \delta(f(\tau))\, \Theta(x_0 - r_0(\tau)) \frac{d}{d\tau} \left[\frac{u^{\beta}(\tau)(x - r(\tau))^{\alpha}}{u(\tau) \cdot (x - r(\tau))} \right].$$

(10.63)

The integral on the right-hand side of the above equation is structurally equivalent to the integral on the right-hand side of Eq. (10.39). The equivalence holds if we interpret $u^{\alpha} = (d/d\tau)x^{\alpha}$ in the same footing with the proper time derivative of

$$\frac{u^{\beta}(x - r)^{\alpha}}{u \cdot (x - r)}.$$

As the integral in Eq. (10.39) ultimately produced the result in Eq. (10.44), using the equivalence of the integrals we can now write the above integration result as

$$\partial^{\alpha} A^{\beta}(x) = \frac{q}{u(\tau_0) \cdot [x - r(\tau_0)]} \frac{d}{d\tau} \left[\frac{u^{\beta}(\tau)(x - r(\tau))^{\alpha}}{u(\tau) \cdot (x - r(\tau))} \right]\Bigg|_{\tau = \tau_0}.$$

(10.64)

From the above result, one can easily write down

$$\partial^\beta A^\alpha(x) = \frac{q}{u(\tau_0) \cdot [x - r(\tau_0)]} \frac{d}{d\tau} \left[\frac{u^\alpha(\tau)(x - r(\tau))^\beta}{u(\tau) \cdot (x - r(\tau))} \right]\Bigg|_{\tau=\tau_0} . \quad (10.65)$$

We see that at the end the importance of retarded time becomes explicit. From the above results, we can now write

$$F^{\alpha\beta}(x) = \frac{q}{u(\tau_0) \cdot [x - r(\tau_0)]} \frac{d}{d\tau} \left[\frac{u^\beta(\tau)(x - r(\tau))^\alpha - u^\alpha(\tau)(x - r(\tau))^\beta}{u(\tau) \cdot (x - r(\tau))} \right]\Bigg|_{\tau=\tau_0} , \quad (10.66)$$

giving us the form of the field strength tensor for fields produced by an arbitrarily moving point charge. The right-hand side has to be calculated in the retarded time.

10.4.1   Setting up the Stage

Before we embark on a full fledged calculation of the electric and magnetic field components from $F^{\alpha\beta}(x)$ we will like to cast the expression of $F^{\alpha\beta}(x)$ into a particular form which will help us to calculate the electromagnetic fields. We know $x_0 - r_0(\tau) = R(\tau)$, $\mathbf{x} - \mathbf{r}(\tau) = \mathbf{R}(\tau)$ and $\hat{\mathbf{n}} = \mathbf{R}/R$. Using these quantities, we can write $(x - r)^\alpha = (R, R\hat{\mathbf{n}})$. From Eq. (10.37), we can write

$$\frac{du^\alpha}{d\tau} = \left(c\gamma^4(\boldsymbol{\beta} \cdot \dot{\boldsymbol{\beta}}), \; c\gamma^4\boldsymbol{\beta}(\boldsymbol{\beta} \cdot \dot{\boldsymbol{\beta}}) + c\gamma^2\dot{\boldsymbol{\beta}} \right) , \quad (10.67)$$

where

$$\dot{\boldsymbol{\beta}} = \frac{d\boldsymbol{\beta}}{dt} , \quad \gamma = \frac{1}{\sqrt{1 - \beta^2}} , \quad \boldsymbol{\beta} = \frac{\mathbf{v}}{c} , \quad \mathbf{v} = \frac{d\mathbf{r}}{dt} .$$

The relation in Eq. (10.67) can be derived by calculating $d\gamma/dt$ and then suitably writing $du^0/d\tau$ and $du^i/d\tau$ in terms of the proper time τ. In our case, $\dot{\boldsymbol{\beta}}$ is in general not zero and the charged particle can accelerate. To tackle the acceleration of the charged particle in the framework of special relativity, we employ two kinds of inertial frames. One frame is called traditionally the laboratory frame where the coordinates are t, x, y, z as we are using. The other frame (or frames) is (are) related to the rest frame of the arbitrarily moving particle. If we fix one coordinate system, for all times, to the moving particle then that frame in general is not an inertial frame as the coordinate system may accelerate with the particle. On the other hand we can have innumerable number of inertial frames, moving with uniform velocities, whose velocity at any instant t coincides with the velocity of the arbitrarily moving particle. In this case, instantaneously we always have an inertial frame which acts as the rest frame of the arbitrarily moving particle. The laboratory frame and the instantaneous rest frames are related by Lorentz transformations and the proper time τ for the charged particle is defined in such instantaneous rest frames. We see special relativity can in general handle accelerated motion using only inertial frames. A brief discussion regarding this issue was presented in Sect. 10.2.

Next we start to calculate the various terms which appear in the expression of the field strength tensor. First we see that

$$\frac{d}{d\tau}[u \cdot (x - r)] = -u_\alpha \frac{dr^\alpha}{d\tau} + (x - r)_\alpha \frac{du^\alpha}{d\tau} = -c^2 + (x - r)_\lambda \frac{du^\lambda}{d\tau},$$

$$(10.68)$$

where we know $dr^\alpha/d\tau = u^\alpha$ and $u^\alpha u_\alpha = c^2$. In these calculations, we will always use $dx^\alpha/d\tau = 0$ as the field point does not depend upon the proper time of the source. Next we have

$$\frac{d}{d\tau}\left[\frac{u^\beta (x - r)^\alpha}{u \cdot (x - r)} \right] = -\frac{u^\beta (x - r)^\alpha}{[u \cdot (x - r)]^2} \frac{d}{d\tau}[u \cdot (x - r)] + \frac{1}{u \cdot (x - r)} \frac{d}{d\tau}[u^\beta (x - r)^\alpha].$$

To evaluate the above expression, we have to calculate

$$\frac{d}{d\tau}[u^\beta (x - r)^\alpha] = -u^\alpha u^\beta + (x - r)^\alpha \frac{du^\beta}{d\tau}.$$

$$(10.69)$$

From this result and the result in Eq. (10.68), we have

$$\frac{d}{d\tau}\left[\frac{u^\beta (x - r)^\alpha}{u \cdot (x - r)} \right] = \frac{u^\beta (x - r)^\alpha}{[u \cdot (x - r)]^2} \left[c^2 - (x - r)_\lambda \frac{du^\lambda}{d\tau} \right]$$
$$- \frac{1}{u \cdot (x - r)} \left[u^\alpha u^\beta - (x - r)^\alpha \frac{du^\beta}{d\tau} \right].$$

$$(10.70)$$

To calculate $F^{\alpha\beta}$ from Eq. (10.66) we can now write

$$\frac{d}{d\tau}\left[\frac{u^\beta (x - r)^\alpha - u^\alpha (x - r)^\beta}{u \cdot (x - r)} \right] = \frac{c^2 - (x - r)_\lambda \frac{du^\lambda}{d\tau}}{[u \cdot (x - r)]^2} \left[u^\beta (x - r)^\alpha - u^\alpha (x - r)^\beta \right]$$
$$+ \frac{1}{u \cdot (x - r)} \left[(x - r)^\alpha \frac{du^\beta}{d\tau} - (x - r)^\beta \frac{du^\alpha}{d\tau} \right].$$

$$(10.71)$$

For later use we define the second-rank tensor $F_p^{\alpha\beta}$ as

$$F_P^{\alpha\beta} \equiv \left\{ \frac{c^2 - (x-r)_\lambda \frac{du^\lambda}{d\tau}}{[u \cdot (x-r)]^2} \right\} \left[u^\beta (x-r)^\alpha - u^\alpha (x-r)^\beta \right]$$

$$+ \frac{1}{u \cdot (x-r)} \left[(x-r)^\alpha \frac{du^\beta}{d\tau} - (x-r)^\beta \frac{du^\alpha}{d\tau} \right]. \qquad (10.72)$$

From Eqs. (10.66), (10.71) and the above equation, we see that the final expression for the field strength tensor is obtained as

$$F^{\alpha\beta}(x) = \left[\frac{q}{u \cdot (x-r)} F_P^{\alpha\beta} \right]_{\text{ret}}. \qquad (10.73)$$

Primarily knowing $F_P^{\alpha\beta}$ will yield the electric and magnetic field values from the Liénard-Wiechert potentials. It must be noted that although we did not start from the gauge field expression in Eq. (10.44) to get $F^{\alpha\beta}$, nonetheless the above field strength tensor corresponds to the same physical situation which gave birth to the gauge field solution in Eq. (10.44).

10.4.2   The Calculation for the Electric Field

To calculate the electric field from the Liénard-Wiechert potential formulation, we require to calculate F_P^{i0} as $E^i = F^{i0}$. Here, i runs from 1 to 3. To find out the expression for F_P^{i0} we first start with

$$(x-r)_\lambda \frac{du^\lambda}{d\tau} = (x-r)_0 \frac{du^0}{d\tau} - (\mathbf{x} - \mathbf{r}) \cdot \frac{d\mathbf{u}}{d\tau}$$
$$= c\gamma^4(\boldsymbol{\beta} \cdot \dot{\boldsymbol{\beta}})R - R\hat{\mathbf{n}} \cdot [c\gamma^4 \boldsymbol{\beta}(\boldsymbol{\beta} \cdot \dot{\boldsymbol{\beta}}) + c\gamma^2 \dot{\boldsymbol{\beta}}]. \qquad (10.74)$$

Then we have

$$c^2 - (x-r)_\lambda \frac{du^\lambda}{d\tau} = c^2 - c\gamma^4 R(\boldsymbol{\beta} \cdot \dot{\boldsymbol{\beta}}) + Rc\gamma^4(\hat{\mathbf{n}} \cdot \boldsymbol{\beta})(\boldsymbol{\beta} \cdot \dot{\boldsymbol{\beta}})$$
$$+ Rc\gamma^2(\hat{\mathbf{n}} \cdot \dot{\boldsymbol{\beta}}). \qquad (10.75)$$

We have

$$u^0(x-r)^i - u^i(x-r)^0 = \gamma cRn^i - \gamma c\beta^i R = R\gamma c(n^i - \beta^i), \qquad (10.76)$$

where n^i is the ith component of the unit vector $\hat{\mathbf{n}}$. The other term which we require to calculate F_P^{i0} is

$$(x-r)^i \frac{du^0}{d\tau} - (x-r)^0 \frac{du^i}{d\tau} = Rn^i c\gamma^4(\boldsymbol{\beta} \cdot \dot{\boldsymbol{\beta}}) - R[c\gamma^4\beta^i(\boldsymbol{\beta} \cdot \dot{\boldsymbol{\beta}}) + c\gamma^2\dot{\beta}^i]$$

$$= Rc\gamma^4 n^i(\boldsymbol{\beta} \cdot \dot{\boldsymbol{\beta}}) - Rc\gamma^4\beta^i(\boldsymbol{\beta} \cdot \dot{\boldsymbol{\beta}}) - Rc\gamma^2\dot{\beta}^i. \tag{10.77}$$

Using the above results in Eq. (10.72), we can write F_P^{i0} as

$$F_P^{i0} = \left[\frac{c^2 - c\gamma^4 R(\boldsymbol{\beta} \cdot \dot{\boldsymbol{\beta}}) + Rc\gamma^4(\hat{\mathbf{n}} \cdot \boldsymbol{\beta})(\boldsymbol{\beta} \cdot \dot{\boldsymbol{\beta}}) + Rc\gamma^2(\hat{\mathbf{n}} \cdot \dot{\boldsymbol{\beta}})}{\gamma c R(1 - \boldsymbol{\beta} \cdot \hat{\mathbf{n}})^2} \right] (n^i - \beta^i)$$

$$+ \frac{\gamma^3 n^i(\boldsymbol{\beta} \cdot \dot{\boldsymbol{\beta}}) - \gamma^3\beta^i(\boldsymbol{\beta} \cdot \dot{\boldsymbol{\beta}}) - \gamma\dot{\beta}^i}{(1 - \boldsymbol{\beta} \cdot \hat{\mathbf{n}})}, \tag{10.78}$$

where we have also used the relation in Eq. (10.48). We will now write the above equation in a particular form which will help us later when we will write down the field strength tensor. We write F_P^{i0} as

$$F_P^{i0} = \frac{c(n^i - \beta^i)}{\gamma R(1 - \boldsymbol{\beta} \cdot \hat{\mathbf{n}})^2} + \left[\frac{\gamma^3(\hat{\mathbf{n}} \cdot \boldsymbol{\beta})(\boldsymbol{\beta} \cdot \dot{\boldsymbol{\beta}}) - \gamma^3(\boldsymbol{\beta} \cdot \dot{\boldsymbol{\beta}}) + \gamma(\hat{\mathbf{n}} \cdot \dot{\boldsymbol{\beta}})}{(1 - \boldsymbol{\beta} \cdot \hat{\mathbf{n}})^2} \right] (n^i - \beta^i)$$

$$+ \frac{\gamma^3(n^i - \beta^i)(\boldsymbol{\beta} \cdot \dot{\boldsymbol{\beta}}) - \gamma\dot{\beta}^i}{(1 - \boldsymbol{\beta} \cdot \hat{\mathbf{n}})}. \tag{10.79}$$

In the above equation, we do not touch the first term on the right-hand side but add the second and the third terms on the right-hand side. The addition of the second and third terms gives

$$\frac{\gamma(\hat{\mathbf{n}} \cdot \dot{\boldsymbol{\beta}})(n^i - \beta^i) - \gamma\dot{\beta}^i[1 - (\hat{\mathbf{n}} \cdot \boldsymbol{\beta})]}{(1 - \boldsymbol{\beta} \cdot \hat{\mathbf{n}})^2},$$

which can be written in compact way using a 3-vector identity. We note that

$$\hat{\mathbf{n}} \times [(\hat{\mathbf{n}} - \boldsymbol{\beta}) \times \dot{\boldsymbol{\beta}}] = (\hat{\mathbf{n}} - \boldsymbol{\beta})(\hat{\mathbf{n}} \cdot \dot{\boldsymbol{\beta}}) - \dot{\boldsymbol{\beta}}[\hat{\mathbf{n}} \cdot (\hat{\mathbf{n}} - \boldsymbol{\beta})]$$

$$= (\hat{\mathbf{n}} - \boldsymbol{\beta})(\hat{\mathbf{n}} \cdot \dot{\boldsymbol{\beta}}) - \dot{\boldsymbol{\beta}}[1 - (\hat{\mathbf{n}} \cdot \boldsymbol{\beta})].$$

Using the above relation, we can now write

$$F_P^{i0} = \frac{\gamma \left\{ \hat{\mathbf{n}} \times [(\hat{\mathbf{n}} - \boldsymbol{\beta}) \times \dot{\boldsymbol{\beta}}] \right\}^i}{(1 - \boldsymbol{\beta} \cdot \hat{\mathbf{n}})^2} + \frac{c(\hat{\mathbf{n}} - \boldsymbol{\beta})^i}{\gamma R(1 - \boldsymbol{\beta} \cdot \hat{\mathbf{n}})^2}, \tag{10.80}$$

where the i superscript specifies the ith component of the 3-vectors. From Eq. (10.73), we can now write the expression for the electric field produced by arbitrary motion of a point charge as

$$\mathbf{E}(x) = q \left[\frac{(\hat{\mathbf{n}} - \boldsymbol{\beta})}{\gamma^2 (1 - \boldsymbol{\beta} \cdot \hat{\mathbf{n}})^3 R^2} \right]_{\text{ret}} + \frac{q}{c} \left\{ \frac{\hat{\mathbf{n}} \times \left[(\hat{\mathbf{n}} - \boldsymbol{\beta}) \times \dot{\boldsymbol{\beta}} \right]}{(1 - \boldsymbol{\beta} \cdot \hat{\mathbf{n}})^3 R} \right\}_{\text{ret}} . \quad (10.81)$$

We see that the electric field produced by an arbitrarily moving point charge has distinctly two parts. One of the parts is directly related to the acceleration $\dot{\boldsymbol{\beta}}$ of the particle. If both acceleration and velocity of the charged particle are zero, the above formula reproduces the electric field due to a static charge. If the charge is moving with uniform velocity then only the first term on the right-hand side of the above equation remains. Traditionally, the first term on the right-hand side of the above equation is called the velocity field. As the second term on the right-hand side only affects the electric field when the charged particle accelerates, traditionally the second term on the right-hand side of the above equation is called the acceleration field. The electric field depends upon the velocity field and the acceleration field calculated in the retarded time.

10.4.3   The Calculation for the Magnetic Field

To calculate the magnetic field from the Liénard-Wiechert potential formulation we require to calculate F_P^{ij}, where $i \neq j$ and i, j runs from 1 to 3. In the present case

$$u^j (x - r)^i - u^i (x - r)^j = \gamma c \beta^j R n^i - \gamma c \beta^i R n^j = R \gamma c (n^i \beta^j - n^j \beta^i) , \quad (10.82)$$

and

$$(x - r)^i \frac{du^j}{d\tau} - (x - r)^j \frac{du^i}{d\tau} = R n^i [c\gamma^4 \beta^j (\boldsymbol{\beta} \cdot \dot{\boldsymbol{\beta}}) + c\gamma^2 \dot{\beta}^j] - R n^j [c\gamma^4 \beta^i (\boldsymbol{\beta} \cdot \dot{\boldsymbol{\beta}}) + c\gamma^2 \dot{\beta}^i] ,$$

which can also be written as

$$(x - r)^i \frac{du^j}{d\tau} - (x - r)^j \frac{du^i}{d\tau} = R c\gamma^2 \left[n^i \dot{\beta}^j - n^j \dot{\beta}^i + \gamma^2 (\boldsymbol{\beta} \cdot \dot{\boldsymbol{\beta}})(n^i \beta^j - n^j \beta^i) \right] . \quad (10.83)$$

Using these relations in Eq. (10.72), we can write F_P^{ij} as

$$F_P^{ij} = \left[\frac{c^2 - c\gamma^4 R (\boldsymbol{\beta} \cdot \dot{\boldsymbol{\beta}}) + R c\gamma^4 (\hat{\mathbf{n}} \cdot \boldsymbol{\beta})(\boldsymbol{\beta} \cdot \dot{\boldsymbol{\beta}}) + R c\gamma^2 (\hat{\mathbf{n}} \cdot \dot{\boldsymbol{\beta}})}{\gamma c R (1 - \boldsymbol{\beta} \cdot \hat{\mathbf{n}})^2} \right] (n^i \beta^j - n^j \beta^i)$$

$$+ \frac{\gamma \left[n^i \dot{\beta}^j - n^j \dot{\beta}^i + \gamma^2 (\boldsymbol{\beta} \cdot \dot{\boldsymbol{\beta}})(n^i \beta^j - n^j \beta^i) \right]}{(1 - \boldsymbol{\beta} \cdot \hat{\mathbf{n}})} . \quad (10.84)$$

To proceed further we define a 3-vector $\tilde{\mathbf{E}}$, whose components are F_P^{i0} as given in Eq. (10.78). This 3-vector can be written as

$$\tilde{\mathbf{E}} = \left[\frac{c^2 - c\gamma^4 R(\boldsymbol{\beta} \cdot \dot{\boldsymbol{\beta}}) + Rc\gamma^4(\hat{\mathbf{n}} \cdot \boldsymbol{\beta})(\boldsymbol{\beta} \cdot \dot{\boldsymbol{\beta}}) + Rc\gamma^2(\hat{\mathbf{n}} \cdot \dot{\boldsymbol{\beta}})}{\gamma c R(1 - \boldsymbol{\beta} \cdot \hat{\mathbf{n}})^2}\right](\hat{\mathbf{n}} - \boldsymbol{\beta})$$
$$+ \frac{\gamma^3(\hat{\mathbf{n}} - \boldsymbol{\beta})(\boldsymbol{\beta} \cdot \dot{\boldsymbol{\beta}}) - \gamma\dot{\boldsymbol{\beta}}}{(1 - \boldsymbol{\beta} \cdot \hat{\mathbf{n}})}, \tag{10.85}$$

and from Eq. (10.73) one understands that the electric field in Eq. (10.81) is obtained as

$$\mathbf{E}(x) = \left[\frac{q}{u \cdot (x - r)}\tilde{\mathbf{E}}(x)\right]_{\text{ret}}. \tag{10.86}$$

Next we define a magnetic field as

$$\tilde{B}_k \equiv -\frac{1}{2}\epsilon_{kij} F_P^{ij}, \tag{10.87}$$

which when combined with Eq. (10.73) shows that the actual magnetic field produced by an arbitrarily moving point charge will be

$$B_k(x) = \left[\frac{q}{u \cdot (x - r)}\tilde{B}_k(x)\right]_{\text{ret}}. \tag{10.88}$$

The effective magnetic field expression in Eq. (10.87) is a generalization of the relation in Eq. (7.11). The minus sign present in the right-hand side of Eq. (10.87) is required as we are using the Minkowski metric $\eta^{\mu\nu}$ to raise or lower tensor indices. Using the expression for F_P^{ij} in Eq. (10.84) we can write, using 3-vector notation,

$$\tilde{\mathbf{B}} = -\left[\frac{c^2 - c\gamma^4 R(\boldsymbol{\beta} \cdot \dot{\boldsymbol{\beta}}) + Rc\gamma^4(\hat{\mathbf{n}} \cdot \boldsymbol{\beta})(\boldsymbol{\beta} \cdot \dot{\boldsymbol{\beta}}) + Rc\gamma^2(\hat{\mathbf{n}} \cdot \dot{\boldsymbol{\beta}})}{\gamma c R(1 - \boldsymbol{\beta} \cdot \hat{\mathbf{n}})^2}\right](\hat{\mathbf{n}} \times \boldsymbol{\beta})$$
$$- \frac{\gamma\left[(\hat{\mathbf{n}} \times \dot{\boldsymbol{\beta}}) + \gamma^2(\boldsymbol{\beta} \cdot \dot{\boldsymbol{\beta}})(\hat{\mathbf{n}} \times \boldsymbol{\beta})\right]}{(1 - \boldsymbol{\beta} \cdot \hat{\mathbf{n}})}, \tag{10.89}$$

which can also be written as

$$\tilde{\mathbf{B}} = \left[\frac{c^2 - c\gamma^4 R(\boldsymbol{\beta} \cdot \dot{\boldsymbol{\beta}}) + Rc\gamma^4(\hat{\mathbf{n}} \cdot \boldsymbol{\beta})(\boldsymbol{\beta} \cdot \dot{\boldsymbol{\beta}}) + Rc\gamma^2(\hat{\mathbf{n}} \cdot \dot{\boldsymbol{\beta}})}{\gamma c R(1 - \boldsymbol{\beta} \cdot \hat{\mathbf{n}})^2}\right][\hat{\mathbf{n}} \times (\hat{\mathbf{n}} - \boldsymbol{\beta})]$$
$$+ \frac{\gamma^3[\hat{\mathbf{n}} \times (\hat{\mathbf{n}} - \boldsymbol{\beta})](\boldsymbol{\beta} \cdot \dot{\boldsymbol{\beta}}) - \gamma(\hat{\mathbf{n}} \times \dot{\boldsymbol{\beta}})}{(1 - \boldsymbol{\beta} \cdot \hat{\mathbf{n}})}. \tag{10.90}$$

From the expression of $\tilde{\mathbf{E}}$ in Eq. (10.85) and the above expression, we see that we can write

$$\tilde{\mathbf{B}} = \hat{\mathbf{n}} \times \tilde{\mathbf{E}}. \qquad (10.91)$$

Replacing the value of $\tilde{\mathbf{B}}$ from the above result in Eq. (10.88) and then using Eq. (10.86) gives us the retarded magnetic field produced by an arbitrarily moving point charge as

$$\mathbf{B}(x) = \left[\hat{\mathbf{n}}\right]_{\text{ret}} \times \mathbf{E}(x). \qquad (10.92)$$

Together with the expression of the electric field, in Eq. (10.81), and the above magnetic field we have the complete electromagnetic field structure which is produced by an arbitrarily moving point charge. These are the electric and magnetic fields produced from the Liénard-Wiechert potentials. In the above expression of the magnetic field, one has to replace the right-hand side in Eq. (10.81) for $\mathbf{E}(x)$. It is seen that the electric field and the magnetic field, at the field point $\mathbf{x}$ at time t, are mutually perpendicular.

10.4.4   The Nature of the Fields

Once we know the electric field and the magnetic field produced by an arbitrarily moving point charge we can calculate the rate of energy flowing out in the form of radiation energy. To figure out the rate at which radiation energy flows out we require to find out the Poynting vector

$$\mathbf{S} = \frac{c}{4\pi} \mathbf{E} \times \mathbf{B},$$

where the electric field and the magnetic field are given in Eqs. (10.81) and (10.92). The radiation emitted by the moving charge at time t' reaches a distance $R(t')$ in time t as shown in Fig. 10.8. If we draw a sphere, centered at the position of the charge at time t', of radius $R(t')$ then an infinitesimal surface area on this sphere will be $R^2 \sin\theta d\theta d\phi$. We have assumed the coordinate system has the origin at the center of the sphere and the polar and the azimuthal angles are suitably defined in that coordinate system. Energy, stored in the fields, incident on this small area per unit time will be $|\mathbf{S}|R^2 \sin\theta d\theta d\phi$. The total energy flowing through the spherical surface with radius R at time t will simply be

$$\int_{\theta=0}^{\pi} \int_{\phi=0}^{2\pi} |\mathbf{S}(t)| R^2 \sin\theta d\theta d\phi.$$

From the nature of the electric and magnetic fields produced by an arbitrarily moving point charge, as given in Eqs. (10.81) and (10.92), we can write the form of $|\mathbf{S}|$ as

$$|\mathbf{S}(t)| = [\cdots]_{\text{ret}} \frac{1}{R^4(t')} + [\cdots]_{\text{ret}} \frac{1}{R^3(t')} + [\cdots]_{\text{ret}} \frac{1}{R^2(t')}, \qquad (10.93)$$

where we have only shown the $R(t')$ dependence of $|\mathbf{S}|(t)$ and $t' = t - (R/c)$. The boxes with the dots inside contain the other terms which define the Poynting vector at time t. The first term on the right-hand side of the above equation, which is proportional to R^{-4}, comes purely from the velocity field part. The second term on the right-hand side, proportional to R^{-3}, comes from the cross term of the velocity field and acceleration field. The last term on the right-hand side of the above equation, proportional to R^{-2}, comes purely from the acceleration field. When $R \to \infty$, it is seen that only the last term on the right-hand side of the expression for $|\mathbf{S}(t)|$ contributes to the energy flow integral. The other terms in $|\mathbf{S}(t)|$ do not contribute to energy flow at large distance from the point charge. Radiation is traditionally defined as flow of energy from a local source to infinite distance, in the absence of any material medium. We see that such a flow is only possible from the acceleration field term in Eq. (10.81).

10.4.5   Electric and Magnetic Fields from an Uniformly Moving Charged Particle in the Liénard-Wiechert Formalism

In Chap. 7, Sect. 7.4.1 we discussed the electromagnetic field produced by a uniformly moving point charge, moving along the x-axis with velocity v. The field point is at P along the positive y-axis. The distance of P from origin is b. The calculations presented in Sect. 7.4.1 did not use retarded time formalism. In this subsection, we will show that the Liénard-Wiechert formalism can reproduce the same old results when the point charge moves with uniform velocity v along the x-axis. Figure 10.9 captures the essence of the retarded time formalism in this case.

Field produced by the moving charge is measured at point P at time t when the position of the charge is at Q as shown in Fig. 10.9. The retarded position of the charge, at time t', is Q'. The signal emitted by the moving charge at Q' reaches point P after a time R/c during which the charge has traveled a distance $(R/c)v = \beta R$. If $Q'Q$ denote the length of the line segment whose endpoints are at Q' and Q we have $Q'Q = \beta R$. From the figure it is clear that $QO = vt$. At $t = 0$ the particle will move from the negative side of the x-axis to the positive side. From the geometry of Fig. 10.9, we have

$$Q'N = Q'Q \cos\theta = \beta R \cos\theta = R\boldsymbol{\beta} \cdot \hat{\mathbf{n}},$$

and consequently

Fig. 10.9 A point charge
moves with uniform velocity
v along the x-axis. The field
point is at P. The
electromagnetic field is
measured at time t at P. The
position of the charge at
retarded time t' is shown as
Q'. At time t the charge is at
Q and its distance to P at
time t is $r = R(t)$. The
distance of Q' to P at time t'
is $R(t')$

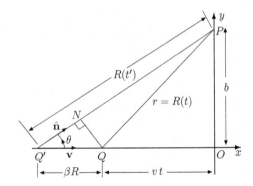

$$PN = PQ' - Q'N = R(1 - \boldsymbol{\beta} \cdot \hat{\mathbf{n}}).$$

As $QN = Q'Q \sin\theta = \beta R \sin\theta$ and $(PN)^2 = r^2 - (QN)^2$ in triangle QNP we have[2]

$$\left[(1 - \boldsymbol{\beta} \cdot \hat{\mathbf{n}}) R \right]^2 = r^2 - \beta^2 R^2 \sin^2\theta . \tag{10.94}$$

From the geometry of triangle QPO we have $r^2 = b^2 + v^2 t^2$ where $b = R \sin\theta$. Using these results in the last equation, we have

$$\left[(1 - \boldsymbol{\beta} \cdot \hat{\mathbf{n}}) R \right]^2 = b^2 + v^2 t^2 - \beta^2 b^2 = b^2 (1 - \beta^2) + v^2 t^2$$

$$= \frac{1}{\gamma^2} (b^2 + \gamma^2 v^2 t^2) . \tag{10.95}$$

Suppose we look at the expression of E_y, in the K frame, in Sect. 7.4.1 given as

$$E_y = \frac{q\gamma b}{(b^2 + \gamma^2 v^2 t^2)^{3/2}} .$$

We can write the expression of E_y using the relation in Eq. (10.95) as

$$E_y = q \left[\frac{b/R}{\gamma^2 (1 - \boldsymbol{\beta} \cdot \hat{\mathbf{n}})^3 R^2} \right]_{\text{ret}} . \tag{10.96}$$

We can compare this result to the velocity field coming from the Liénard-Wiechert formalism as given in Eq. (10.81). The pure velocity field is

---

[2] In keeping with the notation in Chap. 7, Sect. 7.4.1 we are using r here to specify the distance of the point charge to the field point at time t. In this particular exercise r does not specify the world line of the charged particle.

$$\mathbf{E}(x) = q \left[\frac{(\hat{\mathbf{n}} - \boldsymbol{\beta})}{\gamma^2 (1 - \boldsymbol{\beta} \cdot \hat{\mathbf{n}})^3 R^2} \right]_{\mathrm{ret}}, \tag{10.97}$$

where $\boldsymbol{\beta} = \beta\hat{\mathbf{i}}$. We see that the y-component of the above electric field is exactly the same as given in Eq. (10.96) as $n_y = \sin\theta = b/R$. In a similar way, the reader can easily verify that the expression of $\mathbf{E}(x)$ in Eq. (10.97) can reproduce the forms of E_x and E_z in K frame in Sect. 7.4.1.

The discussion presented above clearly shows that the Liénard-Wiechert formalism can reproduce the standard old result related to the electromagnetic field of a uniformly moving point charge. The matching was successful after we casted the old result in terms of retarded variables. The above workout shows clearly how the retarded formalism actually work in a simple case. Next we address the historically important problem related to radiation from an accelerated charge which is moving non-relativistically.

10.5   Power Radiated by an Accelerated Charge Moving Non-relativistically: Larmor's Formula

In this section, we will be dealing with radiation from an accelerated point charge. As radiation can only happen from the acceleration field in the expression for the electric field in Eq. (10.81), we write the relevant fields as

$$\mathbf{E}(x) = \frac{q}{c} \left\{ \frac{\hat{\mathbf{n}} \times \left[(\hat{\mathbf{n}} - \boldsymbol{\beta}) \times \dot{\boldsymbol{\beta}} \right]}{(1 - \boldsymbol{\beta} \cdot \hat{\mathbf{n}})^3 R} \right\}_{\mathrm{ret}}, \quad \mathbf{B}(x) = \left[\hat{\mathbf{n}} \right]_{\mathrm{ret}} \times \mathbf{E}(x). \tag{10.98}$$

It must be noted that the velocity field always remains a part of the electric field but we neglect that part here as the velocity field do not contribute to radiation. In the non-relativistic limit

$$\frac{\mathbf{v}}{c} \equiv \boldsymbol{\beta} \to 0, \tag{10.99}$$

but there is a finite acceleration, implying

$$\dot{\boldsymbol{\beta}} \neq 0. \tag{10.100}$$

In our subsequent calculations, we will systematically neglect terms containing $\boldsymbol{\beta}$ but will retain terms which depend on $\dot{\boldsymbol{\beta}}$. Moreover, in the non-relativistic limit the concept of retarded time looses its significance as although $t' = t - (R/c)$ but as $v \ll c$, the particle moves negligible distance during the time R/c. As a result the position of the moving particle at t and t' essentially coincides. The magnitude c is higher than any other signal velocity in the system and we can neglect the time period R/c. Consequently in the discussion of Larmor's formula we do not care about the

Fig. 10.10 The charge q at time t is shown on its trajectory. The sphere with radius R is drawn with center at the position of the charge at time t. The unit vector $\hat{\mathbf{n}}$ points in the outward direction as shown. The area element, $R^2 d\Omega$ as shown, subtends a solid angle $d\Omega$ at the origin

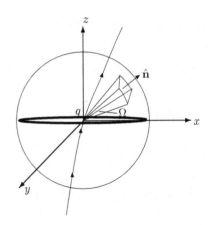

retarded time formalism and write the relevant fields simply as

$$\mathbf{E}(x) = \frac{q}{c}\left[\frac{\hat{\mathbf{n}} \times (\hat{\mathbf{n}} \times \dot{\boldsymbol{\beta}})}{R}\right], \quad \mathbf{B}(x) = \hat{\mathbf{n}} \times \mathbf{E}(x). \tag{10.101}$$

We can write the electric field also as

$$\mathbf{E}(x) = \frac{q}{c}\left[\frac{\hat{\mathbf{n}}(\hat{\mathbf{n}} \cdot \dot{\boldsymbol{\beta}}) - \dot{\boldsymbol{\beta}}}{R}\right], \tag{10.102}$$

which shows that $\hat{\mathbf{n}} \cdot \mathbf{E} = 0$ and $\mathbf{E}$ is polarized in the plane containing $\hat{\mathbf{n}}$ and $\dot{\boldsymbol{\beta}}$. Poynting vector in the present case is

$$\mathbf{S} = \frac{c}{4\pi}\left[\mathbf{E} \times (\hat{\mathbf{n}} \times \mathbf{E})\right] = \frac{c}{4\pi}|\mathbf{E}^2|\hat{\mathbf{n}}. \tag{10.103}$$

Choosing any time t we can specify the position of the charged particle on its trajectory. Taking the origin of the coordinate system in a particular position on the trajectory of the particle at time t we can draw a sphere about the origin whose radius is R. In Fig. 10.10 such a situation is shown. If the field point is at any point on the surface of the sphere then the vector $\hat{\mathbf{n}}$ points outward from the surface as shown in Fig. 10.10. Taking an infinitesimal area $d\mathbf{a} = R^2 d\Omega\,\hat{\mathbf{n}}$ around the field point on the sphere we can write the expression of power emitted across the infinitesimal area as $dP = \mathbf{S} \cdot d\mathbf{a} = R^2|\mathbf{S}|d\Omega$. As a consequence

$$\frac{dP}{d\Omega} = \frac{c}{4\pi}|R\mathbf{E}|^2 = \frac{q^2}{4\pi c}\left|\hat{\mathbf{n}} \times (\hat{\mathbf{n}} \times \dot{\boldsymbol{\beta}})\right|^2, \tag{10.104}$$

Fig. 10.11 The angle
between $\hat{\mathbf{n}}$ and $\dot{\boldsymbol{\beta}}$ at time t is
Θ as shown in the figure.
The size of the vectors are
magnified for the sake of
illustration

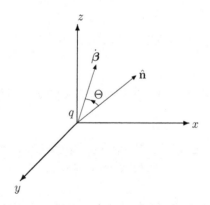

which gives the differential rate of power emitted, at time t, through a solid angle
$d\Omega$ about the origin. If the angle between $\hat{\mathbf{n}}$ and $\dot{\boldsymbol{\beta}}$ is assumed to be Θ, as shown in
Fig. 10.11, we have

$$\hat{\mathbf{n}} \times \dot{\boldsymbol{\beta}} = \dot{\boldsymbol{\beta}} \times (\text{a unit vector perp. to } \hat{\mathbf{n}} \text{ and } \dot{\boldsymbol{\beta}}),$$

which implies

$$\left| \hat{\mathbf{n}} \times (\hat{\mathbf{n}} \times \dot{\boldsymbol{\beta}}) \right|^2 = \dot{\beta}^2 \sin^2 \Theta.$$

Using the above result, we can write the differential power radiated by the non-
relativistic accelerated charge at time t as

$$\frac{dP}{d\Omega} = \frac{q^2}{4\pi c^3} |\dot{\mathbf{v}}|^2 \sin^2 \Theta. \tag{10.105}$$

The above formula shows that the differential power does not actually depend upon
the sign of $\dot{\mathbf{v}}$, it is the same for acceleration and deceleration as long as both have the
same magnitude. The total power radiated by the accelerated charge at time t is

$$P = \int \frac{dP}{d\Omega} d\Omega. \tag{10.106}$$

To do the above integral we orient our coordinate system in such a way that $\dot{\mathbf{v}}$ at
time t coincides with the positive direction of z-axis and consequently Θ becomes
the polar angle specifying $\hat{\mathbf{n}}$. Strictly speaking P is the energy incident per unit time
on a spherical surface whose radius is R at time t. As energy is conserved P is also
the energy flowing out of the charged particle per unit time during its accelerated
motion. The integration is straightforward,

$$P = \frac{q^2}{4\pi c^3} \int_{\Theta=0}^{\pi} \int_{\phi=0}^{2\pi} |\dot{\mathbf{v}}|^2 \sin^3 \Theta \, d\Theta \, d\phi = \frac{q^2}{2c^3} |\dot{\mathbf{v}}|^2 \int_{\Theta=0}^{\pi} \sin^3 \Theta \, d\Theta,$$

which yields the famous Larmor power formula as

$$P = \frac{2}{3}\frac{q^2}{c^3}\,|\dot{\mathbf{v}}|^2\;.$$
(10.107)

The power does not depend upon the sign of $\dot{\mathbf{v}}$ or the sign of the charge.

10.5.1   Power Radiated by a System of Accelerated Charged Particles Moving Non-relativistically

A system of accelerated charged particles will also radiate power. We can neatly write down the expression of this power using the Larmor formula when the charges move non-relativistically. Suppose we have N number of charged particles, with charge q_i, which are accelerating. The dipole moment of the system will then be

$$\mathbf{p} = \sum_{i=1}^{N} q_i \mathbf{x}_i\;.$$
(10.108)

To write the dipole moment we use the same vector as we use in describing the 3-momentum of a particle. We hope this will not produce any confusion as the meaning of the symbols become clear from the context. In the present case, we can use the previous calculation for a single particle and write:

$$\mathbf{E}(x) = \sum_{i=1}^{N} \mathbf{E}_i(x)\,, \quad \mathbf{B}(x) = \sum_{i=1}^{N} \mathbf{B}_i(x)\,,$$
(10.109)

where

$$\mathbf{E}_i(x) = \frac{q_i}{c}\left[\frac{\hat{\mathbf{n}}_i \times (\hat{\mathbf{n}}_i \times \dot{\boldsymbol{\beta}}_i)}{R_i}\right]\,, \quad \mathbf{B}_i(x) = \hat{\mathbf{n}}_i \times \mathbf{E}_i(x)\,.$$
(10.110)

The Poynting vector in the present case is

$$\mathbf{S} = \frac{c}{4\pi}\sum_{i=1}^{N}\left[\mathbf{E}_i \times (\hat{\mathbf{n}}_i \times \mathbf{E}_i)\right] = \frac{c}{4\pi}\sum_{i=1}^{N}|\mathbf{E}_i^2|\hat{\mathbf{n}}_i\;.$$
(10.111)

The rest of the calculation is straightforward giving the total power emitted

$$P = \frac{2}{3c^3} \sum_{i=1}^{N} q_i^2 \, |\dot{\mathbf{v}}_i|^2 = \frac{2}{3c^3} \ddot{\mathbf{p}}^2 \,. \tag{10.112}$$

The above formula gives the power radiated by a system of slowly moving charged particles. The result is expressed in terms of the square of the second-order time derivative of the dipole moment of the system of charges.

10.6 Relativistic Generalization of Larmor's Formula

Till now we have been working with the non-relativistic form of Larmor's formula, in the present subsection we will like to generalize the result and produce a covariant formula for total power emitted by an arbitrarily accelerating point charge. To proceed in this direction we will assume that there are two frames of reference, one is the laboratory frame or the traditional K frame as discussed in Chap. 6 and the other one is the instantaneous inertial rest frame of the charged particle, we will call it the K' frame. The instantaneous rest frame of the particle at time t, in the laboratory frame, is a uniformly moving inertial frame whose velocity coincides with the particle's velocity at time t and the charged particle is at the origin of such an instantaneous rest frame at time t. There are a plethora of inertial frames moving with different velocities which serve as the rest frame for the accelerated particle as time changes continuously. An instantaneous rest frame for the charged particle is shown in Fig. 10.12. The field point is shown as the point with spacetime coordinates $(t', \mathbf{x}')$. We assume that the charged particle is moving with velocity $\mathbf{v}$ at time t in the K frame. The reference frame K' is also moving with uniform velocity $\mathbf{v}$ and the charged particle is at the origin of the primed frame at time t. The charged particle has some acceleration in the laboratory frame and consequently the charged particle should also have some acceleration in the instantaneous K' inertial frame. From this discussion, we see that

Fig. 10.12 The field point in the K' frame is $(t', \mathbf{x}')$, as shown. The K' coordinate system has its origin at the position of the charge on its trajectory at time t. Here primes do not denote source points, the prime symbol specifies coordinates in K' frame

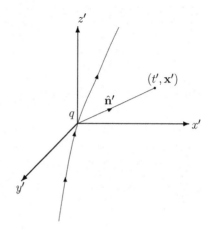

in the frame K' the 3-velocity of the charged particle is zero and 3-acceleration of the charged particle is non-zero. Denoting the 3-velocity parameter of the charged particle, in the K' frame, as $\boldsymbol{\beta}'$ we have in the primed frame

$$\boldsymbol{\beta}' = 0, \quad \dot{\boldsymbol{\beta}}' \neq 0, \quad \boldsymbol{\beta}' \equiv \frac{\mathbf{v}'}{c},$$

where $\mathbf{v}'$ is the 3-velocity of the charged particle in the primed frame. As because the charged particle is accelerating in the laboratory frame it must also be accelerating with respect to the inertial frame K', which is moving with a constant velocity parameter $\boldsymbol{\beta} = \mathbf{v}/c$ with respect to the laboratory frame. All the discussion about radiation from non-relativistic accelerated charged particle can now be applied in the K' frame. The acceleration fields responsible for radiation in the K' frame are obtained from our earlier discussion as

$$\mathbf{E}'(x) = \frac{q}{c} \left[\frac{\hat{\mathbf{n}}' \times (\hat{\mathbf{n}}' \times \dot{\boldsymbol{\beta}}')}{R} \right], \quad \mathbf{B}' = \hat{\mathbf{n}}' \times \mathbf{E}'. \tag{10.113}$$

The linear momentum density of the fields, in the K' frame is then written as

$$\mathbf{g}' = \frac{1}{4\pi c} (\mathbf{E}' \times \mathbf{B}') = \frac{1}{4\pi c} \left[\mathbf{E}' \times (\hat{\mathbf{n}}' \times \mathbf{E}') \right] = \frac{1}{4\pi c} |\mathbf{E}'|^2 \hat{\mathbf{n}}'.$$

Using the expression of the electric field, we can write the above equation as

$$\mathbf{g}' = \frac{q^2}{4\pi c^3 R'^2} \left| \hat{\mathbf{n}}' \times (\hat{\mathbf{n}}' \times \boldsymbol{\beta}') \right|^2 \hat{\mathbf{n}}', \tag{10.114}$$

which shows that for a fixed R' we have $\mathbf{g}'(\hat{\mathbf{n}}') = -\mathbf{g}'(-\hat{\mathbf{n}}')$. As the total linear 3-momentum component of the fields at any time t' is given by $P'^i_{\text{field}} = \int \mathbf{g}'^i \, d^3x'$ in the K' frame, we see that

$$P'^i_{\text{field}} = 0. \tag{10.115}$$

The 4-momentum of the radiation field in the K' frame, at any time t', is related to

$$\mathcal{M}'^\mu = \int \Theta'^{0\mu} \, d^3x', \tag{10.116}$$

which is a well-defined quantity assuming that $\Theta'^{0\mu}$ is well localized about the charge at any instant of time. We can also formally write the 4-momentum of the electromagnetic fields in the K' frame as

$$\mathcal{M}'^\mu = \left(E'_{\text{field}}, \, c P'^i_{\text{field}} \right) \tag{10.117}$$

where E'_{field} is the total energy of the fields due to radiation and P'^i_{field} is the total linear 3-momentum of the radiation fields. As $\mathcal{M}'^\mu$ is a 4-vector, if we want to write the 4-momentum of the radiation fields in K frame then we have to Lorentz transform $\mathcal{M}'^\mu$ to the laboratory frame. Suppose

$$\mathcal{M}^\mu = \left(E_{\text{field}}, c P^i_{\text{field}} \right) , \tag{10.118}$$

is the 4-momentum of the radiation fields in the laboratory frame then we must have

$$E_{\text{field}} = \gamma \left(E'_{\text{field}} + c \sum_i \beta^i P'^i_{\text{field}} \right) = \gamma E'_{\text{field}} , \tag{10.119}$$

as $\mathcal{M}'^i = c P'^i_{\text{field}} = 0$. Here

$$\gamma = \frac{1}{\sqrt{1 - \beta^2}} .$$

In this case, $\boldsymbol{\beta}$ makes an arbitrary angle with the common $x - x'$ direction. The coordinate times in the two frames are related as

$$t = \gamma t' . \tag{10.120}$$

Suppose the amount of energy radiated away by the accelerated charge in a very small time interval t (where time starts at 0 and so $\delta t = t$) in the K frame is E_{field}, and the corresponding quantities in the K' frame are E'_{field} and t'. From this quantities, we can now define the power emitted by the charge in the two frames. The power emitted, in the form of radiation, in frame K' is $P' = E'_{\text{field}}/t'$. Similarly, the power emitted in the K frame is given as $P = E_{\text{field}}/t$. From Eqs. (10.119) and (10.120), we see that

$$P' = \frac{E'_{\text{field}}}{t'} = \frac{E_{\text{field}}}{t} = P , \tag{10.121}$$

which shows that the radiation power emitted by an accelerated charge is Lorentz invariant.

As we know radiation power is Lorentz invariant we can now try to write the most general Lorentz invariant formula for radiation power, due to an accelerated charge. Our strategy can only be successful if the Lorentz invariant form of radiation power reproduces Larmor formula in the non-relativistic case as we already know the exact non-relativistic result. Suppose we successfully write down a Lorentz invariant form of radiation power which reproduces Larmor formula in the appropriate limits, then we can be sure that the form of power we have written down is the proper covariant generalization of radiation power.

To write down the most general Lorentz invariant form of radiation power in the K frame, due to an accelerated charge, we must notice that only $\boldsymbol{\beta}$ and $\dot{\boldsymbol{\beta}}$ can appear in that formula, except other dimensional constants and R. This is because the expressions of the electric and magnetic field only depend upon these vectors and R. The unit vector $\hat{\mathbf{n}}$ gets integrated out in the final expression of power. In the non-relativistic limit, Larmor formula gives the expression of radiated power as

$$P = \frac{2}{3}\frac{q^2}{c^3}|\dot{\mathbf{v}}|^2 .$$

Using the fact that the 3-momentum of the accelerating particle is given by $\mathbf{p} = m\mathbf{v}$, we can write the above equation as

$$P = \frac{2}{3}\frac{q^2}{m^2c^3}\left(\frac{d\mathbf{p}}{dt} \cdot \frac{d\mathbf{p}}{dt}\right) .$$

This equation now can be generalized to give us the desired Lorentz invariant form of power as

$$P = -\frac{2}{3}\frac{q^2}{m^2c^3}\left(\frac{dp^\alpha}{d\tau}\frac{dp_\alpha}{d\tau}\right) , \qquad (10.122)$$

where $dt = \gamma d\tau$ and

$$p^\alpha = m\gamma(c, \mathbf{v}) = \left(\frac{E}{c}, \mathbf{p}\right) ,$$

E being the energy of the particle. In our case we have

$$\gamma = \frac{1}{\sqrt{1-\beta^2}} , \quad \boldsymbol{\beta} = \frac{\mathbf{v}}{c} ,$$

where $\mathbf{v}$ is the velocity of the particle in K frame.

Next we show that the formula in Eq. (10.122) properly reproduces the Larmor formula in the non-relativistic limit. To show this we note

$$\frac{dp^\alpha}{d\tau}\frac{dp_\alpha}{d\tau} = \gamma^2\left(\frac{dp^\alpha}{dt}\frac{dp_\alpha}{dt}\right) = \gamma^2\left[\frac{1}{c^2}\left(\frac{dE}{dt}\right)^2 - \left(\frac{d\mathbf{p}}{dt}\right)^2\right]$$

$$= -\gamma^2\left(\frac{d\mathbf{p}}{dt}\right)^2 + \frac{\gamma^2}{c^2}\left(\frac{dE}{dt}\right)^2 .$$

As $E = mc^2\gamma$ we have

$$\frac{dE}{dt} = mc^2\frac{d\gamma}{dt} = mc^2\frac{\boldsymbol{\beta}\cdot\dot{\boldsymbol{\beta}}}{(1-\beta)^{3/2}} = m\gamma^3(\mathbf{v}\cdot\dot{\mathbf{v}}) .$$

In the non-relativistic limit as $|\mathbf{v}| \to 0$, $\gamma \to 1$ we see

$$\frac{dp^{\alpha}}{d\tau}\frac{dp_{\alpha}}{d\tau} \rightarrow -\left(\frac{d\mathbf{p}}{dt}\right)^2 . \tag{10.123}$$

Using the above result in Eq. (10.122), we see that we get back Larmor's formula in the non-relativistic limit. As a consequence we now claim that the formula, as given in Eq. (10.122), is the proper Lorentz invariant expression for power for radiation fields where radiation is emitted from an arbitrarily accelerating charged particle.

10.6.1 Radiation from a Linearly Accelerated Charge

We will now discuss the power emitted by a linearly accelerated charge. In the present case, we have $\boldsymbol{\beta} \parallel \dot{\boldsymbol{\beta}}$. The Lorentz invariant power is given by Eq. (10.122) as

$$\begin{aligned} P &= -\frac{2}{3}\frac{q^2}{m^2c^3}\left(\frac{dp^{\alpha}}{d\tau}\frac{dp_{\alpha}}{d\tau}\right) \\ &= \frac{2}{3}\frac{q^2}{m^2c^3}\left[\left(\frac{d\mathbf{p}}{d\tau}\right)^2 - \frac{1}{c^2}\left(\frac{dE}{d\tau}\right)^2\right], \end{aligned} \tag{10.124}$$

as $p^0 = E/c$. We know

$$\frac{dE}{d\tau} = m\gamma^2 \boldsymbol{\beta}\cdot\dot{\boldsymbol{\beta}}c^2 ,$$

and

$$\begin{aligned} \frac{d\mathbf{p}}{dt} = m\frac{d(\gamma\mathbf{v})}{dt} &= m\mathbf{v}\frac{d\gamma}{dt} + m\gamma\dot{\mathbf{v}} = mc\gamma^3\boldsymbol{\beta}(\boldsymbol{\beta}\cdot\dot{\boldsymbol{\beta}}) + m\gamma c\dot{\boldsymbol{\beta}} \\ &= m\gamma c\left[\gamma^2\boldsymbol{\beta}(\boldsymbol{\beta}\cdot\dot{\boldsymbol{\beta}}) + \dot{\boldsymbol{\beta}}\right] . \end{aligned}$$

Suppose velocity and acceleration of the charged particle are both along the x-axis, then

$$\boldsymbol{\beta} = \beta\,\hat{\mathbf{i}}, \qquad \dot{\boldsymbol{\beta}} = \dot{\beta}\,\hat{\mathbf{i}},$$

where $\hat{\mathbf{i}}$ is the unit vector along the x-axis. Using the above results we can write

$$\frac{d\mathbf{p}}{d\tau} = mc\left[\gamma^2\beta^2\dot{\beta} + \dot{\beta}\right]\hat{\mathbf{i}} = mc\dot{\beta}(1 + \gamma^2\beta^2)\hat{\mathbf{i}} = mc\gamma^2\dot{\beta}\,\hat{\mathbf{i}}. \tag{10.125}$$

In our present case

$$\frac{dE}{d\tau} = mc^2\gamma^2\beta\dot{\beta} .$$

These results show that for the case of linear acceleration

$$\beta^2 \left(\frac{d\mathbf{p}}{d\tau} \right)^2 = \frac{1}{c^2} \left(\frac{dE}{d\tau} \right)^2 .$$

We can now write Eq. (10.124) as

$$P = \frac{2}{3} \frac{q^2}{m^2 c^3} \left[\left(\frac{d\mathbf{p}}{d\tau} \right)^2 - \beta^2 \left(\frac{d\mathbf{p}}{d\tau} \right)^2 \right] = \frac{2}{3} \frac{q^2}{m^2 c^3} \left(\frac{d\mathbf{p}}{d\tau} \right)^2 (1 - \beta^2) .$$

Using the relation $dt = \gamma d\tau$, we have

$$P = \frac{2}{3} \frac{q^2}{m^2 c^3} \left(\frac{d\mathbf{p}}{dt} \right)^2 , \tag{10.126}$$

where $\mathbf{p} = m\gamma \mathbf{v}$. This is the expression for power emitted by a linearly accelerated point charge.

Linear acceleration of charged particles takes place in particle accelerators. In particle accelerators, one applies electric field to accelerate charged particles. The electric field supplies energy to the particle and as a result of which the charged particle accelerates. As the charged particle accelerates linearly it radiates electromagnetic energy. There is a balance between the rate at which it is energized and the power emitted in radiation. To understand more about this balance we can compare the emitted power and the rate at which the charged particle is energized. Due to the action of the electric field along the x-axis suppose the particle moves from x to $x + dx$ in the time interval bounded by t and $t + dt$. The force acting on the particle during this time is assumed to be constant, and is denoted by F along x-axis. In this infinitesimal time interval, the particle gains $dE = Fdx$ amount of energy where $F = dp/dt$ and $p = |\mathbf{p}|$. From these relations, we can write

$$\frac{dp}{dt} = F = \frac{dE}{dx} ,$$

using which we can now write the expression of power emitted by the charged particle as

$$P = \frac{2}{3} \frac{q^2}{m^2 c^3} \left(\frac{dE}{dx} \right)^2 . \tag{10.127}$$

The emitted power depends on the mass of the charged particle and $(dE/dx)^2$ but not on the total energy E at anytime. We can now compare the expression of P and the rate at which the charge is energized, i.e., dE/dt. We first note that

$$\frac{dE}{dt} = F \frac{dx}{dt} = Fv = \frac{dE}{dx} v .$$

Using this relation, we can now write

$$\frac{P}{(dE/dt)} = \frac{2}{3}\frac{q^2}{m^2c^3}\left(\frac{dE}{dx}\right)^2\left(\frac{1}{v\frac{dE}{dx}}\right) = \frac{2}{3}\frac{q^2}{m^2c^3}\frac{1}{v}\left(\frac{dE}{dx}\right),$$

which in the relativistic limit $(v \to c)$ becomes

$$\frac{P}{(dE/dt)} \sim \frac{2}{3}\frac{q^2}{m^2c^4}\left(\frac{dE}{dx}\right) = \frac{2}{3}\frac{(q^2/mc^2)}{mc^2}\left(\frac{dE}{dx}\right). \qquad (10.128)$$

The reason for writing the above formula in the way it is represented will become clear once we note the physical dimension of the quantities appearing on the rightmost part of the above equation. We know $P/(dE/dt)$ is dimensionless and consequently the right hand side of the above equation should also be dimensionless. The quantity dE/dx has dimension of energy by length. We note mc^2 has the dimension of energy, whereas q^2/mc^2 has the dimension of length. The dimensionless nature of the right-hand side of the above equation becomes manifest when we write it in the way it is represented. From the above equation, we see that $P/(dE/dt)$ becomes appreciable when

$$\frac{dE}{dx} \sim \frac{mc^2}{(q^2/mc^2)}. \qquad (10.129)$$

The above equation shows that when the gain in energy of the accelerating charged particle is mc^2 while traversing a distance q^2/mc^2, the radiated power becomes approximately of the same order of energy gain by the particle per unit time. In C.G.S. units the mass of an electron $m \sim 9 \times 10^{-28}$ gm, $c \sim 3 \times 10^{10}$ cm s^{-1} and $q \sim 4.8 \times 10^{-10}$ e.s.u.. Using these numbers one can easily calculate

$$\frac{q^2}{mc^2} \sim 2.8 \times 10^{-13} \text{ cm}.$$

It is also known that $mc^2 \sim 0.5$ MeV, where we deliberately use the S.I. unit of energy. Using the above numbers we get

$$\frac{mc^2}{(q^2/mc^2)} \sim 10^{14} \text{ MeV m}^{-1}.$$

This is a huge number and in reality the rate of energy gain by accelerated charged particles per unit length remains much smaller than this number and as a result we do not have to bother about the effects of radiation in linear accelerators.

10.6.2   Radiation from a Charged Particle in Circular Motion

Here, we discuss the case of radiation from a charged particle undergoing circular motion. We assume a charged particle to be moving in a circular trajectory with a fixed center. As in circular motion the centripetal acceleration is perpendicular to the direction of the linear velocity of the particle so in this case the acceleration vector is perpendicular to the velocity vector, $\boldsymbol{\beta} \perp \dot{\boldsymbol{\beta}}$. We assume from the start that the amount of radiation emitted by the charge is not destabilizing the orbit. If we ignore radiation completely then the energy of the particle does not change in one full revolution. Taking radiation into account this statement does not remain true, the energy of the particle changes in one full revolution. The 3-momentum, $\mathbf{p}$, of the moving charged particle is rotating rapidly as the particle moves fast on a circular trajectory. For a rotating vector $\mathbf{p}$, rotating with angular velocity ω, we know

$$\left| \frac{d\mathbf{p}}{dt} \right| = \omega |\mathbf{p}| , \tag{10.130}$$

where ω is assumed to be a constant. For the circular motion of the charged particle, we assume

$$\left| \frac{d\mathbf{p}}{d\tau} \right| = \gamma \omega |\mathbf{p}| \gg \frac{1}{c} \frac{dE}{d\tau} , \tag{10.131}$$

where τ is the proper time of the charged particle and E is its energy. The above statement implies that the rate of change of the 3-momentum is much more compared to the rate of change of E/c on the trajectory. This assumption is sensible because in the opposite limit the particle should have emitted so much energy that it could not have maintained a circular trajectory. We are assuming proper circular trajectories of the particle and as a result the above assumption is an important one. We know the power emitted by a charged particle moving in a circular orbit is

$$P = -\frac{2}{3} \frac{q^2}{m^2 c^3} \left(\frac{dp^\alpha}{d\tau} \frac{dp_\alpha}{d\tau} \right) = -\frac{2}{3} \frac{q^2}{M^2 c^3} \left[\frac{1}{c^2} \left(\frac{dE}{d\tau} \right)^2 - \left(\frac{d\mathbf{p}}{d\tau} \right)^2 \right] .$$

Using the assumption stated in Eq. (10.131), the above equation becomes

$$P \simeq \frac{2}{3} \frac{q^2}{m^2 c^3} \left(\frac{d\mathbf{p}}{d\tau} \right)^2 . \tag{10.132}$$

In a circular orbit with radius ρ, we know that the linear velocity of the charged particle $c\beta = \omega \rho$. This gives $|\mathbf{p}| = mc\gamma\beta$. Using Eq. (10.131), we can write the last equation as

$$P \simeq \frac{2}{3} \frac{q^2}{m^2 c^3} \gamma^2 \omega^2 |\mathbf{p}|^2 ,$$

where one can plug in the expression of $|\mathbf{p}|$ to obtain the final expression of power emitted by a charged particle in circular motion as

$$P \simeq \frac{2}{3} \left(\frac{q^2 c}{\rho^2} \right) \gamma^4 \beta^4 . \tag{10.133}$$

From this formula one can approximately calculate the energy emitted δE, in the form of radiation, in one full revolution of the charged particle. If the time period of one full revolution for the particle is T then we know $\delta E = TP$ where P is the power emitted by the moving particle. In the present case

$$T = \frac{2\pi \rho}{\beta c} ,$$

and hence the radiative energy loss per revolution is given by

$$\delta E = \frac{2\pi \rho}{\beta c} P = \frac{4\pi}{3} \frac{q^2}{\rho} \beta^3 \gamma^4 . \tag{10.134}$$

As the energy of the particle is $E = mc^2 \gamma$, we can write the above formula as

$$\delta E = \frac{4\pi}{3} \frac{q^2}{\rho} \beta^3 \left(\frac{E}{mc^2} \right)^4 , \tag{10.135}$$

which gives δE as a function of E for a charged particle undergoing circular motion. The energy loss formula yields important information about circular motion of charged particles in particle accelerators as Synchrotron or the Betatron. We see that in one full revolution the radiated energy from the charged particle $\delta E \propto \gamma^4$. For near relativistic velocities, this factor γ^4 can be high. In general, the electrons are forced to move in circular orbits in the accelerators by magnetic fields. The magnetic field does not inject energy to the radiating particle. Unlike linear accelerators, where electric fields can continuously inject energy and keep the accelerating particle steady, in circular accelerators one has to frequently turn on radio frequency oscillators which pump electrical energy to the moving charged particles to re-energize them.

We can also write the expression for power, as given in Eq. (10.132), as

$$P \simeq \frac{2}{3} \frac{q^2}{m^2 c^3} \gamma^2 \left(\frac{d\mathbf{p}}{dt} \right)^2 . \tag{10.136}$$

If we now compare the above result with the corresponding expression for power emitted by a linearly accelerated charged particle, whose expression is given in

Eq. (10.126), we see that the ratio of power emitted by a charged particle in circular motion is roughly γ^2 times the power emitted by a linearly accelerated charged particle. This result shows that for the ultrarelativistic case the power emitted in circular motion is much more than the power emitted in the case of linear acceleration.

10.7   Angular Distribution of Radiation Emitted by the Accelerated Charge

We know the expressions for the radiation fields produced by a charged particle in arbitrary accelerated motion from Sect. 10.5. The expressions for the electric and magnetic fields are given in Eq. (10.98). From the form of the fields we can calculate the Poynting vector, for radiation from accelerated charge, as

$$\mathbf{S} = \frac{c}{4\pi} \left[\mathbf{E} \times \mathbf{B}\right]_{\text{ret}} = \frac{c}{4\pi} \left[|\mathbf{E}|^2 \hat{\mathbf{n}}\right]_{\text{ret}} . \tag{10.137}$$

In writing the above expression, we have used the fact that $\hat{\mathbf{n}} \cdot \mathbf{E} = 0$. The above expression for the Poynting vector specifies the energy per unit time flowing normally through unit area positioned at $\mathbf{x}$ at time t. If the particle is moving with relativistic speed then the right-hand side of the above equation has to be calculated at the retarded time $t' = t - R(t)/c$. Here, t' is the coordinate time when the source charge emitted radiation and t is the time when that radiation reached point $\mathbf{x}$. From Eq. (10.98) and the last equation we can write

$$[\mathbf{S} \cdot \hat{\mathbf{n}}]_{\text{ret}} = \frac{q^2}{4\pi c} \left\{ \frac{1}{R^2} \left| \frac{\hat{\mathbf{n}} \times \left[(\hat{\mathbf{n}} - \boldsymbol{\beta}) \times \dot{\boldsymbol{\beta}}\right]}{(1 - \boldsymbol{\beta} \cdot \hat{\mathbf{n}})^3} \right|^2 \right\}_{\text{ret}}, \tag{10.138}$$

which is calculated at retarded time. Now suppose we want to calculate the energy emitted per unit area E_A, along the direction $\hat{\mathbf{n}}$, during a finite period of acceleration from the point charge where the time period stretched from $t_1' = T_1$ to $t_2' = T_2$. We can then write

$$E_A = \int_{t_1 = T_1 + \frac{R(T_1)}{c}}^{t_2 = T_2 + \frac{R(T_2)}{c}} [\mathbf{S} \cdot \hat{\mathbf{n}}]_{\text{ret}} \, dt \tag{10.139}$$

where the integration is over the coordinate time of the field point as ultimately the electric and magnetic fields are functions of t. We can write the above integral in the retarded time t' by using the relation

$$t = t' + \frac{R(t')}{c}, \quad \text{or} \quad ct = ct' + R(t').$$

From the above relation, we get

$$c\frac{dt}{dt'} = c + \frac{dR(t')}{dt'}, \tag{10.140}$$

which gives the derivative of the filed coordinate time with respect to the retarded time. Here $\mathbf{R}(t') = \mathbf{x} - \mathbf{r}(t')$. As $R^2 = \mathbf{R}^2$ we have

$$R\frac{dR}{dt'} = \mathbf{R} \cdot \frac{d\mathbf{R}}{dt'} = -\mathbf{R} \cdot \frac{d\mathbf{r}(t')}{dt'} = -\mathbf{R} \cdot \mathbf{v}(t'),$$

where $\mathbf{v}(t') = d\mathbf{r}(t')/dt'$. From the above result, we see that

$$\frac{dR}{dt'} = -\hat{\mathbf{n}} \cdot \mathbf{v}(t'), \tag{10.141}$$

where $\hat{\mathbf{n}} = \mathbf{R}/R$. Using the above relation, we can now write Eq. (10.140) as

$$\frac{dt}{dt'} = 1 - \boldsymbol{\beta} \cdot \hat{\mathbf{n}}. \tag{10.142}$$

Now we can write Eq. (10.139) as

$$E_A = \int_{t_1'=T_1}^{t_2'=T_2} (\mathbf{S} \cdot \hat{\mathbf{n}}) \frac{dt}{dt'} dt' = \int_{t_1'=T_1}^{t_2'=T_2} \left[(\mathbf{S} \cdot \hat{\mathbf{n}})(1 - \boldsymbol{\beta} \cdot \hat{\mathbf{n}}) \right] dt', \tag{10.143}$$

where the integration is now on retarded time. From the integrand we see that $\left[(\mathbf{S} \cdot \hat{\mathbf{n}})(1 - \boldsymbol{\beta} \cdot \hat{\mathbf{n}}) \right]$ is the energy emitted per unit time by the charged particle along a unit area, whose normal is along $\hat{\mathbf{n}}$ direction, during time T_1 to T_2.

The energy passing through area element $R^2 d\Omega$, whose normal direction is specified by $\hat{\mathbf{n}}$, per unit time is then given by $\left[(\mathbf{S} \cdot \hat{\mathbf{n}})(1 - \boldsymbol{\beta} \cdot \hat{\mathbf{n}}) \right] R^2 d\Omega$. Here, the infinitesimal area is around the field point which is R distance away from the charge. By definition the infinitesimal power emitted by the charged particle is then given by

$$dP = \left[(\mathbf{S} \cdot \hat{\mathbf{n}})(1 - \boldsymbol{\beta} \cdot \hat{\mathbf{n}}) \right] R^2 d\Omega,$$

which immediately gives

$$\frac{dP}{d\Omega} = R^2 (\mathbf{S} \cdot \hat{\mathbf{n}})(1 - \boldsymbol{\beta} \cdot \hat{\mathbf{n}}). \tag{10.144}$$

Both sides of the above equation are now calculated in the retarded time, or the source time, and so we do not use the restarted time symbol anymore. Using the expression of $(\mathbf{S} \cdot \hat{\mathbf{n}})$, we can write the above expression as

$$\frac{dP}{d\Omega} = \frac{q^2}{4\pi c} \frac{\left|\hat{\mathbf{n}} \times \left[(\hat{\mathbf{n}} - \boldsymbol{\beta}) \times \dot{\boldsymbol{\beta}}\right]\right|^2}{(1 - \boldsymbol{\beta} \cdot \hat{\mathbf{n}})^5}. \tag{10.145}$$

We will use this formula to investigate the direction dependence of power emission from an accelerated charged particle.

Once we have established the general expression of the differential power emitted by an arbitrarily moving point charge we want to see how that expression can be modified for the particular case of linearly accelerated motion. In the present case as because $\boldsymbol{\beta}$ and $\dot{\boldsymbol{\beta}}$ are parallel we have $(\hat{\mathbf{n}} - \boldsymbol{\beta}) \times \dot{\boldsymbol{\beta}} = \hat{\mathbf{n}} \times \dot{\boldsymbol{\beta}}$. Using this result, we can write

$$\frac{dP}{d\Omega} = \frac{q^2}{4\pi c} \frac{\left|\hat{\mathbf{n}} \times (\hat{\mathbf{n}} \times \dot{\boldsymbol{\beta}})\right|^2}{(1 - \boldsymbol{\beta} \cdot \hat{\mathbf{n}})^5}. \tag{10.146}$$

If the angle between $\boldsymbol{\beta}$ (or $\dot{\boldsymbol{\beta}}$) and $\hat{\mathbf{n}}$ is Θ, we have

$$\left|\hat{\mathbf{n}} \times (\hat{\mathbf{n}} \times \dot{\boldsymbol{\beta}})\right|^2 = \dot{\beta}^2 \sin^2 \Theta = \frac{\dot{v}^2 \sin^2 \Theta}{c^2},$$

and as a result we can write

$$\frac{dP}{d\Omega} = \frac{q^2 \dot{v}^2}{4\pi c^3} \frac{\sin^2 \Theta}{(1 - \beta \cos \Theta)^5}. \tag{10.147}$$

When $\beta \to 0$ the above formula reproduces the differential power corresponding to the Larmor formula as given in Eq. (10.105). We can now try to find out the extremum of the differential power. To do that we have to solve the equation

$$\frac{d}{d\Theta}\left[\frac{\sin^2 \Theta}{(1 - \beta \cos \Theta)^5}\right] = 0, \tag{10.148}$$

for some value or values of Θ. The above equation reduces to

$$3\beta \cos^2 \Theta + 2\cos \Theta - 5\beta = 0,$$

which has roots

$$\cos \Theta = \frac{1}{3\beta}\left[-1 \pm \sqrt{1 + 15\beta^2}\right]. \tag{10.149}$$

The above equation shows that there can be two extrema for the differential power. A little bit of calculation shows that in reality only one extremum exists. To see this point we take the solution with minus sign before the square root on the right-hand side and assume $\beta \sim 1$. In this case, we get

Fig. 10.13 Rescaled
differential power plotted
with respect to Θ for
$\beta = 0.8$. The value of $\Theta_{\max}$
is slightly greater than 0.3
(around 17°) as shown

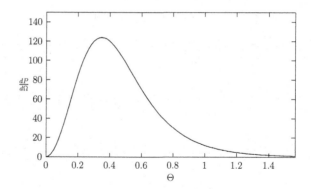

$$\cos \Theta \sim -\frac{5}{3},$$

which is an impossible solution as $\cos \Theta > -1$. Consequently, we only have one extremum for the differential power and that is obtained by choosing the plus sign before the square root in the expression for $\cos \Theta$.

Till now we have been talking about the extremum of $dP/d\Omega$ without mentioning whether that extremum is a maximum or a minimum. To analytically answer this question one has to find the second-order derivative of $dP/d\Omega$ with respect to Θ. This calculation is cumbersome and we do not proceed in that line. Here we use the graphical method to see the nature of the extremum. It turns out that the only extremum of the differential power is a maximum. This fact is shown in Fig. 10.13 where we plot a rescaled version of $dP/d\Omega$ with respect to Θ for $\beta = 0.8$. In the plot, the differential power is scaled down by a factor $q^2\dot{v}^2/4\pi c^3$. This scaling is unimportant here as we are plotting the differential power with respect to Θ at any time instant. The plot clearly shows the maximum at a Θ value somewhere slightly greater than 0.3. We have only plotted the differential power for Θ values ranging from 0 to $\pi/2$ as the only maximum lies in that part. From now on we will call the Θ value, for which the maximum of differential power is obtained, as $\Theta_{\max}$. In our case we have specified the angles in radians, in conventional units $\Theta_{\max} \sim 17°$.

From what we have discussed so far we can write

$$\Theta_{\max} = \cos^{-1}\left[\frac{1}{3\beta}\left(\sqrt{1 + 15\beta^2} - 1\right)\right]. \tag{10.150}$$

As $\gamma^2 = 1/(1 - \beta^2)$, we have

$$\beta^2 = 1 - \frac{1}{\gamma^2}.$$

For large γ, corresponding to $\beta \sim 1$, the above relation gives

$$\beta \simeq 1 - \frac{1}{2\gamma^2}, \quad \text{and} \quad \frac{1}{\beta} \simeq 1 + \frac{1}{2\gamma^2}.$$

Using these above relations, we see that

$$
\frac{1}{3\beta}\left(\sqrt{1+15\beta^2}-1\right) = \frac{1}{3}\left(1+\frac{1}{2\gamma^2}\right)\left[\left(1+15-\frac{15}{\gamma^2}\right)^{1/2}-1\right]
$$

$$
= \frac{1}{3}\left(1+\frac{1}{2\gamma^2}\right)\left\{\left[16\left(1-\frac{15}{16\gamma^2}\right)\right]^{1/2}-1\right\}
$$

$$
= \frac{1}{3}\left(1+\frac{1}{2\gamma^2}\right)\left[4\left(1-\frac{15}{32\gamma^2}\right)-1\right],
$$

where we have assumed a large γ value. Doing the above calculation up to order of $1/\gamma^2$ we obtain

$$
\frac{1}{3\beta}\left(\sqrt{1+15\beta^2}-1\right) \simeq 1-\frac{1}{8\gamma^2},
$$

which says that in the ultrarelativistic limit

$$
\Theta_{\max} \simeq \cos^{-1}\left(1-\frac{1}{8\gamma^2}\right). \tag{10.151}
$$

Using the series expansion of the cosine function up to second order in the (small) angular variable, we get

$$
\cos\Theta_{\max} = 1-\frac{1}{8\gamma^2} = 1-\frac{\Theta_{\max}^2}{2},
$$

yielding

$$
\Theta_{\max} \simeq \pm\frac{1}{2\gamma}, \quad \text{(in radians)} \tag{10.152}
$$

for the ultrarelativistic case. This is in general a small angle. For ultrarelativistic motion of the charged particle, the differential power will be peaked along the direction of motion and the direction along which the differential power peaks will make a very small angle with the direction of motion of the charged particle.

The way differential power is emitted by the charged particle is shown in the polar plot in Fig. 10.14. In this plot, the variation $dP/d\Omega$ with respect to Θ has been shown. Each point on the curve has coordinates $(dP/d\Omega, \Theta)$. The angle Θ is measured with respect to the velocity direction. To figure out the differential power in a particular Θ direction one may draw a line connecting the origin and a particular point on the curve, where the line makes angle Θ with the velocity direction. The length of the line corresponds to $dP/d\Omega$ for that particular Θ. In Fig. 10.14, we have rescaled the differential power by the factor $q^2\dot{v}^2/4\pi c^3$ and consequently the polar plot is more symbolic in nature. The actual values of the differential power for various angles

Fig. 10.14 A representative polar plot showing the variation of differential power emitted by the moving charge, when $\beta = 0.8$, with respect to angle Θ. In this plot, we have used a rescaled form of $dP/d\Omega$

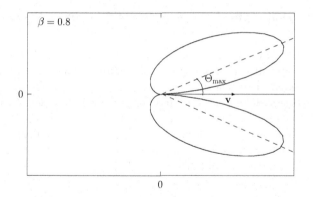

may be more or less than the values one gets by looking at the above figure. The plot clearly shows the two possible values of Θ_{max} for a particular β (or γ).

Knowing that Θ is in general very small, for ultrarelativistic motion, one may directly write the differential power expression in terms of Θ. In this case, we have

$$\frac{dP}{d\Omega} = \frac{q^2 \dot{v}^2}{4\pi c^3} \frac{\sin^2 \Theta}{(1 - \beta \cos \Theta)^5} = \frac{q^2 \dot{v}^2}{4\pi c^3} \frac{\Theta^2}{\left(1 - \beta + \frac{\beta \Theta^2}{2}\right)^5}$$

$$= \frac{q^2 \dot{v}^2}{4\pi c^3} \frac{32\Theta^2}{\left[2(1 - \beta) + \beta\Theta^2\right]^5} . \tag{10.153}$$

Using the fact $1 - \beta \simeq 1/2\gamma^2$ and $\beta \sim 1$ (both of which are consistent for large γ) in the denominator, we ultimately get

$$\frac{dP}{d\Omega} = \frac{8q^2 \dot{v}^2}{\pi c^3} \gamma^8 \frac{(\gamma \Theta)^2}{\left[1 + (\gamma \Theta)^2\right]^5} . \tag{10.154}$$

This expression is easier to handle as it is an algebraic function of Θ.

In a similar way, one can find out the Θ dependence of differential power for the case where velocity and acceleration are perpendicular to each other. In this book, we do not discuss further as the new calculations will not bring in any new conceptual points.

Before we end this chapter we want to present a brief application of the Larmor formula and show how one can use it to get a very useful quantity called the Thomson scattering cross-section.

10.8   The Thomson Scattering Cross-Section

In astrophysics one often finds electromagnetic radiation to be incident on free electrons. Particularly in the physics of the cosmic microwave background radiation this
kind of light matter interaction plays a very important part. Here we discuss the
case where a plane electromagnetic wave is incident on a free charged particle with
charge q and mass m. The electromagnetic wave is composed of an electric field and
a magnetic field, which are oscillating in perpendicular directions. We assume that
the free charge is moving very slowly compared to the velocity of light. In that case
from the form of Lorentz force law

$$\mathbf{F} = q \left(\mathbf{E} + \frac{1}{c} \mathbf{v} \times \mathbf{B} \right) ,$$

we see that only the electric field will impart appreciable amount of force on the
charged particle as $v/c \sim 0$. As a result of this the charged particle will be accelerated
and emit radiation. The emitted radiation will not be flowing in the same direction
as that of the incident radiation. We can interpret this chain of events as a scattering
process between electromagnetic wave and a charged particle. The charged particle
scatters the incident electromagnetic wave into a different direction.

We assume the incident electric field is of the form

$$\mathbf{E}(\mathbf{x}, t) = E_0 \boldsymbol{\epsilon} e^{i(\mathbf{k} \cdot \mathbf{x} - \omega t)} = \boldsymbol{\epsilon} E(\mathbf{x}) e^{-i\omega t} ,$$

where $\boldsymbol{\epsilon}$ specifies the polarization of the incident wave. Here $E(\mathbf{x}) \equiv E_0 e^{i\mathbf{k} \cdot \mathbf{x}}$. The
differential power emitted in solid angle $d\Omega$ is given by the Larmor formula. When
the charge moves very slowly we know

$$\frac{dP}{d\Omega} = \frac{q^2}{4\pi c^3} |\dot{\mathbf{v}}|^2 \sin^2 \Theta ,$$

where Θ is the angle between the observation direction and the acceleration vector.

Figure 10.15 shows the geometry of the scattering phenomenon. The linear polar-
ization basis vectors $\boldsymbol{\epsilon}_1$ and $\boldsymbol{\epsilon}_2$ are fixed along the x and y axes. These vectors are
chosen to be orthonormal. The charged particle is assumed to be at the origin and
the incident wave is traveling along the z axis as shown. The electric field is in the
$x - y$ plane. The charged particle is accelerated along the electric field direction in
the $x - y$ plane. The wave is scattered along direction $\hat{\mathbf{n}}$. The angle between $\hat{\mathbf{n}}$ and
the acceleration vector is Θ. From Lorentz force law, we can write

$$m\dot{\mathbf{v}} = q E(\mathbf{x}) \boldsymbol{\epsilon} e^{-i\omega t} ,$$

which gives

Fig. 10.15 The charge is at
the origin. Incident wave
propagates along z-direction.
Electromagnetic wave is
polarized along the $x - y$
plane. The polarization
vector is shown making an
angle ψ with the x-axis. The
wave is scattered along the
direction $\hat{\mathbf{n}}$

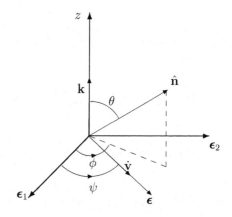

$$\dot{\mathbf{v}} = \epsilon \frac{q E(\mathbf{x})}{m} e^{-i\omega t} \,, \tag{10.155}$$

which shows that $\dot{\mathbf{v}}$ is an oscillating function of time. As the acceleration of the
particle is oscillating in time the differential power is also an oscillating function of
time. Moreover $\dot{\mathbf{v}}$ is also a function of ψ as

$$\epsilon = \cos \psi \epsilon_1 + \sin \psi \epsilon_2 \,. \tag{10.156}$$

To find out the scattering cross-section of electromagnetic waves by the charged
particle we will like to take a time average of the differential power emitted in an
oscillation cycle. This amounts to saying that we look at the average power emitted,
in a certain solid angle, in one cycle of the oscillating electric field. We do not
resolve events in time period less than a cycle of oscillation and consequently this
procedure is meaningful in our calculation. As $\dot{\mathbf{v}}$ is a function of time and Θ is the
angle between $\dot{\mathbf{v}}$ and the fixed direction of observation we actually require the time
average of $|\dot{\mathbf{v}}|^2 \sin^2 \Theta$ appearing in the expression of differential power. We can write

$$\langle |\dot{\mathbf{v}}|^2 \sin^2 \Theta \rangle = \langle |\dot{\mathbf{v}}|^2 (1 - \cos^2 \Theta) \rangle = \langle |\dot{\mathbf{v}}|^2 - (\dot{\mathbf{v}} \cdot \mathbf{n})^2 \rangle \,.$$

In our case $\mathbf{n} = \sin \theta \cos \phi \hat{\mathbf{x}} + \sin \theta \sin \phi \hat{\mathbf{y}} + \cos \theta \hat{\mathbf{z}}$. As a consequence of this

$$\dot{\mathbf{v}} \cdot \mathbf{n} = \sin \theta \cos(\psi - \phi) \frac{q E(\mathbf{x})}{m} e^{-i\omega t} \,.$$

If we have two complex functions $f(\mathbf{x}, t) = \tilde{f}(\mathbf{x}) e^{-i\omega t}$ and $g(\mathbf{x}, t) = \tilde{g}(\mathbf{x}) e^{-i\omega t}$ then
the time average of the product of the functions is given by

$$\langle f(\mathbf{x}, t) g(\mathbf{x}, t) \rangle = \frac{1}{2} \mathrm{Re} \left[\tilde{f}(\mathbf{x}) \tilde{g}^*(\mathbf{x}) \right] \,. \tag{10.157}$$

Here, we are assuming that in general $\tilde{f}(\mathbf{x})$ and $\tilde{g}(\mathbf{x})$ are complex functions as $\tilde{f}(\mathbf{x}) = f_0 e^{-i\phi_f}$ and $\tilde{g}(\mathbf{x}) = g_0 e^{-i\phi_g}$ where the amplitudes f_0, g_0 and the phases ϕ_f, ϕ_g are assumed to be real functions of $\mathbf{x}$. In the present case, the time average of the product of the functions stands for

$$\langle f(\mathbf{x}, t) g(\mathbf{x}, t) \rangle = \frac{1}{T} \int_0^T f_0 g_0 \cos(\omega t + \phi_f) \cos(\omega t + \phi_g)\, dt \,,$$

where $T = 2\pi/\omega$ is the time period of oscillation. This average is taken over the real parts of the oscillating functions $f(\mathbf{x}, t)$ and $g(\mathbf{x}, t)$. In a similar way for two vectors $\mathbf{a}(\mathbf{x}, t) = \tilde{\mathbf{a}}(\mathbf{x}) e^{-i\omega t}$ and $\mathbf{b}(\mathbf{x}, t) = \tilde{\mathbf{b}}(\mathbf{x}) e^{-i\omega t}$, where $\tilde{\mathbf{a}}(\mathbf{x}) = \mathbf{a}_0 e^{-i\phi_a}$ and $\tilde{\mathbf{b}}(\mathbf{x}) = \mathbf{b}_0 e^{-i\phi_b}$ we can define the time average of the inner product of the vectors. In this case, the time average of the scalar product of these vectors turns out to be

$$\langle \mathbf{a}(\mathbf{x}, t) \cdot \mathbf{b}(\mathbf{x}, t) \rangle = \frac{1}{2} \mathrm{Re} \left[\tilde{\mathbf{a}}(\mathbf{x}) \cdot \tilde{\mathbf{b}}^*(\mathbf{x}) \right]. \tag{10.158}$$

Using these results we get

$$\langle |\dot{\mathbf{v}}|^2 \sin^2 \Theta \rangle = \frac{q^2 E_0^2}{2m^2} \left[1 - \sin^2 \theta \cos^2(\psi - \phi) \right]. \tag{10.159}$$

The average differential power can now be written as

$$\left\langle \frac{dP}{d\Omega} \right\rangle = \frac{q^4 E_0^2}{8\pi m^2 c^3} \left[1 - \sin^2 \theta \cos^2(\psi - \phi) \right]. \tag{10.160}$$

Now we have all the ingredients to calculate the differential scattering cross-section for Thomson scattering.

The differential scattering cross-section is defined to be the ratio between the average differential power radiated in a certain direction and the amount of energy incident on the particle per unit time per unit area. One can write

$$\frac{d\sigma}{d\Omega} = \frac{\text{Average energy radiated in solid angle } d\Omega \text{ per unit time}}{\text{Average incident energy per unit area per unit time}},$$

where the average incident energy per unit area per unit time is the time average of the Poynting vector for the plane electromagnetic wave. From Chap. 4, Eq. (4.50), we know that the time averaged Poynting vector for the plane wave in free space is

$$\langle \mathbf{S} \rangle = \frac{c}{8\pi} E_0^2 \,.$$

Using this result, we can write the Thomson scattering differential cross-section as:

$$\frac{d\sigma}{d\Omega} = \left(\frac{q^2}{mc^2}\right)^2 \left[1 - \sin^2\theta \cos^2(\psi - \phi)\right].$$ (10.161)

This result gives us the Thomson scattering cross-section when the initial plane wave is linearly polarized in a particular direction. For unpolarized initial wave one has to average out the above result on the polarization direction ψ. For unpolarized incident wave, the differential scattering cross-section is

$$\frac{d\sigma}{d\Omega} = \left(\frac{q^2}{mc^2}\right)^2 \left\{1 - \sin^2\theta \left[\frac{1}{2\pi}\int_{\psi=0}^{2\pi}\cos^2(\psi - \phi)\,d\psi\right]\right\}.$$ (10.162)

The integral can easily be done and one gets the final result as

$$\frac{d\sigma}{d\Omega} = \frac{1}{2}\left(\frac{q^2}{mc^2}\right)^2 (1 + \cos^2\theta).$$ (10.163)

To find the cross-section, we integrate the above result with respect to solid angle and get

$$\sigma_{\text{T}} = \frac{1}{2}\left(\frac{q^2}{mc^2}\right)^2 \int_{\phi=0}^{2\pi}\int_{\theta=0}^{\pi}(1 + \cos^2\theta)\sin\theta\,d\theta\,d\phi = \frac{8\pi}{3}\left(\frac{q^2}{mc^2}\right)^2,$$

(10.164)

giving us the total Thomson scattering cross-section. In our previous discussion, we found that for the electron $q^2/mc^2 = 2.8 \times 10^{-13}$cm. Using this number we get $\sigma_{\text{T}} \sim 0.66 \times 10^{-24}$ cm^2.

In our analysis of electromagnetic wave scattering from a charged particle, we have used classical concepts. Here, we have assumed that the electron accepts energy from the incident radiation and then it re-radiates it in a different direction. If the frequency of the incident radiation is high then the incident radiation cannot be treated as a pure classical wave, one has to apply quantum mechanical laws in that regime as high-frequency photons not only imparts energy to the free charge but also imparts 3-momentum to the charge. Consequently we see that when the initial momentum of the photon, $\hbar\omega/c$, is comparable with mc, the natural momentum scale attached with the charged particle, we expect the charged particle to show quantum mechanical recoil and in that regime the above theory will not be able to predict the scattering cross-section. The quantum mechanical treatment of the above scattering process is formally given by the theory of Compton scattering.

10.9 Problems

1. Derive the form of $du^\alpha/d\tau$ as given in Eq. (10.67).
2. From the relation given in Eq. (10.64) figure out what will be $\partial_\alpha A^\beta(x)$. Once you know $\partial_\alpha A^\beta(x)$, then calculate $\partial_\alpha A^\alpha(x)$. Can you justify the result?
3. Derive the expression of $\tilde{\mathbf{B}}$ in Eq. (10.89) from the expression for F_P^{ij} in Eq. (10.84). Then show that Eq. (10.90) naturally follows from Eq. (10.89).
4. Show that the expression of $\mathbf{E}(x)$ in Eq. (10.97) can reproduce the forms of E_x and E_z in K frame in Sect. 7.4.1.
5. Derive the time average given in Eq. (10.157).
6. Derive the form of the gauge potentials using the advanced Green function.
7. First convince yourself that Eq. (10.1) remains the same if you interchange $\mathbf{x}$ with $-\mathbf{x}$ and t with $-t$. Then look at the expression of the Liénard-Wiechert potentials and try to verify whether they are invariant under such transformations. You will see that the gauge fields are symmetric under the above spatial transformation but there is a problem with time replacements. You see an interesting phenomenon, although the equation you started admitted the operations of interchanging $\mathbf{x}$ with $-\mathbf{x}$ and t with $-t$ the solution to that equation has lost the time reversal property. What can be the reason for this lost symmetry in the solution? Once you get the answer to the last question, can you now prescribe a method in which the gauge solutions will also have the symmetries of the mother equation?
8. From the Liénard-Wiechert potentials we have calculated the electric and magnetic fields. We know which part of those fields specifies radiation. The part which specifies radiation does not apparently come as an oscillating field with some specified frequency ω. In our previous study of electromagnetic waves we expect the radiation to be like spherical waves or plane waves, with some wave number and frequency. Still we claim that those potentials here specify radiation. Is there anything wrong or are we missing something? Try to think about this problem and convince yourself that nothing is wrong here.
9. Suppose we have a classical Hydrogen atom where the classical electron rotates in a circular orbit around the proton, the interaction is of the Coulomb type. Take the radius of the orbit to be a_0. As the electron is in a rotational motion, we know it is accelerating and hence radiating. Assuming that the electron is always in a nearly circular orbit and the rate of radiation is sufficiently well approximated by the classical, non-relativistic Larmor formula, find out the time when the electron will spiral into the origin. Here, you know the time rate of energy change of the electron from the Larmor formula. You can also find out the total energy of the electron in a central force field and take a time derivative of it. Equating those two expressions of rate of change of energy will give you a differential equation in terms of the radial coordinate. Integrate the last equation to find out the result. Assume that the primary acceleration of the electron is the centripetal acceleration.
10. You have calculated the power radiated by a charged particle in linear acceleration and in circular motion. It is known that the power radiated by the charged particle

in circular motion is more compared to the power emitted by the same particle in linear motion, particularly this difference is appreciable in extreme relativistic motion. Using this information can you figure out the power emitted by an arbitrarily moving accelerated charged particle in the extreme relativistic limit? You may actually ignore the parallel component of the acceleration (parallel to velocity) and assume that the instantaneous centripetal acceleration is completely given by the acceleration which is perpendicular to the velocity direction (for the calculation of power). Now you can yourself figure out how the situation can be handled.

Chapter 11
Radiation Reaction in Brief

In this chapter, we will discuss briefly about radiation reaction. Radiation reaction is a very important topic and many prominent physicists have thought about radiation reaction in some detail. The basic idea about radiation reaction was implicitly present in the discussion of radiation by an arbitrarily moving point charge. We will see how the idea of radiation reaction takes shape in an independent way in this Chapter. In our brief discussion on radiation reaction, in this chapter, we will not use the covariant language of electrodynamics. We will try to explain the main points using a Newtonian approach. Once the reader understands the basic physics of radiation reaction he/she can open any other book on covariant formulation of electrodynamics to understand the covariant approach. We will start with the topic of Abraham-Lorentz equation of motion of a charged particle.

11.1 Abraham-Lorentz Equation of Motion

While we were discussing radiation by an arbitrarily moving charged particle in the last chapter, we saw that for radiation we require the charged particle to accelerate. A particle can accelerate only when some external force $\mathbf{F}_{ext}$ acts on it for some finite time. In reality as the external force accelerates the charged particle, the particle starts to radiate electromagnetic radiation. As electromagnetic radiation carries away energy from the charged particle, it is natural to think that the charged particle loses energy while accelerating. As the charged particle loses energy due to radiation, the velocity of the charged particle will decrease. The particle velocity cannot increase while radiation happens because in that case we will have a runaway situation where more radiation will produce more acceleration and that will produce more radiation and so on. From this discussion, we can infer that an accelerated charge which is radiating loses energy and this energy loss produces a deceleration.

© Springer Nature Singapore Pte Ltd. 2021
K. Bhattacharya and S. Mukhopadhyay, *Introduction to Advanced Electrodynamics*,
https://doi.org/10.1007/978-981-16-7802-8_11

The decelerating effect on the charge can be assumed to be produced by some force acting on the charge $\mathbf{F}_{rad}$. This force $\mathbf{F}_{rad}$ is called the force due to radiation reaction. In the Newtonian paradigm, acceleration and deceleration can only be produced by external forces on the particle. In the presence of electromagnetic radiation by an accelerating charge, we see that we have two kinds of forces acting on the charged particle, one is $\mathbf{F}_{ext}$ and the other is $\mathbf{F}_{rad}$. Before we include the effect of radiation reaction we briefly specify the equation of motion of the charged particle ignoring radiation. If we ignore radiation we are also ignoring radiation reaction force. In this case, the equation of motion for the charged particle is simply

$$m\dot{\mathbf{v}} = \mathbf{F}_{ext} \, , \qquad\qquad (11.1)$$

where m is the mass of the charged particle and $\mathbf{v}$ is the velocity of the particle. This equation of motion of the charged particle does not take care of any radiation. Next we want to see how this equation of motion for the charged particle is modified when we introduce electromagnetic radiation from the accelerated charge.

In the non-relativistic limit ($\beta \to 0$), we know that as the particle accelerates it radiates energy at the rate given by Larmor formula:

$$P = \frac{2}{3}\frac{q^2}{c^3} \, |\dot{\mathbf{v}}|^2 \, .$$

This energy emission must modify the mechanics of the moving particle and as we can take care of radiation by a radiation reaction force, the modified equation of motion for the charged particle becomes

$$m\dot{\mathbf{v}} = \mathbf{F}_{ext} + \mathbf{F}_{rad} \, , \qquad\qquad (11.2)$$

where we do not know the form of $\mathbf{F}_{rad}$ but we have to find out its form as we go along. The form of $\mathbf{F}_{ext}$ is supposed to be known. To deduce the form of $\mathbf{F}_{rad}$ we assume that

1. $\mathbf{F}_{rad}$ vanishes when $\dot{\mathbf{v}} = 0$ for an extended time period. This is a natural assumption as when the particle does not accelerate there is no radiation and in the absence of radiation there cannot be any radiation reaction.
2. The radiation reaction force $\mathbf{F}_{rad}$ is proportional to the square of the charge, q^2. This is also a straightforward assumption as the Larmor formula shows power is proportional to q^2 so the radiation energy per unit time does not depend on the sign of the charge.

We will use these two assumptions to deduce the form of $\mathbf{F}_{rad}$.

Suppose the accelerating charged particle is observed during a time interval stretching from t_1 to t_2. During this time the force $\mathbf{F}_{rad}$ does work due to which the particle loses energy in the form of radiation. The work done by $\mathbf{F}_{rad}$ during this time is given by

$$\int_{t_1}^{t_2} \mathbf{F}_{\text{rad}} \cdot \mathbf{v}\, dt = -\int_{t_1}^{t_2} P(t)\, dt = -\frac{2}{3}\frac{q^2}{c^3}\int_{t_1}^{t_2} \dot{\mathbf{v}} \cdot \dot{\mathbf{v}}\, dt\,, \tag{11.3}$$

where $P(t)$ is the Larmor power expression. One can easily do the above integral and obtain

$$\int_{t_1}^{t_2} \mathbf{F}_{\text{rad}} \cdot \mathbf{v}\, dt = -\frac{2}{3}\frac{q^2}{c^3}\left[(\mathbf{v} \cdot \dot{\mathbf{v}})\Big|_{t_1}^{t_2} - \int_{t_1}^{t_2} \ddot{\mathbf{v}} \cdot \mathbf{v}\, dt\right]. \tag{11.4}$$

This result cannot be evaluated unless we know more information about the time dependence of the velocity vector of the charged particle. We can proceed if we assume some simple features about the nature of motion. One of the simplest cases where we can actually omit the first term on the right-hand side of the above integral is related to periodic motion of the particle. If the time period is $T = t_2 - t_1$ then we know that the velocity and acceleration of the particle at t_2 remains the same as the corresponding terms at time t_1. In this case

$$(\mathbf{v} \cdot \dot{\mathbf{v}})\Big|_{t_1}^{t_2} = 0\,.$$

On the other hand if at t_1 or t_2 the situation is such that the velocity of the particle becomes perpendicular to the acceleration vector, even then the above condition remains valid and one can simply write

$$\int_{t_1}^{t_2} \mathbf{F}_{\text{rad}} \cdot \mathbf{v}\, dt = \int_{t_1}^{t_2}\left(\frac{2}{3}\frac{q^2}{c^3}\ddot{\mathbf{v}} \cdot \mathbf{v}\right) dt\,. \tag{11.5}$$

The above equation is valid for any periodic, non-relativistic motion of the charged particle. We can write the above equation as

$$\int_{t_1}^{t_2}\left(\mathbf{F}_{\text{rad}} - \frac{2}{3}\frac{q^2}{c^3}\ddot{\mathbf{v}}\right) \cdot \mathbf{v}\, dt = 0\,. \tag{11.6}$$

For the integral to vanish for all pertinent cases in general the integrand has to vanish. As in the general case the velocity of the particle in motion is non-zero we must have

$$\mathbf{F}_{\text{rad}} = \frac{2}{3}\frac{q^2}{c^3}\ddot{\mathbf{v}}\,. \quad (|\mathbf{v}| \ll c) \tag{11.7}$$

It can be seen that the radiation reaction force vanishes if $\dot{\mathbf{v}}$ vanishes for an extended time period as in that case $\ddot{\mathbf{v}}$ also vanishes. The force is also proportional to q^2. Perhaps this is the first time you are seeing that a force depending on rate of change of acceleration. The above relation holds in the non-relativistic limit. In classical mechanics, we are accustomed to see forces depending on acceleration, here the force of radiation reaction depends on the third time derivative of the position vector. We will see later that this fact makes radiation reaction phenomenon a bit esoteric.

From the previous chapter, we already know that q^2/mc^2 has the dimension of length. From this information we can now define a quantity τ with the dimension of time as

$$\tau \equiv \frac{2}{3} \frac{q^2}{mc^3}.$$ (11.8)

To get some understanding about the magnitude of τ we can evaluate τ for the electron. From the previous chapter, we know for the electron $\frac{q^2}{mc^2} \sim 2.8 \times 10^{-13}$ cm. From this estimation we can easily see that

$$\tau \simeq 10^{-24}\,\text{s},$$ (11.9)

for the electron. This number shows that in general the quantity τ is very small. This quantity τ plays a very important role in the dynamics of the accelerating particle which radiates. Using τ we can now write the equation of motion for the radiating charged particle

$$m\dot{\mathbf{v}} = \mathbf{F}_{\text{ext}} + \frac{2}{3}\frac{q^2}{c^3}\ddot{\mathbf{v}},$$

as

$$m(\dot{\mathbf{v}} - \tau\ddot{\mathbf{v}}) = \mathbf{F}_{\text{ext}}.$$ (11.10)

This equation is traditionally called the Abraham-Lorentz equation of motion. The reader must note that the above equation is a higher derivative (involving three time derivatives of position vector) equation of motion. This higher derivative term produces some of the peculiar features related with radiation.

To see the peculiar nature of the Abraham-Lorentz equation we see that this equation predicts radiation even when there is no external force $\mathbf{F}_{\text{ext}}$. Suppose we set $\mathbf{F}_{\text{ext}} = 0$, then the Abraham-Lorentz equation gives

$$\dot{\mathbf{v}} = \tau\ddot{\mathbf{v}}.$$ (11.11)

If this equation had only one solution, $\dot{\mathbf{v}} = 0$ and $\ddot{\mathbf{v}} = 0$ then everything should have been fine. The above solution says that the charged particle is at rest or moves with uniform velocity such that there is no electromagnetic radiation. In absence of any external force, this is a fine solution. The difficulty is that this equation does have another nontrivial solution as

$$\dot{\mathbf{v}} = \mathbf{a}_0\, e^{t/\tau},$$ (11.12)

where $\mathbf{a}_0$ is the acceleration of the particle at $t = 0$. This solution grows indefinitely with time. This kind of a solution is generally called a runaway solution which wildly increases without bound. Moreover, this solution is problematic in another sense, this solution does not satisfy the property $\mathbf{v} \cdot \dot{\mathbf{v}} = 0$ at any time.

Some of the readers may think that because τ is too small a number, we may interpret $-m\tau\ddot{\mathbf{v}}$ term as a perturbation term on the equation

$$m\dot{\mathbf{v}} = \mathbf{F}_{\text{ext}} \, ,$$

to get realistic solutions of Eq. (11.10). If this scheme succeeds then it will reduce most of the difficulties in our case as in general we know how to solve for particle motions in an external well-behaved force field. In the perturbative approach, the effect of radiation will only produce minor deviations from the path of the unperturbed solution and all the problems of runaway solution will be tamed as the perturbed solutions will only slightly differ from the motion of a particle in an external force. This scheme unfortunately cannot be applied in our case as the perturbation term $-m\tau\ddot{\mathbf{v}}$ contains a higher order derivative of the velocity of the particle. To solve for the perturbations, we then require an extra initial condition for the problem. Standard perturbation theories do not allow us to include extra initial conditions, the unperturbed and perturbed solutions are assumed to satisfy the same initial conditions. As a consequence we see the higher derivative term, how small it may be, cannot be interpreted as a perturbative term. The Abraham-Lorentz equation has to be solved as an independent and complete equation with its pathologies.

To tackle the problem of the runaway solution of the Abraham-Lorentz equation, one can transform the above equation into an integro-differential equation. In the next section, we will see how the integro-differential equation tries to tackle the problem of the runaway solution.

11.2 The Integro-Differential Equation from Abraham-Lorentz Equation

To proceed with the integro-differential approach, we point out another difficulty of the runaway solution of the Abraham-Lorentz equation. This difficulty is related to the charge-less limit of the solution. If the charge of the particle vanishes we expect the runaway solution to convert to the standard solution of a particle moving in the absence of any external force. The solution should be $\dot{\mathbf{v}} = 0$. We see from the expression of τ, as $q = 0$, τ vanishes and from the runaway solution we see in that limit the acceleration does not vanish but it diverges. This is another problematic point of the Abraham-Lorentz equation.

To tackle this problem, we try to reformulate the radiation reaction formalism in a slightly different way. We suppose that the external force only depends on time so that we can write the basic equation of motion as

$$m(\dot{\mathbf{v}} - \tau\ddot{\mathbf{v}}) = \mathbf{F}_{\text{ext}}(t) \, . \tag{11.13}$$

From our previous understanding of the solution of the above equation in the absence of any external force we can now propose a solution as

$$\dot{\mathbf{v}} = e^{t/\tau}\,\mathbf{u}(t)\,, \tag{11.14}$$

where $\mathbf{u}(t)$ is some 3-vector which only depends on time. In our discussion, this vector $\mathbf{u}(t)$ will be used as an auxiliary variable, we will only use it in transit and it will never appear in the final equation for acceleration of the particle. Differentiating the above equation once with respect to time we get

$$\ddot{\mathbf{v}} = \frac{1}{\tau}\dot{\mathbf{v}} + e^{t/\tau}\,\dot{\mathbf{u}}\,,$$

which can be rearranged as

$$\dot{\mathbf{v}} - \tau\ddot{\mathbf{v}} = -\tau e^{t/\tau}\,\dot{\mathbf{u}}\,.$$

Using Eq. (11.13), the above equation transforms to

$$m\dot{\mathbf{u}} = -\frac{1}{\tau}e^{-t/\tau}\mathbf{F}_{\text{ext}}(t)\,. \tag{11.15}$$

This is a differential equation of the auxiliary vector $\mathbf{u}$. Finally to get rid of the auxiliary variable we integrate both sides of the above equation from some time t_0 to t. We assume the acceleration of the charged particle to be momentarily zero at t_0 or $\mathbf{u}(t_0) = 0$. Integrating the above equation, we get

$$m\int_{t_0}^{t}\dot{\mathbf{u}}\,dt = -\frac{1}{\tau}\int_{t_0}^{t}e^{-t'/\tau}\,\mathbf{F}_{\text{ext}}(t')\,dt'\,. \tag{11.16}$$

This equation yields

$$m\mathbf{u}(t) = me^{-t/\tau}\,\dot{\mathbf{v}}(t) = -\frac{1}{\tau}\int_{t_0}^{t}e^{-t'/\tau}\,\mathbf{F}_{\text{ext}}(t')\,dt'\,,$$

where we have used the relation in Eq. (11.14). From this equation, we can now write the desired integro-differential equation as

$$m\,\dot{\mathbf{v}}(t) = -\frac{e^{t/\tau}}{\tau}\int_{t_0}^{t}e^{-t'/\tau}\,\mathbf{F}_{\text{ext}}(t')\,dt'\,. \tag{11.17}$$

This is the integro-differential equation which corresponds to the Abraham-Lorentz equation. We will see that this equation does address the problem of the runaways solution in a positive way. Although it can address some difficulties of radiation reaction mechanism, the above form of the equation is not intuitively very appealing. We see from the above equation that the acceleration of the particle at any time t

is simply not proportional to the external force acting on it at time t. Actually the acceleration of the particle at time t depends on the future $(t_0 > t)$ or history $(t_0 < t)$ of the particle as an integration on the applied force is involved. To be specific, the above equation violates causality. Later we will show that the problem of causality can be addressed by taking quantum effects into consideration. These ideas are certainly strange and do not match our idea of a classical Newtonian system. It shows that radiation by itself is an interesting topic and to understand it we require various counter-intuitive concepts.

Next we see how the integro-differential equation takes care of the runaway solution. To do so we write the last equation as

$$m\,\dot{\mathbf{v}}(t) = \frac{1}{\tau} \int_t^{t_0} e^{-(t'-t)/\tau}\, \mathbf{F}_{\text{ext}}(t')\, dt' . \tag{11.18}$$

Without any loss of generality we can assume that $t_0 > t$. In general τ is a very small quantity, we have seen that for the electron $\tau \sim 10^{-24}$s. For other heavier charged particles τ will be lesser than this value. Due to the smallness of τ we see that the factor $e^{-(t'-t)/\tau}$ inside the integrand is appreciable only when $t' - t < \tau$. On the other hand when $t' - t > \tau$ the exponential factor in the integrand does not contribute appreciably. Utilizing this fact we define a parameter

$$s \equiv \frac{t' - t}{\tau} . \tag{11.19}$$

At $t' = t$ we have $s = 0$ and when $t' = t_0 \gg \tau$ we have a finite s which is positive as $t_0 > t$. Using the variable s we can now write the integro-differential equation as

$$m\,\dot{\mathbf{v}}(t) = \int_0^{\infty} e^{-s}\, \mathbf{F}_{\text{ext}}(t + \tau s)\, ds , \tag{11.20}$$

where we have pushed the upper limit of the integration to infinity as for values of $s \gg 1$ the integrand does not contribute. Assuming $\mathbf{F}_{\text{ext}}(t + \tau s)$ an analytical function we can Taylor expand it around t as

$$\mathbf{F}_{\text{ext}}(t + \tau s) = \sum_{n=0}^{\infty} \frac{(\tau s)^n}{n!} \frac{d^n \mathbf{F}_{\text{ext}}(t)}{dt^n} . \tag{11.21}$$

We can now use this expansion of the external force inside the last integral and get

$$m\,\dot{\mathbf{v}}(t) = \sum_{n=0}^{\infty} \frac{\tau^n}{n!} \left[\int_0^{\infty} e^{-s}\, s^n\, ds \right] \frac{d^n \mathbf{F}_{\text{ext}}(t)}{dt^n} . \tag{11.22}$$

From the definition of the Gamma function, we know

$$\int_0^\infty e^{-x} x^{n-1}\, dx = \Gamma(n) = (n-1)!\,,\qquad(11.23)$$

for a real variable x, we have

$$m\,\dot{\mathbf{v}}(t) = \sum_{n=0}^\infty \tau^n \frac{d^n \mathbf{F}_{\text{ext}}(t)}{dt^n}\,.\qquad(11.24)$$

We will now see that this equation can address many of the old problems of the Abraham-Lorentz equation.

From the above equation, we see that

1. the right-hand side is a series in τ^n where $\tau \propto q^2$. From this fact it can easily be seen that as $q=0$ the right-hand side vanishes and we get $m\,\dot{\mathbf{v}}(t) = 0$ for a free particle as expected. This implies uncharged particles will never be accelerated under the external force and there will be no radiation.
2. If $q \neq 0$ but the external force vanishes, i.e., $\mathbf{F}_{\text{ext}}(t) = 0$ then we see that $m\,\dot{\mathbf{v}}(t) = 0$ as expected. If there is no external force to accelerate the charged particle, the particle moves like a free particle and does not radiate. The issue about the runaway solution does not arise in this formulation.

We see that Eq. (11.24) does address some of the most difficult issues related with the Abraham-Lorentz equation. Although this form of the equation does address some of the problems it also opens up a different problem related to causality, a topic which we briefly discussed before.

The problem about the acausal nature of radiation reaction shows up in this formulation as

$$m\,\dot{\mathbf{v}}(t) = \int_0^\infty e^{-s}\,\mathbf{F}_{\text{ext}}(t + \tau s)\, ds\,,$$

which shows that the acceleration at time t depends on the force at some future time $t + \tau s$ where s can be a large number. This fact makes the law of motion acausal. We will see here that the difficulty with acausal behavior can be addressed by invoking the quantum nature of interactions. Quantum corrections may help us to avoid the acausal nature of the problem. To be more specific we expand the right-hand side of Eq. (11.24) in a series as

$$m\,\dot{\mathbf{v}}(t) = \mathbf{F}_{\text{ext}}(t) + \tau \frac{d\mathbf{F}_{\text{ext}}(t)}{dt} + \tau^2 \frac{d^2 \mathbf{F}_{\text{ext}}(t)}{dt^2} + \cdots,\qquad(11.25)$$

where $\tau \sim 10^{-24}$ s for electron. For other heavier charged particles, carrying electronic unit of charge, the corresponding τ will be smaller. Acausality arises from terms as $\tau\,(d\mathbf{F}_{\text{ext}}/dt)$, $\tau^2\,(d^2\mathbf{F}_{\text{ext}}/dt^2)$ and other terms in the series. All these terms are sensitive to the change in $\mathbf{F}_{\text{ext}}(t)$ or its derivatives in a time period of τ. We see that

$$\tau \frac{d\mathbf{F}_{\text{ext}}(t)}{dt} \sim \mathbf{F}_{\text{ext}}(t + \tau) - \mathbf{F}_{\text{ext}}(t)\,,$$

where we have assumed the derivative is approximately constant during τ. In a similar way, we can represent the higher order terms in the Taylor expansion as differences in the derivatives of the external force during τ. As τ is too small a time period, how much can a force originating from macroscopic agents change during that time? To feel the smallness of τ we must first give you an estimate of the classical electron radius. In reality, the reader is now familiar with the classical electron radius

$$r_e = \frac{q^2}{mc^2} \sim 2.8 \times 10^{-13}\,\text{cm}\,. \tag{11.26}$$

This radius is obtained by calculating the electrostatic self-energy of the electron (assuming it to be a sphere with radius r_e and uniform charge density) and equating it with the relativistic rest energy mc^2. The reader must note that r_e in reality does not say that the electron is actually a sphere with a certain radius, it only gives a benchmark value of length associated with the electron. We know from quantum mechanics that the electron does not have any specific shape, it is described by a probability density function. From the classical electron radius, we see that light travels this distance in r_e/c seconds. It can easily be seen that $\tau \sim r_e/c$, which implies that τ is so small that it is the time in which light traverses the classical electron radius.

In reality, for classical forces, it is extremely difficult or impossible to vary in such small time scale or the corresponding length scale. As a result of this, the quantities responsible for the acausal behavior in Eq. (11.25) do not contribute appreciably and causality is maintained. In quantum mechanics we have some uncertainty relations. A commonly used uncertainty relation involves the uncertainty in energy and uncertainty in time. If in a quantum system the uncertainty in energy measurement is ΔE and the uncertainty in time measurement is Δt then we know

$$\Delta E\, \Delta t \geq \hbar\,, \tag{11.27}$$

where $\hbar = 6.626 \times 10^{-29}\,\text{erg s}$ is the reduced Planck constant. If the uncertainty in energy measurement, for a fixed time uncertainty, is of the order of the rest energy of any particle then there is a possibility of particle creation in that time window Δt. Suppose we are observing an electron in its trajectory for a very small amount of time Δt then there is a chance we will discover that the electron is clouded by virtual partners, the partners are electron-positron pairs. The positron is the antiparticle of an electron, it has same mass and spin as the electron but opposite electric charge. The particle cloud around the real electron is virtual in the sense that the energy of the particles constituting it are uncertain as these particles appear as bubbles of energy uncertainty. These virtual particles do not have any exact energy, they have uncertain energy. To find the time scale in which we expect to see the cloud of virtual particles around the electron we set $\Delta E \sim mc^2$, where $m \sim 9 \times 10^{-28}\,\text{gm}$ is the electronic mass. The time uncertainty corresponding to these ΔE is

$$\Delta t \sim \frac{\hbar}{mc^2} = O(100)\tau\,, \tag{11.28}$$

where $O(100)$ is a number of the order of 10^2 and $\tau \sim 10^{-24}\,\mathrm{s}$. From the above equation, we see that if Δt is smaller than $10^2\tau$ then the energy uncertainty will be such that many virtual electron-positron pairs will arise around the real electron and instead of an electron there will be a charged cloud. From the above description, it becomes clear that any description of the theory in time scales less than $10^2\tau$ will require quantum mechanical corrections. This shows that our classical description of the electron ceases when we reach timescales less than $10^2\tau$ and consequently we will never be able to predict classically about forces and their variation in a time scale around τ. As the acausal behavior is expected to appear in time scales of the order of τ it can be predicted that quantum mechanical corrections will set in around those time scales and make the theory well behaved. In this book, we will not discuss the actual nature of the quantum corrections as it is beyond the scope of this book.

11.3   Problems

1. Suppose a charged particle is moving in central force field where the potential is $V(r)$. You may assume that it is moving in a closed, periodic orbit. In absence of radiation reaction, the energy and angular momentum of the particle remains constant. If we introduce radiation reaction force then the energy and angular momentum will not be constants. Find out the rate at which the energy and angular momentum of the particle decays in presence of radiation. As both energy and angular momentum cyclically varies in one cycle (assuming the radiative loss to be small) of the orbit you may actually find out the time average of the decay rates in one cycle of the orbit.

2. Let us take the model of the Lorentz atom, where a charged particle is bound by a one-dimensional linear restoring force with force constant $k = m\omega_0^2$. In absence of radiation damping, the charged particle oscillates with constant amplitude at the characteristic frequency ω_0. Introduce the effect of radiation due to the acceleration of the charge and write down the Abraham-Lorentz equation of motion in the integro-differential form for such a system. Solve it with a trial solution as $x(t) = x_0 e^{-\alpha t}$. Find out the form of α. In your solution, you should be careful to avoid the runaway solution.

3. Let us try to solve a higher derivative differential equation to see the general nature of these equations. The equation is (in one dimension)

$$\dddot{x} = \lambda \ddot{x} + \beta \delta(t) \,,$$

where λ and β are some constants with proper dimensions. Verify that the last equation has a solution

$$x(t) = A e^{t/\lambda} + Bt + C \,,$$

in regions where $t \neq 0$. Here A, B, C are constants.

In such situations, one matches the solution in region $t < 0$ with the solution at $t > 0$ keeping in mind that the Dirac delta function is there, which will contribute maximally at $t = 0$. To make life simple you can assume that the particle is at rest prior to $t = 0$. We know $x(t)$ is continuous at $t = 0$. What can you derive from this information?

Convince yourself that $\text{Disc} \left[x(t) \big|_{t=0} \right] = 0$. To show that $\dot{x}(t)$ is really continuous at $t = 0$ you can assume the opposite statement and come to a contradiction. If $\dot{x}(t)$ was discontinuous at $t = 0$ then in general one expects $\ddot{x}(t) \propto \delta(t)$. From this fact you can calculate the third derivative of $x(t)$. Then show that the original differential equation is not obtained in this assumption.

To obtain more information on the matching condition you can now integrate the original differential equation, with respect to time, from $-\epsilon$ to ϵ where ϵ is a very small time interval. Once you do all these you will have all the conditions to match the solutions in the two regions.

In the solutions, you will still have the exponential part. This part looks like the runaway solution. To understand why we call it the runaway solution you can see that in the original differential equation the source term is the Delta function term, which does not act at $t > 0$, but still there is fair amount of acceleration (exponentially increasing $\ddot{x}$). Can you eliminate this runaway solution?

To eliminate the runaway solution you may relax the condition that the particle was at rest in the region $t < 0$. Impose the initial conditions at $t \ll 0$ the particle was at $x = 0$. Also assume that the coefficient for $e^{t/\lambda}$ in the region $t > 0$ is zero, eliminating the runaway solution. You can now find out the full solution.

4. From the form of the Liénard-Wiechert potentials, as given in Eq. (10.52), can you say what are the gauge fields on the charge (when $R = 0$)? You used retarded Green function to derive the expression of the fields. Suppose a charge is accelerating and loosing energy as radiation, can you think of the time-reversed process here? Does the field expressions, you have calculated before, admit this time reversal symmetry? If the answer is no, then what can you do to make the above process time reversal symmetric?

Chapter 12
Cherenkov Radiation

In this chapter, we will discuss the interesting topic of Cherenkov radiation. Cherenkov radiation is emitted by a charged particle, like electron, which moves with a uniform velocity greater than the velocity of light inside a dielectric medium. Motion with a uniform velocity may be an approximation but the theory can be constructed in a relatively simple manner using this assumption. Superluminal velocity inside a dielectric medium is not a problem as the velocity of light decreases from its value in vacuum inside the medium. In reality, the complete description of the process is too complicated, and we will try to explain as much as possible using our classical description of electrodynamics inside a dielectric medium. The dielectric medium is supposed to be dispersive and non-magnetic. In a non-magnetic medium the magnetization is assumed to be zero. As the charge moves relatively fast, we will use the conventions of a relativistic system although the whole description of Cherenkov radiation is not written in a Lorentz covariant formalism. The reason we will not use a fully Lorentz covariant language is related to the fact that we will exclusively work in the rest frame of the dielectric medium. The results we obtain will be true only in the rest frame of the dielectric. In a fully covariant picture, the medium can move with a uniform 3-velocity. It is quite formidable a task to produce a fully covariant theory of Cherenkov radiation and is certainly beyond the scope of this book. The lack of Lorentz covariance does not make the present analysis less appealing because Cherenkov radiation is generally observed in media which are static in the laboratory.

The topic of energy loss by a moving charged particle inside a material medium is a widely studied subject. Bohr initiated such studies in the first half of the twentieth century. These energy loss mechanism calculations were based on the interaction of the moving charged particle with the electrons attached to atoms of the medium. In plasmas, the theory is a bit different as in plasmas the electrons inside the medium have more freedom to move about. The charged particle while passing through the medium deposit energy on the electrons of the medium and the total energy loss per unit length (of the moving particle) can be calculated by an incoherent addition of

© Springer Nature Singapore Pte Ltd. 2021
K. Bhattacharya and S. Mukhopadhyay, *Introduction to Advanced Electrodynamics*,
https://doi.org/10.1007/978-981-16-7802-8_12

the individual energy loss due to each binary encounter. These calculations did not predict the existence of any kind of radiation inside the medium.

In this chapter, we will not give a historic overview of the energy loss mechanism of moving charged particles inside dielectric medium. We will directly discuss Cherenkov radiation. As we go along, it will be seen that the origin of Cherenkov radiation is completely different from the origin of radiation from accelerated charges in vacuum. Cherenkov radiation is a complex phenomenon which originates from a moving charge inside a medium; in this case, the process of radiation involves the dielectric medium as a whole. This radiation arises due to complex interaction between the fast-moving charged particle and the polarizable medium constituents. In our approach we will not directly work with atoms and molecules which have electrons attached with them, we will treat the medium as a dielectric whose electromagnetic properties are guided by the dielectric constant. The charge distribution and other properties of the medium constituents will be taken care of by the form of the dielectric constant which specifies a macroscopic medium. Cherenkov radiation happens when the moving charged particle interacts coherently with the whole medium and consequently this phenomenon is complex in nature. To formulate the basic theory of Cherenkov radiation, we will first discuss the electromagnetic fields produced by a fast-moving charged particle inside a dielectric medium.

12.1 Electric and Magnetic Field Due to a Uniformly Moving Charged Particle Inside Dielectric Medium

Let us once again write down Maxwell's equation in a material medium

$$\nabla \cdot \mathbf{D} = 4\pi\rho, \quad \nabla \times \mathbf{H} - \frac{1}{c}\frac{\partial \mathbf{D}}{\partial t} = \frac{4\pi}{c}\mathbf{J}, \qquad (12.1)$$

$$\nabla \cdot \mathbf{B} = 0, \quad \nabla \times \mathbf{E} + \frac{1}{c}\frac{\partial \mathbf{B}}{\partial t} = 0, \qquad (12.2)$$

where $\mathbf{D} = \mathbf{E} + 4\pi\mathbf{P}$ is the electric displacement vector inside the medium and $\mathbf{P}$ is dielectric polarization vector. The magnetic field $\mathbf{H} = \mathbf{B} - 4\pi\mathbf{M}$ where $\mathbf{M}$ is magnetization. In our case, we are assuming a non-magnetic medium and consequently $\mathbf{M} = 0$ and $\mathbf{B} = \mathbf{H}$. In a dispersive medium, the Fourier transformed relation between the electric displacement and electric field is

$$\mathbf{D}(\mathbf{k}, \omega) = \epsilon(\omega)\mathbf{E}(\mathbf{k}, \omega), \qquad (12.3)$$

where the Fourier transforms are defined as

$$\mathbf{D}(\mathbf{x}, t) = \frac{1}{(2\pi)^2} \int_{-\infty}^{\infty} d^3k \int_{-\infty}^{\infty} d\omega \, \mathbf{D}(\mathbf{k}, \omega) \, e^{i(\mathbf{k}\cdot\mathbf{x}-\omega t)} \, ,$$

$$\mathbf{E}(\mathbf{x}, t) = \frac{1}{(2\pi)^2} \int_{-\infty}^{\infty} d^3k \int_{-\infty}^{\infty} d\omega \, \mathbf{E}(\mathbf{k}, \omega) \, e^{i(\mathbf{k}\cdot\mathbf{x}-\omega t)} \, .$$

All other fields are Fourier transformed in the same manner. Here, $\epsilon(\omega)$ is the dielectric constant inside the medium. As the material medium is assumed to be non-magnetic we assume the magnetic permeability to be unity.

The gauge fields are still defined via the relations

$$\mathbf{B} = \nabla \times \mathbf{A}, \quad \mathbf{E} = -\nabla\Phi - \frac{1}{c}\frac{\partial \mathbf{A}}{\partial t}, \tag{12.4}$$

where Φ is the electromagnetic scalar potential and $\mathbf{A}$ is the vector potential. Later we will cast all the relevant equations in terms of the gauge fields. In terms of the gauge fields, the inhomogeneous Maxwell equations can be written without any cross product and we will be able to write down the formal solutions for $\mathbf{A}$ and Φ in a simple straightforward way. Once we know the forms of the solutions for the gauge potentials, we can find out the electromagnetic field via the equations. To solve the relevant Maxwell equation, we have to know about the source contribution which comes from a fast-moving particle with charge Zq and mass m. Here, q can be taken as the electronic charge and Z is a positive number. The charge is supposed to be moving through the medium with a uniform 3-velocity $\mathbf{v}$.

As the fast-moving charge moves through the medium it produces electric and magnetic field around it. The electric field can polarize some part of the dielectric medium around the charge. The rapid fluctuations in induced polarization may cause energy to spread out further away from the charge producing a kind of electromagnetic radiation inside the medium. We will later see that this radiation can happen only when the velocity of the charged particle inside the medium exceeds the velocity of light inside the medium.

In this section and henceforth we will primarily work in the Fourier space. Our convention regarding Fourier transform was described previously in Sect. 6.8 of Chap. 6 and we will follow that convention explicitly. In the Fourier transformed space, we have

$$\mathbf{E}(\mathbf{k}, \omega) = -i\mathbf{k}\Phi(\mathbf{k}, \omega) + \frac{i\omega}{c}\mathbf{A}(\mathbf{k}, \omega), \quad \mathbf{B}(\mathbf{k}, \omega) = i\mathbf{k} \times \mathbf{A}(\mathbf{k}, \omega), \tag{12.5}$$

which are the Fourier transforms of the expressions in Eq. (12.4). The Fourier transformed version of the first inhomogeneous Maxwell equation in a medium is

$$i\mathbf{k} \cdot \mathbf{D}(\mathbf{k}, \omega) = 4\pi\rho(\mathbf{k}, \omega) \, .$$

Using the relation between $\mathbf{D}(\mathbf{k}, \omega)$ and $\mathbf{E}(\mathbf{k}, \omega)$, we can write the above equation as

$$i\mathbf{k} \cdot \left[-i\mathbf{k}\Phi(\mathbf{k}, \omega) + \frac{i\omega}{c}\mathbf{A}(\mathbf{k}, \omega) \right] = \frac{4\pi\rho(\mathbf{k}, \omega)}{\epsilon(\omega)},$$

which can be finally transformed into

$$k^2\Phi(\mathbf{k}, \omega) - \frac{\omega}{c}\mathbf{k} \cdot \mathbf{A}(\mathbf{k}, \omega) = \frac{4\pi\rho(\mathbf{k}, \omega)}{\epsilon(\omega)}. \tag{12.6}$$

Similarly, we can write the Fourier transform of the other inhomogeneous Maxwell equation, in a medium, as

$$i\mathbf{k} \times [i\mathbf{k} \times \mathbf{A}(\mathbf{k}, \omega)] + \frac{i\omega}{c}\epsilon(\omega)\left[-i\mathbf{k}\Phi(\mathbf{k}, \omega) + \frac{i\omega}{c}\mathbf{A}(\mathbf{k}, \omega) \right] = \frac{4\pi}{c}\mathbf{J}(\mathbf{k}, \omega).$$

After doing some basic algebra we can write the above equation as

$$i\mathbf{k}\left[i\mathbf{k} \cdot \mathbf{A}(\mathbf{k}, \omega) - \frac{i\omega}{c}\epsilon(\omega)\Phi(\mathbf{k}, \omega) \right] + \left[k^2 - \frac{\omega^2}{c^2}\epsilon(\omega) \right]\mathbf{A}(\mathbf{k}, \omega) = \frac{4\pi}{c}\mathbf{J}(\mathbf{k}, \omega). \tag{12.7}$$

We know that the solutions Φ and $\mathbf{A}$ by solving the equations in Eq. (12.6) and the above equation are not unique. There can be other solutions related to the gauge potentials via gauge transformations. To eliminate the arbitrariness of the gauge field solutions we have to choose some restricting condition. In microscopic electrodynamics, the restricting condition was the Lorenz gauge condition given by

$$\frac{1}{c}\frac{\partial\Phi(\mathbf{x}, t)}{\partial t} + \nabla \cdot \mathbf{A}(\mathbf{x}, t) = 0.$$

The Fourier transformed form of the above condition is

$$\mathbf{k} \cdot \mathbf{A}(\mathbf{k}, \omega) - \frac{\omega}{c}\Phi(\mathbf{k}, \omega) = 0. \tag{12.8}$$

In the present case, we have to generalize the Lorenz gauge condition inside a dielectric medium. In this case, it turns out to be easier to generalize the condition in the Fourier transformed form as the dielectric constant is naturally defined as a function of angular frequency. Modification of the above equation due to the dielectric medium must involve some medium dependent function. The only medium dependent function in a dielectric is the dielectric constant. This gives us a hint that Eq. (12.8) must be modified in such a manner that in a medium it has some $\epsilon(\omega)$ dependence and the form of the modified expression must reduce to microscopic version in the absence of the medium. Here, we propose a simple modification of the gauge condition with which we will work in this chapter on Cherenkov radiation. The modification of the Fourier transformed version of the Lorenz gauge condition is given as

$$\mathbf{k} \cdot \mathbf{A}(\mathbf{k}, \omega) - \frac{\omega}{c}\epsilon(\omega)\Phi(\mathbf{k}, \omega) = 0. \tag{12.9}$$

This equation reduces to Eq. (12.8) in the absence of the medium where $\epsilon(\omega) = 1$. Using this condition, we can write Eqs. (12.7) and (12.6) as

$$\left[k^2 - \frac{\omega^2}{c^2}\epsilon(\omega) \right] \mathbf{A}(\mathbf{k}, \omega) = \frac{4\pi}{c}\mathbf{J}(\mathbf{k}, \omega), \tag{12.10}$$

$$\left[k^2 - \frac{\omega^2}{c^2}\epsilon(\omega) \right] \Phi(\mathbf{k}, \omega) = \frac{4\pi\rho(\mathbf{k}, \omega)}{\epsilon(\omega)}. \tag{12.11}$$

These are the gauge restricted form of the inhomogeneous Maxwell equations inside a dielectric medium. To solve these equations, we require to know the form of the source charge density and source current density. The charge density corresponding to a charge Zq moving with uniform velocity $\mathbf{v}$ is

$$\rho(\mathbf{x}, t) = Zq\,\delta^3(\mathbf{x} - \mathbf{v}t). \tag{12.12}$$

In our convention

$$
\begin{aligned}
(2\pi)^2\rho(\mathbf{k}, \omega) &= \int_{-\infty}^{\infty} d^3x \int_{-\infty}^{\infty} dt\, e^{-i(\mathbf{k}\cdot\mathbf{x}-\omega t)}\,\rho(\mathbf{x}, t) \\
&= Zq \int_{-\infty}^{\infty} d^3x \int_{-\infty}^{\infty} dt\, e^{-i(\mathbf{k}\cdot\mathbf{x}-\omega t)}\,\delta^3(\mathbf{x} - \mathbf{v}t) \\
&= Zq \int_{-\infty}^{\infty} dt\, e^{-i(\mathbf{k}\cdot\mathbf{v}-\omega)t}.
\end{aligned}
$$

We can also write the last step of the above calculation as

$$
\begin{aligned}
(2\pi)^2\rho(\mathbf{k}, \omega) &= Zq(2\pi)\left[\frac{1}{2\pi} \int_{-\infty}^{\infty} dt\, e^{-i(\mathbf{k}\cdot\mathbf{v}-\omega)t} \right] \\
&= Zq(2\pi)\delta(\omega - \mathbf{k}\cdot\mathbf{v}).
\end{aligned} \tag{12.13}
$$

Using this information, we can now write the Fourier transformed charge density as

$$\rho(\mathbf{k}, \omega) = \frac{Zq}{2\pi}\delta(\omega - \mathbf{k}\cdot\mathbf{v}). \tag{12.14}$$

The 3-current density is given by $\mathbf{J}(\mathbf{x}, t) = Zq\mathbf{v}\,\delta^3(\mathbf{x} - \mathbf{v}t)$. As the velocity vector is a constant, we have

$$\mathbf{J}(\mathbf{k}, t) = \mathbf{v}\,\rho(\mathbf{k}, \omega). \tag{12.15}$$

The above relations give us the Fourier transforms of the source charge density and current.

From the expression of the scalar electrodynamic potential in Eq. (12.11), we can now write

$$\Phi(\mathbf{k}, \omega) = \frac{2Zq}{\epsilon(\omega)} \frac{\delta(\omega - \mathbf{k} \cdot \mathbf{v})}{\left[k^2 - \frac{\omega^2}{c^2}\epsilon(\omega)\right]} . \tag{12.16}$$

The form of the vector potential is obtained from Eq. (12.10) as

$$\mathbf{A}(\mathbf{k}, \omega) = \frac{2Zq}{c} \frac{\delta(\omega - \mathbf{k} \cdot \mathbf{v})}{\left[k^2 - \frac{\omega^2}{c^2}\epsilon(\omega)\right]} \mathbf{v} ,$$

which can be compactly written as

$$\mathbf{A}(\mathbf{k}, \omega) = \epsilon(\omega)\frac{\mathbf{v}}{c}\Phi(\mathbf{k}, \omega) . \tag{12.17}$$

Now we have solved the Maxwell equations for the gauge fields. The next step is to figure out the physical electromagnetic fields from these gauge field solutions. To write the Fourier transforms for the electric and magnetic fields induced by the uniformly moving charge inside the medium we will use Eq. (12.5). The electric and magnetic fields in the present case are

$$\mathbf{E}(\mathbf{k}, \omega) = i\left[\frac{\omega\epsilon(\omega)}{c}\beta - \mathbf{k}\right]\Phi(\mathbf{k}, \omega), \quad \mathbf{B}(\mathbf{k}, \omega) = i\epsilon(\omega)(\mathbf{k} \times \beta)\Phi(\mathbf{k}, \omega),$$

$$\tag{12.18}$$

where as usual $\beta = \mathbf{v}/c$ and we know the expression of the scalar potential. Using the above field expressions, we can now calculate the energy dissipated by a uniformly moving charged particle inside a dielectric medium. To find out the energy dissipated by the charged particle we will use the above expression of the fields and calculate the Poynting vector. The expression of the Poynting vector will have to be calculated assuming the charged particle is moving uniformly and it covers infinite distance inside the medium. All these assumptions are not practical but turn out to be immensely helpful for the theoretical calculation for the emitted power. We will point out the various assumptions we will use and will also discuss the limitations of the assumptions.

Fig. 12.1 Picture showing a charged particle, inside the dielectric, moving with uniform velocity **v** along the x-direction. The electric field and magnetic field is calculated at point P whose coordinates are $(0, b, 0)$

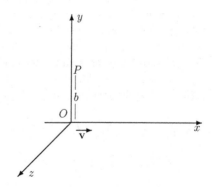

12.2 Energy Loss by the Charged Particle Moving Inside the Dielectric Medium

In this section, we will first calculate the electric and magnetic fields at point P due to the motion of a charged particle moving with uniform velocity **v** along the x-axis. The charged particle is supposed to be moving inside the dielectric medium. As there is an axial symmetry of the situation, where the symmetry axis is along the velocity vector, we can chose the point P along any transverse direction to the axis of symmetry and find the fields there. As shown in Fig. 12.1, we assume the point P to be lying along the y-direction and has coordinates $(0, b, 0)$. We will like to find out the electric and magnetic fields due to the moving charge at point P. Here, we are assuming that the charge is uniformly moving along the x-axis for an infinite time and the system is infinitely long. In reality, this assumption is never realized but compared to the size of the charged particle the size of the macroscopic dielectric is practically infinite. We have to understand this assumption about the indefinite existence of the particle trajectory in this light. Moreover in reality it is difficult for the particle to maintain uniform velocity. Most likely the velocity of the particle will change. We are assuming that the charged particle is not ionizing the medium heavily and consequently the particle does not lose much energy. The uniformity of motion is an assumption which works finely in the case of Cherenkov radiation and using this assumption we will be able to figure out the energy loss mechanism in an intuitive way.

12.2.1 Calculation of the Fields

First we calculate the electric field at point P in Fig. 12.1 due to motion of the charge. We will calculate the Fourier transform of the field and consequently we will be working in the frequency domain. One has to be careful that the fields calculated in such a way cannot be interpreted as the field produced at any particular time. To know the field at P at any particular time one has to Fourier transform the results we obtain. As the coordinate of P is $(0, b, 0)$ the electric field at P is $\mathbf{E}(b\hat{\mathbf{j}}, \omega)$ where $\hat{\mathbf{j}}$ is the unit vector along the y-axis. Another point must be noted here. As the fields $\mathbf{E}(\mathbf{x}, t)$ and $\mathbf{B}(\mathbf{x}, t)$ are real fields there Fourier transforms $\mathbf{E}(\mathbf{k}, \omega)$ and $\mathbf{B}(\mathbf{k}, \omega)$ are not necessarily real. All of the quantities, as the Fourier transform of the fields and $\rho(\mathbf{k}, \omega)$, $\mathbf{J}(\mathbf{k}, \omega)$ are complex. With this brief introduction we write

$$\mathbf{E}(b\hat{\mathbf{j}}, \omega) = \frac{1}{(2\pi)^{3/2}} \int_{-\infty}^{\infty} d^3k \, \mathbf{E}(\mathbf{k}, \omega) \, e^{i\mathbf{k}\cdot\mathbf{x}} = \frac{1}{(2\pi)^{3/2}} \int_{-\infty}^{\infty} d^3k \, \mathbf{E}(\mathbf{k}, \omega) \, e^{ibk_y} ,$$

(12.19)

where we are using the conventions about Fourier transforms set in Chap. 6. Specifically the x-component of the electric filed $\mathbf{E}(\mathbf{k}, \omega)$ is obtained from Eq. (12.18), and it is

$$E_x(\mathbf{k}, \omega) = \frac{2Ziq}{\epsilon(\omega)} \left[\frac{\omega\epsilon(\omega)}{c^2} v - k_x \right] \frac{\delta(\omega - k_x v)}{\left[k^2 - \frac{\omega^2}{c^2}\epsilon(\omega) \right]} ,$$

(12.20)

where we only use the expression $v = |\mathbf{v}|$ in the above equation as the velocity vector has only the x-component in our case. From the last two equations, we can now write

$$E_x(b\hat{\mathbf{j}}, \omega) = \frac{2Ziq}{\epsilon(\omega)} \frac{1}{(2\pi)^{3/2}} \int_{-\infty}^{\infty} d^3k \left[\frac{\omega\epsilon(\omega)}{c^2} v - k_x \right] \frac{\delta(\omega - k_x v)}{\left[k^2 - \frac{\omega^2}{c^2}\epsilon(\omega) \right]} e^{ibk_y} .$$

(12.21)

We can write the integrals separately as

$$E_x(b\hat{\mathbf{j}}, \omega) = \frac{2Ziq}{\epsilon(\omega)} \frac{1}{(2\pi)^{3/2}} \int_{-\infty}^{\infty} dk_z \int_{-\infty}^{\infty} dk_y \, e^{ibk_y} \int_{-\infty}^{\infty} dk_x \left[\frac{\omega\epsilon(\omega)}{c^2} v - k_x \right]$$
$$\times \frac{\delta(\omega - k_x v)}{k_x^2 + (k_y^2 + k_z^2) - \frac{\omega^2}{c^2}\epsilon(\omega)} .$$

(12.22)

We have written the integrals in such a way so that the reader must understand that we will do the k_x integral first. The k_x integral is relatively easier due to the existence of the Dirac-delta function. We first note that

$$\delta(\omega - k_x v) = \delta\left(v \left(k_x - \frac{\omega}{v} \right) \right) = \frac{1}{v} \delta\left(k_x - \frac{\omega}{v} \right) ,$$

where v is the magnitude of the 3-velocity along the positive x-axis and is always positive. Using this expression of the Dirac-delta function, we can write the last integral form of the x-component of the electric field as

$$E_x(b\hat{\mathbf{j}}, \omega) = \frac{2Ziq}{\upsilon\epsilon(\omega)} \frac{1}{(2\pi)^{3/2}} \int_{-\infty}^{\infty} dk_z \int_{-\infty}^{\infty} dk_y\, e^{ibk_y} \int_{-\infty}^{\infty} dk_x \left[\frac{\omega\epsilon(\omega)}{c^2}\upsilon - k_x\right]$$

$$\times \frac{\delta\left(k_x - \frac{\omega}{\upsilon}\right)}{k_x^2 + (k_y^2 + k_z^2) - \frac{\omega^2}{c^2}\epsilon(\omega)}, \tag{12.23}$$

where the k_x integration can be easily done

$$E_x(b\hat{\mathbf{j}}, \omega) = \frac{2Ziq}{\upsilon\epsilon(\omega)} \frac{1}{(2\pi)^{3/2}} \int_{-\infty}^{\infty} dk_z \int_{-\infty}^{\infty} dk_y\, e^{ibk_y} \left[\frac{\omega\epsilon(\omega)}{c^2}\upsilon - \frac{\omega}{\upsilon}\right]$$

$$\times \frac{1}{k_y^2 + k_z^2 + \left[\frac{\omega^2}{\upsilon^2} - \frac{\omega^2}{c^2}\epsilon(\omega)\right]}. \tag{12.24}$$

Before we proceed further we define a quantity called λ as

$$\lambda^2 \equiv \frac{\omega^2}{\upsilon^2} - \frac{\omega^2}{c^2}\epsilon(\omega), \tag{12.25}$$

which shows λ in general is complex and has the dimension of inverse length. If $\upsilon < c/\sqrt{\epsilon}$ then λ is real on the other hand if $\upsilon > c/\sqrt{\epsilon}$ we have an imaginary λ. One should note that in a non-magnetized dielectric $c/\sqrt{\epsilon}$ is the phase velocity of electromagnetic radiation inside the medium. We see that as long as the velocity of the charged particle remains greater then the phase velocity of electromagnetic radiation inside the medium λ remains imaginary. Using the above definition, we can now write

$$E_x(b\hat{\mathbf{j}}, \omega) = -\frac{2Ziq\omega}{\upsilon^2(2\pi)^{3/2}} \left[\frac{1}{\epsilon(\omega)} - \beta^2\right] \int_{-\infty}^{\infty} dk_y\, e^{ibk_y} \int_{-\infty}^{\infty} \frac{dk_z}{k_z^2 + (k_y^2 + \lambda^2)}. \tag{12.26}$$

We can easily do the k_z integral as we know for any constant a

$$\int_{-\infty}^{\infty} \frac{dx}{x^2 + a^2} = \frac{1}{a}\tan^{-1}\frac{x}{a}\Big|_{-\infty}^{\infty}, \tag{12.27}$$

and consequently

$$\int_{-\infty}^{\infty} \frac{dk_z}{k_z^2 + (k_y^2 + \lambda^2)} = \frac{\pi}{(\lambda^2 + k_y^2)^{1/2}}.$$

Using this standard result, we get

$$E_x(b\hat{\mathbf{j}}, \omega) = -\frac{(2\pi)Ziq\omega}{\upsilon^2(2\pi)^{3/2}} \left[\frac{1}{\epsilon(\omega)} - \beta^2\right] \int_{-\infty}^{\infty} dk_y \frac{e^{ibk_y}}{(\lambda^2 + k_y^2)^{1/2}}. \tag{12.28}$$

The k_y integral can be done by using the integral representation of the zeroth order modified Bessel function.

The modified Bessel functions are solutions of Bessel's equation

$$z^2 \frac{d^2 w}{dz^2} + z \frac{dw}{dz} + (z^2 - \alpha^2)w = 0 \,, \tag{12.29}$$

when z is replaced by iz. Here z is a complex number and w is a complex function. The parameter α is also a complex number. The modified Bessel differential equation is

$$z^2 \frac{d^2 w}{dz^2} + z \frac{dw}{dz} - (z^2 + \alpha^2)w = 0 \,, \tag{12.30}$$

whose solutions are traditionally called $I_\alpha(z)$ and $K_\alpha(z)$. The function $I_\alpha(z)$ is called modified Bessel function of the first kind of order α and $K_\alpha(z)$ is modified Bessel function of the second kind of order α. The zeroth order modified Bessel function of the second kind has an integral form

$$\int_{-\infty}^{\infty} \frac{e^{izs}}{(1+s^2)^{1/2}} ds = 2K_0(z) \,, \tag{12.31}$$

where z in general is complex. To evaluate the k_y integral in Eq. (12.28), we will use the above integral form of $K_0(z)$.

We can write the k_y integral in Eq. (12.28) as:

$$\int_{-\infty}^{\infty} dk_y \frac{e^{ibk_y}}{(\lambda^2 + k_y^2)^{1/2}} = \frac{1}{\lambda} \int_{-\infty}^{\infty} dk_y \frac{e^{ibk_y}}{\left[1 + \left(\frac{k_y}{\lambda}\right)^2\right]^{1/2}} \,.$$

Defining a new variable $\kappa = k_y/\lambda$, where we temporarily assume λ to be real, we can convert the last integral as

$$\frac{1}{\lambda} \int_{-\infty}^{\infty} dk_y \frac{e^{ibk_y}}{\left[1 + \left(\frac{k_y}{\lambda}\right)^2\right]^{1/2}} = \int_{-\infty}^{\infty} \frac{e^{ib\lambda\kappa}}{(1 + \kappa^2)^{1/2}} d\kappa \,.$$

From Eq. (12.31) and the above result, we can finally write the k_y integral as

$$\int_{-\infty}^{\infty} dk_y \frac{e^{ibk_y}}{(\lambda^2 + k_y^2)^{1/2}} = \int_{-\infty}^{\infty} \frac{e^{ib\lambda\kappa}}{(1 + \kappa^2)^{1/2}} d\kappa = 2K_0(b\lambda) \,. \tag{12.32}$$

To get the above result we temporarily assumed λ to be real, but the above result is true even when λ is complex. In our case λ is complex and we will use the above result keeping λ to be complex.

Now we have all the ingredients to write down the expression of the electric field, produced by the uniformly moving charge inside the dielectric, in the frequency domain. From the expression of the field in Eq. (12.28) and the result of the k_y integral we can write

$$E_x(b\hat{\mathbf{j}}, \omega) = -\frac{Ziq\omega}{v^2}\sqrt{\frac{2}{\pi}}\left[\frac{1}{\epsilon(\omega)} - \beta^2\right] K_0(\lambda b). \tag{12.33}$$

This expression gives the x-component of the electric field produced at P due to a uniformly moving charged particle along the x-axis. Next we will like to calculate the other components of the electric field.

Next we calculate the z-component of the electric field $\mathbf{E}(b\hat{\mathbf{j}}, \omega)$. The Fourier transform of the z-component of the field is obtained from Eq. (12.18) and it can be written as

$$E_z(\mathbf{k}, \omega) = -ik_z\Phi(\mathbf{k}, \omega) = -\frac{2Ziq}{\epsilon(\omega)}\frac{\delta(\omega - k_x v)}{\left[k^2 - \frac{\omega^2}{c^2}\epsilon(\omega)\right]}k_z. \tag{12.34}$$

From this expression we can now write the z-component of the electric field at P in the frequency domain as

$$\begin{aligned}
E_z(b\hat{\mathbf{j}}, \omega) &= \frac{1}{(2\pi)^{3/2}}\int_{-\infty}^{\infty} d^3k\, E_z(\mathbf{k}, \omega)\, e^{ibk_y} \\
&= -\frac{2Ziq}{\epsilon(\omega)}\frac{1}{(2\pi)^{3/2}}\int_{-\infty}^{\infty} d^3k\, k_z \frac{\delta(\omega - k_x v)}{\left[k_z^2 + k_x^2 + k_y^2 - \frac{\omega^2}{c^2}\epsilon(\omega)\right]}e^{ibk_y}.
\end{aligned} \tag{12.35}$$

From the above equation, we see that the k_z integral yields zero as the integrand is an odd function of k_z. As a result we have

$$E_z(b\hat{\mathbf{j}}, \omega) = 0. \tag{12.36}$$

This shows that in our chosen geometry of the system the z-component of the electric field at P due to the uniformly moving charge vanishes.

The y-component of the electric field $\mathbf{E}(b\hat{\mathbf{j}}, \omega)$ at point P due to the motion of the charge remains to be calculated. From Eq. (12.18) we see that

$$E_y(\mathbf{k}, \omega) = -ik_y\Phi(\mathbf{k}, \omega) = -\frac{2Ziq}{\epsilon(\omega)}\frac{\delta(\omega - k_x v)}{\left[k^2 - \frac{\omega^2}{c^2}\epsilon(\omega)\right]}k_y.$$

From this expression, we can now write the form of the y-component of the electric field at P in the frequency domain as

$$
\begin{aligned}
E_y(b\hat{\mathbf{j}}, \omega) &= \frac{1}{(2\pi)^{3/2}} \int_{-\infty}^{\infty} d^3 k \, E_y(\mathbf{k}, \omega) \, e^{ibk_y} \\
&= -\frac{2Ziq}{\epsilon(\omega)} \frac{1}{(2\pi)^{3/2}} \int_{-\infty}^{\infty} d^3 k \, k_y \frac{\delta(\omega - k_x v)}{\left[k_z^2 + k_x^2 + k_y^2 - \frac{\omega^2}{c^2}\epsilon(\omega) \right]} e^{ibk_y} .
\end{aligned}
$$

From our previous discussion, we can write the above integral as

$$
E_y(b\hat{\mathbf{j}}, \omega) = -\frac{2Ziq}{v\epsilon(\omega)} \frac{1}{(2\pi)^{3/2}} \int_{-\infty}^{\infty} dk_z \int_{-\infty}^{\infty} dk_y \, k_y \, e^{ibk_y} \int_{-\infty}^{\infty} dk_x \frac{\delta(k_x - \frac{\omega}{v})}{\left[k_z^2 + k_x^2 + k_y^2 - \frac{\omega^2}{c^2}\epsilon(\omega) \right]} .
$$

We can immediately integrate out k_x using the Dirac-delta function and get

$$
E_y(b\hat{\mathbf{j}}, \omega) = -\frac{2Ziq}{v\epsilon(\omega)} \frac{1}{(2\pi)^{3/2}} \int_{-\infty}^{\infty} dk_y \, k_y \, e^{ibk_y} \int_{-\infty}^{\infty} \frac{dk_z}{\left[k_z^2 + (k_y^2 + \lambda)^2 \right]} .
$$

(12.37)

The k_z integral can be done using the standard result in Eq. (12.27). After the k_z integral we get

$$
E_y(b\hat{\mathbf{j}}, \omega) = -\frac{(2\pi)Ziq}{v\epsilon(\omega)} \frac{1}{(2\pi)^{3/2}} \int_{-\infty}^{\infty} dk_y \frac{k_y \, e^{ibk_y}}{(\lambda^2 + k_y^2)^{1/2}} .
$$

(12.38)

To evaluate the k_y integral we will again use the properties of the modified Bessel function of the second kind. The modified Bessel function $K_0(z)$ satisfies the property

$$
\frac{dK_\alpha}{dz} = \frac{\alpha}{z} K_\alpha(z) - K_{\alpha+1}(z) ,
$$

(12.39)

from which we get

$$
\frac{dK_0}{dz} = -K_1(z) .
$$

(12.40)

We will use this property to evaluate the k_y integral. From Eq. (12.32), we know

$$
\int_{-\infty}^{\infty} dk_y \frac{e^{ibk_y}}{(\lambda^2 + k_y^2)^{1/2}} = 2K_0(b\lambda) ,
$$

which yields

$$\frac{d}{db} \int_{-\infty}^{\infty} dk_y \, \frac{e^{ibk_y}}{(\lambda^2 + k_y^2)^{1/2}} = 2\lambda K_0'(b\lambda) \,,$$

where $K_0'(b\lambda) = dK_0/d(b\lambda) = -K_1(b\lambda)$ using Eq. (12.40). The differentiation of the integral on the left-hand side produces

$$\int_{-\infty}^{\infty} dk_y \, \frac{k_2 \, e^{ibk_y}}{(\lambda^2 + k_y^2)^{1/2}} = 2i\lambda \, K_1(b\lambda) \,. \tag{12.41}$$

Using this result, we can now write y-component of the electric field at P in the frequency domain, from Eq. (12.38), as

$$E_y(b\hat{\mathbf{j}}, \omega) = \frac{Zq\lambda}{v\epsilon(\omega)} \sqrt{\frac{2}{\pi}} K_1(b\lambda) \,, \tag{12.42}$$

where $K_1(b\lambda)$ is the first-order modified Bessel function of the second kind.

Next we will calculate the magnetic field at point P due to the motion of the uniformly charged particle along the x-axis. Here also we will find the field in the frequency domain. It is known from Eq. (12.18) the total Fourier transform of the magnetic field is

$$\mathbf{B}(\mathbf{k}, \omega) = i\epsilon(\omega)(\mathbf{k} \times \boldsymbol{\beta})\Phi(\mathbf{k}, \omega) \,.$$

As $\boldsymbol{\beta}$ is along the x-axis we can immediately write $B_x(\mathbf{k}, \omega) = 0$ giving

$$B_x(b\hat{\mathbf{j}}, \omega) = 0 \,. \tag{12.43}$$

To find $B_y(b\hat{\mathbf{j}}, \omega)$ we first note

$$B_y(\mathbf{k}, \omega) = i\epsilon(\omega)\beta k_z \Phi(\mathbf{k}, \omega) \,,$$

which predicts

$$\begin{aligned}
B_y(b\hat{\mathbf{j}}, \omega) &= \frac{1}{(2\pi)^{3/2}} \int_{-\infty}^{\infty} d^3k \, B_y(\mathbf{k}, \omega) \, e^{ibk_y} \\
&= \frac{2Ziq\beta}{(2\pi)^{3/2}} \int_{-\infty}^{\infty} d^3k \, k_z \, \frac{\delta(\omega - k_x v)}{\left[k_z^2 + k_x^2 + k_y^2 - \frac{\omega^2}{c^2}\epsilon(\omega) \right]} e^{ibk_y} \,.
\end{aligned}$$

We see the integration above is similar to the one appearing in the expression of $E_z(b\hat{\mathbf{j}}, \omega)$, in Eq. (12.35), and as a result we have

$$B_y(b\hat{\mathbf{j}}, \omega) = 0 \,. \tag{12.44}$$

The above results show that both the x and y-components of the magnetic field, due to the motion of the charge, at P are zero.

At last we calculate the z-component of the magnetic field at P in the frequency domain, due to the uniform motion of charge along the x-axis. In this case

$$B_z(\mathbf{k}, \omega) = -i\epsilon(\omega)k_y \frac{v}{c}\Phi(\mathbf{k}, \omega) = -\frac{2iZq}{c}vk_y \frac{\delta(\omega - k_x v)}{k^2 - \frac{\omega^2}{c^2}\epsilon(\omega)}.$$

From this expression we can now write

$$B_z(b\hat{\mathbf{j}}, \omega) = \epsilon(\omega)\beta E_y(b\hat{\mathbf{j}}, \omega),\qquad(12.45)$$

where the result can be anticipated from the previous calculation of $E_y(b\hat{\mathbf{j}}, \omega)$.

We have calculated all the field components in the frequency domain. Next we will like to calculate the energy loss by the moving charge. The moving charge excites the whole medium in a coherent way and consequently it loses energy.

12.2.2 The Energy Loss Mechanism

To calculate the energy loss mechanism we first assume that the charged particle is moving uniformly, with velocity greater than light in the medium, for an infinite time. The particle is moving along the x-axis, it was at $x \to -\infty$ at $t \to -\infty$ and finally the particle moves to $x \to \infty$ at $t \to \infty$. The reader can easily understand that this is an assumption which requires some explanation. In reality the Cherenkov condition, related to the super-fast motion of the charged particle, only holds for a very brief time period of motion of the charged particle inside the medium. During this time period the charged particle will be able to emit certain number of electromagnetic waves. As because the classical electron radius is very small, of the order of 10^{-13}cm, the brief length scale over which constant superluminal speed is maintained appears to be too long and for all practical purpose we can take this length to be infinite. This heuristic argument shows that our assumption of an infinite length scale for the problem is reasonable. As the moving particle emits radiation it loses energy and after sometime its velocity falls from the superluminal limit and Cherenkov radiation stops.

To calculate the energy loss from such a fast-moving charged particle inside the medium we assume that the charged particle track is surrounded by a hypothetical cylinder whose radius is a. The situation is shown in Fig. 12.2. We assume that the radius of the cylinder is smaller than the distance where the energy from the particle is ultimately deposited but greater than the atomic dimensions. The energy emission process has a cylindrical symmetry about the x-axis. To calculate the energy emitted by the charged particle per unit length we consider a strip, as shown in the figure above, with infinitesimal thickness dx. First we want to find out how much energy

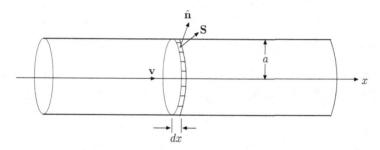

Fig. 12.2 The trajectory of the uniformly moving charged particle inside a hypothetical cylinder surrounding it. The radius of the cylinder is a and it extends infinitely on both sides. A small cross-section of the cylinder is shown whose thickness is dx

flows out through the (infinitesimal) surface of such a strip when the particle moves from one extremity to the other extremity of the cylinder. Energy starts to flow out from the infinitesimal surface as the particle starts its motion from the left end of the cylinder and this energy flow continues as long as the particle moves and reaches the other extreme end of the cylinder.

In an infinitesimal time interval dt, the energy passing through the strip is

$$dE = (2\pi a dx)(\mathbf{S} \cdot \hat{\mathbf{n}})dt ,$$

where $\mathbf{S}$ is the Poynting vector calculated on any point on the strip and $\hat{\mathbf{n}}$ is the unit outward normal at the same point. In Fig. 12.2 the relative position of both the vectors are shown. Due to cylindrical symmetry $\mathbf{S}$ makes the same angle with the unit normal for all points on the strip. Due to cylindrical symmetry of the problem in reality the quantity $\mathbf{S} \cdot \hat{\mathbf{n}}$ remains the same for all points on the strip and as a result we can evaluate the scalar product for any position of the base point (where the vectors are planted) on the strip. Using this information we will calculate $\mathbf{S} \cdot \hat{\mathbf{n}}$ for a base point situated at the topmost position of the strip. For this point $\hat{\mathbf{n}} = \hat{\mathbf{j}}$, where we have assumed that the y-axis lies on the plane of the paper. The Poynting vector is given by

$$\mathbf{S} = \frac{c}{4\pi}\mathbf{E} \times \mathbf{B} = \frac{c}{4\pi}\mathbf{E}(\hat{\mathbf{i}}E_y B_z - \hat{\mathbf{j}}E_x B_z) ,$$

where now we are using the fields in normal coordinate space and time. Later we will connect our calculation with the expression of the fields in frequency domain. If we fix $\hat{\mathbf{n}} = \hat{\mathbf{j}}$ then

$$\mathbf{S}(\mathbf{x}, t) \cdot \hat{\mathbf{n}} = \mathbf{S}(\mathbf{x}, t) \cdot \hat{\mathbf{j}} = -\frac{c}{4\pi}E_x(\mathbf{x}, t)B_z(\mathbf{x}, t) , \qquad (12.46)$$

where we can express $E_x(\mathbf{x}, t)$ and $B_z(\mathbf{x}, t)$ from the expressions of them evaluated in the previous subsection. Using the above expression of the dot product of the two vectors, we can now write

$$dE = -(2\pi a dx)\frac{c}{4\pi}E_x(a\hat{\mathbf{j}}, t)B_z(a\hat{\mathbf{j}}, t)dt = -\frac{c}{2}adx\, E_x(a\hat{\mathbf{j}}, t)B_z(a\hat{\mathbf{j}}, t)dt\,.$$

$$(12.47)$$

The energy deposited by the charged particle on the strip per unit length, during time t to $t + dt$ is then given by

$$\left(\frac{dE}{dx}\right)_{[t,t+dt]} = -\frac{c}{2}a E_x(a\hat{\mathbf{j}}, t)B_z(a\hat{\mathbf{j}}, t)dt\,.$$

$$(12.48)$$

From the above expression we can now write down the energy deposited per unit length by the moving charged particle, for the whole journey, as

$$\frac{dE}{dx} = -\frac{ca}{2}\int_{t=-\infty}^{\infty} E_x(a\hat{\mathbf{j}}, t)B_z(a\hat{\mathbf{j}}, t)dt\,.$$

$$(12.49)$$

We will now express the fields in the frequency domain. To do so we use the Fourier transforms as

$$E_x(t) = \frac{1}{\sqrt{2\pi}}\int_{\omega=-\infty}^{\infty} E_x(\omega)e^{-i\omega t}d\omega\,, \quad B_z(t) = \frac{1}{\sqrt{2\pi}}\int_{\omega'=-\infty}^{\infty} B_z(\omega')e^{-i\omega' t}d\omega'\,,$$

and write

$$\begin{aligned}\frac{dE}{dx} &= -\frac{ca}{4\pi}\int_{t=-\infty}^{\infty}\left[\int_{\omega=-\infty}^{\infty} E_x(a\hat{\mathbf{j}}, \omega)e^{-i\omega t}d\omega\right]\left[\int_{\omega'=-\infty}^{\infty} B_z(a\hat{\mathbf{j}}, \omega')e^{-i\omega' t}d\omega'\right]dt \\ &= -\frac{ca}{2}\int_{\omega=-\infty}^{\infty}\int_{\omega'=-\infty}^{\infty}\left[\frac{1}{2\pi}\int_{t=-\infty}^{\infty} e^{-it(\omega+\omega')}dt\right] E_x(a\hat{\mathbf{j}}, \omega)B_z(a\hat{\mathbf{j}}, \omega')d\omega d\omega' \\ &= -\frac{ca}{2}\int_{\omega=-\infty}^{\infty}\int_{\omega'=-\infty}^{\infty} \delta(\omega+\omega') E_x(a\hat{\mathbf{j}}, \omega)B_z(a\hat{\mathbf{j}}, \omega')d\omega d\omega' \\ &= -\frac{ca}{2}\int_{\omega=-\infty}^{\infty} E_x(a\hat{\mathbf{j}}, \omega)B_z(a\hat{\mathbf{j}}, -\omega)d\omega\,.\end{aligned}$$

$$(12.50)$$

To avoid negative frequency, we write the above expression as

$$\frac{dE}{dx} = -\frac{ca}{2}\left[\int_{\omega=-\infty}^{0} E_x(a\hat{\mathbf{j}}, \omega)B_z(a\hat{\mathbf{j}}, -\omega)d\omega + \int_{\omega=0}^{\infty} E_x(a\hat{\mathbf{j}}, \omega)B_z(a\hat{\mathbf{j}}, -\omega)d\omega\right]\,,$$

where we can now change variable for the first integral and write

$$\frac{dE}{dx} = -\frac{ca}{2}\int_{\omega=0}^{\infty}\left[E_x(a\hat{\mathbf{j}}, -\omega)B_z(a\hat{\mathbf{j}}, \omega) + E_x(a\hat{\mathbf{j}}, \omega)B_z(a\hat{\mathbf{j}}, -\omega)\right]d\omega\,.$$

As $E_x(\mathbf{x}, t)$ and $B_z(\mathbf{x}, t)$ are real functions we must have $B_z^*(a\hat{\mathbf{j}}, \omega) = B_z(a\hat{\mathbf{j}}, -\omega)$ and $E_x^*(a\hat{\mathbf{j}}, \omega) = E_x(a\hat{\mathbf{j}}, -\omega)$. Using these information, we can write

$$\frac{dE}{dx} = -\frac{ca}{2} \int_{\omega=0}^{\infty} \left[E_x^*(a\hat{\mathbf{j}}, \omega) B_z(a\hat{\mathbf{j}}, \omega) + E_x(a\hat{\mathbf{j}}, \omega) B_z^*(a\hat{\mathbf{j}}, \omega) \right] d\omega \,,$$

which can be compactly written as

$$\frac{dE}{dx} = -ca \int_{\omega=0}^{\infty} \mathrm{Re} \left[E_x(a\hat{\mathbf{j}}, \omega) B_z^*(a\hat{\mathbf{j}}, \omega) \right] d\omega \,, \tag{12.51}$$

where $\mathrm{Re}[z]$ stands for the real part of the complex number z. The above equation will be useful for us in deciding the conditions favorable for Cherenkov radiation.

From the discussion in the last section, we can write the relevant field values

$$E_x(a\hat{\mathbf{j}}, \omega) = -\frac{Ziq\omega}{v^2} \sqrt{\frac{2}{\pi}} \left[\frac{1}{\epsilon(\omega)} - \beta^2 \right] K_0(\lambda a) \,, \tag{12.52}$$

$$E_y(a\hat{\mathbf{j}}, \omega) = \frac{Zq\lambda}{v\epsilon(\omega)} \sqrt{\frac{2}{\pi}} K_1(a\lambda) \,, \tag{12.53}$$

$$B_z(a\hat{\mathbf{j}}, \omega) = \epsilon(\omega)\beta E_y(a\hat{\mathbf{j}}, \omega) \,, \tag{12.54}$$

where

$$\lambda^2 = \frac{\omega^2}{v^2} \left[1 - \beta^2 \epsilon(\omega) \right] \,. \tag{12.55}$$

The quantity λ can be complex or pure imaginary. For Cherenkov radiation one must have

$$|\lambda a| \gg 1 \,. \tag{12.56}$$

If this condition is not maintained then the charged particle will lose energy in other ways, it will not radiate inside the medium. Because of the above condition, we require the asymptotic values of the special functions $K_0(\lambda a)$ and $K_1(\lambda a)$.

In general when $|z| \gg 1$ one can expand $K_\alpha(z)$ as

$$K_\alpha(z) \sim \sqrt{\frac{\pi}{2z}} e^{-z} \left(1 + \frac{\mu - 1}{8z} + \cdots \right) \,, \tag{12.57}$$

where $\mu = 4\alpha^2$ and for the above formula to hold true one must have $|\arg(z)| < 3\pi/2$. As we are really interested in the limit when $|z| \gg 1$ we will neglect the second term of the above formula and write

$$K_0 \sim \sqrt{\frac{\pi}{2\lambda a}} e^{-\lambda a}, \quad K_1 \sim \sqrt{\frac{\pi}{2\lambda a}} e^{-\lambda a}, \tag{12.58}$$

which shows that in the asymptotic limit both $K_0(\lambda a)$ and $K_1(\lambda a)$ look the same. Using the above asymptotic limit for the special functions, we can now write

$$E_x(a\hat{\mathbf{j}}, \omega) \sim \frac{Ziq\omega}{c^2} \left[1 - \frac{1}{\beta^2 \epsilon(\omega)} \right] \frac{e^{-\lambda a}}{\sqrt{\lambda a}}, \tag{12.59}$$

$$E_y(a\hat{\mathbf{j}}, \omega) \sim \frac{Zq}{v\epsilon(\omega)} \sqrt{\frac{\lambda}{a}} e^{-\lambda a}, \tag{12.60}$$

from which we immediately get

$$B_z^*(a\hat{\mathbf{j}}, \omega) = \frac{Zq}{c} \sqrt{\frac{\lambda^*}{a}} e^{-\lambda^* a}. \tag{12.61}$$

Using Eq. (12.51), we can write

$$\frac{dE}{dx} = \frac{Z^2 q^2}{c^2} \int_{\omega=0}^{\infty} \text{Re} \left\{ \left(-i\sqrt{\frac{\lambda^*}{\lambda}} \right) \omega \left[1 - \frac{1}{\beta^2 \epsilon(\omega)} \right] e^{-(\lambda+\lambda^*)a} \right\} d\omega, \tag{12.62}$$

which gives the energy emitted by the moving charged particle per unit length when the condition $|\lambda a| \gg 1$ is satisfied. From the above equation we can figure out when Cherenkov radiation will occur. Depending on the nature of λ we will get various conditions, λ can be real, complex or pure imaginary. Throughout our discussion we will assume $\epsilon(\omega)$ to be predominantly real as we are interested in light propagation inside the medium. It is known that Cherenkov radiation produces a faint glow inside the medium and consequently the medium is transparent to the radiation. If on the other hand the imaginary part of the dielectric constant was appreciably large then most of this light energy should have been absorbed or may not have been produced at all.

As the dielectric constant is assumed to be real we see that if λ is also real then the integrand of Eq. (12.62) is purely imaginary whose real part vanishes. We immediately notice that there will be negligible energy emission for the charged particle when λ is real and Cherenkov radiation is impossible in this limit.

One can also have complex λ. From the expression of λ one can see that it can be complex when $\epsilon(\omega)$ is complex. We have assumed the dielectric constant to be predominantly real, but it can have a small imaginary part. If the real part of λ is negative we see that $e^{-(\lambda+\lambda^*)a}$ is an increasing exponential function. As a increases this function increases exponentially and consequently we will get more energy per unit length from the moving particle at large a. In reality, the energy per unit length must not increase for large a, it is physically impossible. To eradicate this impossible situation, we will only take cases where the real part of λ has positive sign. We can always choose the sign of the real part of λ, at least locally around a certain ω, as λ is defined via its square. If the real part of λ is positive then $e^{-(\lambda+\lambda^*)a}$ is a decreasing exponential function of a and hence the energy emitted per unit length decreases exponentially with a. This happens around any ω.

If λ is purely imaginary then $e^{-(\lambda+\lambda^*)a} = 1$ and the exponential factor does not damp the energy emitted per unit length expression. In this limit

$$-i\sqrt{\frac{\lambda^*}{\lambda}} = 1.$$

In this limit the energy emission per unit length by the moving particle becomes independent of a and as a result energy can be deposited far away from the moving particle. This case corresponds to radiation and gives rise to Cherenkov effect. As λ is purely imaginary, we must have

$$\beta^2\epsilon(\omega) > 1,$$

for some real $\epsilon(\omega)$. This condition can also be written as

$$v > \frac{c}{\sqrt{\epsilon(\omega)}}, \qquad (12.63)$$

which shows that for Cherenkov radiation to occur the charged particle must have a speed which is greater than the magnitude of the phase velocity of light in the dielectric medium. In this limit, we can write the expression for energy emission per unit length as

$$\left(\frac{dE}{dx}\right)_{rad} = \frac{Z^2q^2}{c^2} \int_{[\epsilon(\omega)>1/\beta^2]} \omega\left[1 - \frac{1}{\beta^2\epsilon(\omega)}\right] d\omega, \qquad (12.64)$$

where the integral is over the range of ω where the condition $\epsilon(\omega) > 1/\beta^2$ is maintained. The above result is the celebrated Frank-Tamm result. One can see that for purely imaginary λ the electric field and the magnetic field as given in Eqs. (12.59) and (12.60) have oscillatory solutions in space. One must note that this oscillatory solution becomes apparent when $|\lambda a| \gg 1$, as only in this limit the field have the exponentials depending on $-\lambda a$. In this limit, the fields in normal space and time show wave-like oscillations. The quantity $|\lambda|$ behaves like the wave vector although it is represented by the traditional symbol of wavelength.

Cherenkov radiation does not occur for all possible frequencies; there is only a certain band of frequency near the resonant frequency ω_0 of the medium where radiation can occur. The picture of the allowed band is shown in Fig. 12.3. In the figure we see that a dashed horizontal line, whose ordinate is one, is drawn. The real dielectric constant is plotted above this dashed line, the dielectric constant increases with ω till the anomalous dispersion region is reached. The velocity of the particle, assumed to be constant, is slightly less than c and the horizontal line representing $1/\beta^2$ shows the constant nature of the velocity parameter. Initially $\epsilon(\omega)$ is less than $1/\beta^2$ and the Cherenkov radiation condition is not obeyed. As ω increases towards the

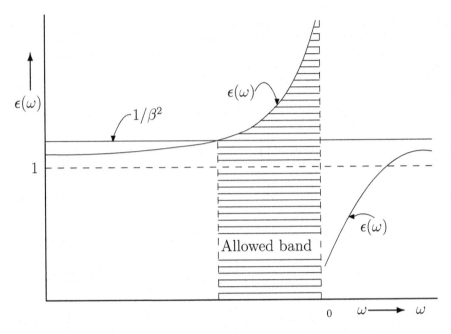

Fig. 12.3 Figure showing the allowed band of frequency where Cherenkov radiation can happen. Here ω_0 is a resonant frequency of the dielectric medium. The velocity of the particle is slightly less than c

resonant frequency ω_0, due to normal dispersion, the dielectric constant increases and reaches $1/\beta^2$. As soon as $\epsilon(\omega) > 1/\beta^2$ the Cherenkov radiation condition is obeyed and the moving particle can emit radiation. Near the resonant frequency ω_0 due to anomalous dispersion effect the dielectric constant decreases rapidly and the Cherenkov radiation condition is violated. As a result of this only a region of frequency is available for Cherenkov radiation. Mostly for electrons this frequency corresponds to blue light in the visible spectrum.

12.3   Properties of Cherenkov Radiation

Next we discuss some properties of Cherenkov radiation. As we have seen that the physical system in which Cherenkov radiation arises has a cylindrical symmetry. Using this symmetry, we can analyze the system in any plane which touches the x-axis in one line. Here, we assume the particle is moving uniformly along the x-axis. In Fig. 12.4 the fields and the Poynting vector are shown. Energy is flowing along the Poynting vector and the observer observes the energy transferred along the direction of the Poynting vector. The whole picture is drawn in the plane defined by the velocity vector and the Poynting vector. This plane happens to be the $x - y$

Fig. 12.4 The electric field, magnetic field and Poynting vector for Cherenkov radiation. The electric field and the Poynting vector are shown in the $x - y$ plane

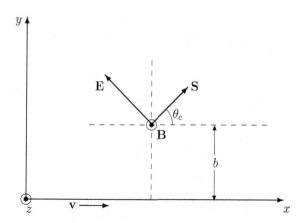

plane in our chosen geometry. In Fig. 12.4 the z-axis and the magnetic field are perpendicular to the plane of the paper, represented by dotted circles. The plane of the picture is defined by the observation direction and the direction of the velocity vector. We have seen before that the electromagnetic fields for a moving charge, in a dielectric medium, is such that the electric field is along the $x - y$ plane and the magnetic field is along the z-axis. This fact shows that the electric and magnetic fields due to the uniformly moving charge are mutually perpendicular to each other. From practical considerations we know that the Poynting vector

$$\mathbf{S} = \frac{c}{4\pi} \mathbf{E} \times \mathbf{B},$$

makes some angle with the axis of the cylinder. Radiation is detected in that angle. This angle is the Cherenkov angle θ_c as shown in the figure. In the figure we see that for the chosen θ_c, the components E_x and E_y both cannot have the same sign. In our chosen geometry, E_x is negative while E_y is positive. From the figure we can now write

$$\tan \theta_c = -\frac{E_x}{E_y}. \tag{12.65}$$

From the expressions of E_x and E_y in Eqs. (12.59) and (12.60) we see that

$$\frac{E_x}{E_y} = \frac{i\omega v}{\lambda c^2} \left[\frac{\beta^2 \epsilon(\omega) - 1}{\beta^2} \right], \tag{12.66}$$

where the expression of λ is given in Eq. (12.55). We will see immediately that choosing

$$\lambda = -\frac{i\omega}{v} \left[\beta^2 \epsilon(\omega) - 1 \right]^{1/2},$$

gives us the physically interesting Cherenkov angle. In this case the Cherenkov angle
has some positive value between zero and ninety degrees. Using the above expression
of λ, we get

$$\frac{E_x}{E_y} = -\left[\beta^2 \epsilon(\omega) - 1\right]^{1/2} ,$$

yielding

$$\tan \theta_c = \left[\beta^2 \epsilon(\omega) - 1\right]^{1/2} . \tag{12.67}$$

The above expression gives a positive θ_c. From the above result, we can write

$$\cos \theta_c = \frac{1}{\beta \sqrt{\epsilon(\omega)}} , \tag{12.68}$$

which is physically relevant because when the Cherenkov radiation condition is
maintained we must have $\beta^2 \epsilon(\omega) > 1$. From the above result, we see that Cherenkov
radiation can happen at some positive angle (with respect to the direction of motion)
less than ninety degrees and this angle is actually dependent on the frequency of
radiation. The result implies that various colors of light will have different Cherenkov
angles for propagation. Moreover Fig. 12.4 shows that Cherenkov radiation is linearly
polarized, the plane of polarization is the plane defined by the observation direction
and the direction of the velocity vector of the moving particle.

Next we try to figure out the wavefront of Cherenkov radiation. As because the
particle which emits Cherenkov radiation is moving faster than the speed of light (in
the dielectric) the wavefront originating from the particle is different from normal
spherical wavefronts. In Fig. 12.5 we show the waves originating from the particle
when the particle is moving faster than light in the medium. The figure shows the
path traversed by the particle in time t. During this time the particle travels a distance
vt. Initially, the particle emits a spherical wave whose radius becomes $ct/\sqrt{\epsilon}$ at time
t. As $v > c/\sqrt{\epsilon}$, the particle has moved outside the spherical wavefront during time
t. While the particle travels it is emitting continuously spherical waves. All these
spherical waves nicely fit in a cone as shown in Fig. 12.5. Such a construction is
natural and one can easily show that all the spherical waves will touch the cone
when the particle has reached one extremity at time t by using properties of similar
triangles. The net effect is that at time t the energy associated with each spherical
wave is concentrated on the surface of the cone and the surface of the cone effectively
acts as the wavefront. Energy is propagated at an angle θ_c, as shown. This kind of
behavior is associated with shock waves and so we call the cone the shock cone
associated with Cherenkov radiation. Although we have drawn a two-dimensional
drawing of the cone in reality all the circles are spheres and the cone is in three-
dimensional space. The cone shown in the figure corresponds to a conical surface in
three-dimensional space.

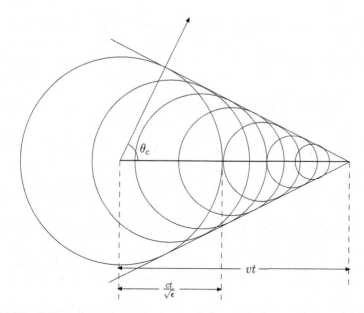

Fig. 12.5 The Cherenkov shock cone

As we know $\theta_c = \theta_c(\omega)$, light of different colors will make slightly different angles with respect to the velocity direction. Measuring this angle one can determine the magnitude as well as direction of the velocity vector if one has a fair estimate of the dielectric constant of the medium. Cherenkov radiation is efficiently used by high-energy physicists in modern neutrino detection experiments. In these experiments the neutrinos (coming from outside) hit a target inside some liquid (generally ultra-pure water), this produces high energy electrons. This high-energy electrons may travel faster than speed of light inside the liquid producing Cherenkov radiation. By observing the radiated light experimentalists can infer about the direction of the incoming neutrino. This technique of detecting neutrino properties using Cherenkov radiation is used in Super Kamioka Neutrino Detection Experiment in Japan.

12.4 Problems

1. Write down briefly on the four ways charged particles lose energy while passing through a material medium. The four specific ways are Ionization, Excitation, Bremsstrahlung and Cherenkov radiation. Can you specify a fundamental difference of energy dissipation by the charged particle between the first three methods and Cherenkov radiation?

2. Closely related to the first question is the point about the distance scales (from the particle's trajectory). The first three modes of energy dissipation are expected to

dominate in the regions around the particle's trajectory whereas in the last case, namely radiation, energy is dissipated in far away regions. From this information can you justify why we model the medium as a dielectric medium instead of a set of atoms or molecules (as is done in the calculation of energy dissipation in the other three cases). Can you now give an intuitive meaning to Eq. (12.56). To answer the last part you may use the fact that in Cherenkov radiation (from electrons) one primarily gets light in the visible spectrum.

3. Write down the oscillatory form of the electric and magnetic fields when the Cherenkov condition is fulfilled.

4. We derived the electric and magnetic fields for a uniformly moving charge in Chap. 7. There we saw that the fields go as $1/r^2$ and consequently these fields do not radiate. It is easy to understand it as a static charge in any frame does not radiate, Lorentz transforming it to another inertial frame moving uniformly will not produce any radiation. Does the same logic hold in presence of a dielectric medium? A static charge inside a dielectric does not radiate, but a uniformly moving charge can. Comparing with the previous case can you find out a loophole in the logic and answer why a uniformly moving charge may radiate inside a static medium?

5. From Fig. 12.5 prove that all the spherical waves will touch the shock cone exactly at one point (at time t) and the wave front will be exactly as shown.

6. Express the solutions of the modified Bessel functions (the modified Bessel functions of the first kind and the second kind) in terms of the Bessel function of the first kind and the Hankel function. From the knowledge of the Bessel functions show that the modified Bessel functions are monotonically increasing or decreasing functions when they are real.

Index

K. Bhattacharya and S. Mukhopadhyay, *Introduction to Advanced Electrodynamics*,
https://doi.org/10.1007/978-981-16-7802-8

Printed in the United States
by Baker & Taylor Publisher Services